CXC Integrated Science

June Mitchelmore
Formerly Education Officer (Science)
Ministry of Education, Kingston, Jamaica

John Phillips
Formerly Science Teacher
Harrison College, Bridgetown, Barbados

John Steward
Formerly Science Adviser
Ministry of Education, Georgetown, Guyana
and Curriculum Consultant, Dominica

CAMBRIDGE UNIVERSITY PRESS
Cambridge, New York, Melbourne, Madrid, Cape Town, Singapore, São Paulo, Delhi

Cambridge University Press
The Water Club, Beach Road, Granger Bay, Cape Town 8005, South Africa

www.cambridge.org
Information on this title: www.cambridge.org/9780521013390

© Cambridge University Press 1986, 2002

This book is in copyright. Subject to statutory exception
and to the provisions of relevant collective licensing agreements,
no reproduction of any part may take place without the written
permission of Cambridge University Press.

First published 1986
Second edition 2002
5th printing 2008

Printed in Dubai by Oriental Press

ISBN   978-0-521-01339-0 paperback

Design and Typesetting: The Nimble Mouse, Cape Town
Illustrators: Karen Ahlschläger, Catherine Crookes, Kay Fish and Robin McBride
....

Cambridge University Press has no responsibility for the persistence or accuracy of URLS
for external or third-party Internet websites referred to in this book, and does not
guarantee that any content on such websites is, or will remain, accurate or appropriate.

....

If you want to know more about this book or any other Cambridge University Press
publication, phone us at (021) 412-7800, fax us at (021) 419-8418 or send an e-mail
to capetown@cambridge.org

Preface

This is the second edition of *CXC Integrated Science*. It completely covers the Caribbean Examination Council's (CXC) new syllabus in Integrated Science (single award) – both Basic and General Proficiency objectives – with the first examination in 2002.

The book is arranged in the form of double page spreads which cover particular groups of specific objectives. Activities, which form an integral part of teaching, are interwoven in the text. Questions on each page challenge the students' understanding, and their answers form a summary of the work covered. The introductory section is an overview of previous work and also outlines some of the practical and study skills which are needed for the course. The remaining four sections of the book cover sections A to D of the syllabus.

> The arrangement of topics will allow teachers to easily find material to arrange into their own teaching schemes.

Where possible we have tried to integrate objectives from different areas of science. We have illustrated particular points with examples from everyday life. The numerous drawings and photographs will help students to visualise different scientific processes and practical procedures.

New features of the second edition include:
- full colour artwork and photographs
- discussion of new topics of particular relevance to the Caribbean
- extension of some sections to allow further development of content
- CXC-style Integrated science questions
- a section on developing the skills needed for school-based assessment (SBA).

We would like to express our thanks to the teachers who have given helpful comments on the book, and to Dr Judith Hardy who commented on the medical aspects. We are also grateful to Betty Macdonald at the CXC office in Barbados, the library staff at the PAHO office, Barbados, and the CUP African Branch for their assistance.

June Mitchelmore
John Phillips
John Steward

Acknowledgements

The publishers would like to thank the following for permission to reproduce their photographs:

© 2002 Mark Van Aardt/Fotozone 7 (top left); © David Simson 136 (top right); © JAMSTEC 264; Altus Pienaar/Fotozone 256; Biophoto Associates 38, 151; Bronwyn Lloyd 35, 56, 57, 61, 172, 209; C James Webb 149; Chad Henning/iAfrika Photos 231; Diane Ross 107; Dr N C Hilyard 116; Eric Miller/iAfrika Photos 298; Fotozone 201; Gallo Images 53, 93 (top), 94, 154 (fly and rat); Geoff Spiby/CPL 12 (bottom), 196 (top left); INPRA 16, 82, 197 (right), 213, 271, 295 (left), 311, 317; Jim Mejuto/Photo Access 185 (bottom); John Phillips 173, 178, 275; Juergen Berger, Max-Planck Institute/SPL 289; June Hassall 7 (top right, bottom left and bottom right), 11, 12 (top), 25, 30, 41, 42, 55, 60, 67, 70, 73, 136 (top left and bottom right), 143, 146, 169, 180, 182 (top), 197 (left), 204, 212, 219, 237, 240, 248, 249, 262, 265, 306, 310, 318; Lester Lefkowitz/Photo Access 185 (top); Mandla Mnyakama/iAfrika Photos 182 (bottom); Manfred Kage/SPL 144 (left); Mark van Aardt/Fotozone 300, 301; Mark Vints 196 (bottom); Mitchell Jones 6; Mr L W Pillar 263; Neil Owen 141; Pauline Rowles 29, 185 (clock), 186, 202, 211, 215, 255, 284; Peter Steyn/Photo Access 241; Photo Access 51, 64, 154 (cockroach and mosquito), 250; Prof. C L Griffiths 144 (right); *Sunday Times* 290; Wendy Lee 59, 93 (bottom), 124, 136 (bottom left), 148, 156, 159, 190, 193, 208, 210, 252, 295 (right).

CPL = Cape Photo Library
SPL = Science Photo Library

Cover photos: June Hassall (top right), Photo Access/Lester Lefkowitz (top left), Photo Access (bottom right), Touchline Media (bottom left).

The illustrations in this edition were based on those done by Marlborough Design, Oxford, United Kingdom for the first edition.

The publishers acknowledge the use of copyright material from the following sources and would like to thank the copyright holders:

The Caribbean Examination Council for the extracts from the CXC syllabus on pages 30–31. Plant diagrams pages 40 (right), 41 and 45 from *Tropical Biological Drawings* (Mitchelmore), Macmillan Education.

Contents

Introduction

Matter, energy and living things
Living and non-living things	6
What are the units of measurement?	8
How are living things built up?	10
Who's who among living things? Parts 1 and 2	12
What is matter made of? Parts 1 and 2	16
What is energy?	20
How can matter change?	22
Why is carbon so important?	24

Studying science
Why study science?	26
How do we use information?	28
Practical work	30
Setting up experiments	32
How do we use equipment and materials?	34
How do we use our results?	36

Living things

Reproduction and growth
How do organisms reproduce?	38
What are flowers?	40
How are fruits and seeds formed?	42
How do plants grow?	44
How do we grow?	46
What changes occur in adolescence?	48
How do we reproduce?	50
What are sexually transmitted diseases?	52
Is there a population problem?	54
How can we control population numbers? Parts 1 and 2	56
Why are you just like you are?	60
What were you like before you were born?	62
Why are pre- and post-natal care important?	64

Food and nutrition
How are plants able to make food?	66
How do substances move?	68
What is a leaf like inside?	70
How is food used by the plant?	72
What are you eating?	74
What are you really eating?	76
What is a balanced diet?	78
Use and abuse of drugs	80
How do we use our teeth?	82
How do we digest our food?	84
How do enzymes help in digestion?	86
How do enzymes work?	88

Transport
How are substances moved around in plants?	90
What is our blood made of?	92
How does our blood circulate?	94
Problems with our circulatory system	96

Respiration
How do we breathe?	98
How are gases exchanged?	100
Problems with our respiratory system	102
What is respiration?	104
What are aerobic and anaerobic respiration?	106

Excretion
How do organisms get rid of waste?	108
What do the kidneys do?	110
How do the kidneys work?	112

Temperature control
Why do temperatures change?	114
Can we control heat transfer?	116
How do living things control their temperature?	118
How do human temperatures vary?	120

Interdependence of life processes
What is the nervous system?	122
What are involuntary and voluntary actions?	124
What is the endocrine system?	126
How are life processes interrelated?	128

Our environment

The physical environment
How do we measure and regulate temperature?	130
How do we use light to see?	132
Why is ventilation important? Parts 1 and 2	134

Water and the aquatic environment
Why is water important for life?	138
How do we obtain the water we need?	140
Would you rather be a fish?	142
Where do fish fit into food chains?	144
Should we try to keep water clean?	146

Health and hygiene
What are parasites?	148
Which parasites might we find in humans?	150
What are the dangers of poor sanitation?	152
How can we control pests?	154
How can we keep our surroundings clean?	156
How can we stop food from spoiling?	158
How can we be protected against disease?	160

Contents

Safety
How can we avoid accidents?	162
Working safely Parts 1 and 2	164
How can we control fires?	168
How do we give first aid?	170

Electricity
Which substances conduct electricity?	172
What are the characteristics of electricity?	174
Is resistance important?	176
How much does electricity cost?	178
How do we use electricity safely? Parts 1 and 2	180
Magnets and electromagnets	184
How do we get and use electricity?	186

Sources of energy
How can we make crude oil useful?	188
How can we get energy from fuels?	190
What are the effects of using cars?	192
How do we use solar energy?	194
How can we make better use of solar energy?	196

Materials
How reactive are some of the metals? Parts 1 and 2	198
Why do we use metals and alloys in the home?	202
Why do we use plastics in the home?	204
What do we use for washing and cleaning?	206
How can we keep ourselves clean?	208
How can we keep household appliances clean?	210
What causes rusting?	212
How can we try to prevent iron from rusting?	214
What happens when we try to dissolve things?	216
What do we mean by hard and soft water?	218
Why are acids and alkalis important?	220
Household chemicals Parts 1 and 2	222

Machines
How do machines make work easier? Parts 1 and 2	226
How efficient are machines?	230

Soils
What is soil?	232
How do soils differ?	234
How can we use our knowledge about soil?	236
How do we get the best out of the soil?	238
How can we conserve the soil?	240

Interdependence of living things
What organisms live in the soil?	242
What organisms live round about us?	244
How are materials recycled in nature?	246
Pollution and conservation	248

Interacting with our environment

Recreation
Why do we need exercise?	250
How can we train our bodies?	252
What kinds of objects float?	254
How can we survive under water?	256
How are objects affected by moving through air?	258
What do you expect from sports equipment?	260

Gathering information
How can we find out about our surroundings?	262
Can waves transfer energy?	264
What are sounds?	266
How do our ears work?	268
How is sound recorded?	270
Can we extend the range of sound?	272
How do we communicate via radio waves?	274
How does light behave?	276
How do our eyes work?	278
Problems with our eyes	280
What is colour?	282
How are images reproduced?	284

Movement

Movement within living things
Why do we need transport systems?	286
How are transport systems protected?	288

Movement on land, in water and in air
How do collisions transfer energy?	290
Does gravity affect balance?	292
Which objects are most stable?	294
How is movement opposed?	296
How do objects and animals fly?	298

Movement in space
How can we travel in space?	300
Where are we in space?	302
How is Earth affected by other bodies?	304
How are organisms affected by the sea?	306

Land and air movements
How does pressure vary?	308
What causes changes in the weather?	310
What are hurricanes?	312
What are earthquakes?	314
What are volcanoes?	316
Living in the Caribbean	318

The electromagnetic spectrum	320
Sample questions in Integrated Science	321
Index	328

Living and non-living things

Look around you in the classroom and outside. Think about what you can see, hear, smell, touch and feel. How are the things similar and different? What are the things made of? How do they work? How can we use them …? These are some of the questions you will be answering as you go through this course.

As you look at the things around you, the most important differences you will notice are those between living and non-living things.

Activity | Living and non-living things

1 Look carefully at the photograph below. Make a list of all the things you can see.

2 Make a table in your notebook with two columns. Use the headings **Living things** and **Non-living things**. Write the names of the things into the correct columns.
3 Discuss your completed table with a friend. Have you put the same things in the same columns? Discuss any differences you notice.

Let's look at the characteristics of living things, which distinguish them from non-living things.

Living things make or eat food (nutrition)

All living things need food. Food is the fuel they use for all their activities.

Plants can make their own food from simple substances: carbon dioxide, water and mineral salts. They are 'self-feeding' or **autotrophic**. They contain **chlorophyll** (a green pigment) and carry out photosynthesis. This means that they use energy from the sun to make their own food (pages 66–7).

Animals cannot make their own food. They need complex compounds that contain energy. They eat plants, or animals which have eaten plants (pages 74–5). They are 'other feeding' or **heterotrophic**.

Living things respire

The food which an organism makes or eats is taken into the cells of its body. As energy is needed, for example for movement or response, some of the food is broken down by respiration to release the trapped energy. At the same time carbon dioxide and water are produced.

All living cells of plants and animals carry out respiration all the time. Respiration is similar to burning: the combination of oxygen with a fuel to release energy. In living things respiration is brought about by special substances called **enzymes** and can happen without high temperatures.

Non-living things like cars and machines cannot make their own food, nor do they go around eating other things! Humans have to put fuel, such as diesel and gasoline, into them. The fuel does not become a part of their 'body', it just stays in the engine. The fuel is burned to release energy for the movement of the car or the work of the machine. This burning involves high temperatures and is not brought about by enzymes.

Living things excrete

A living organism is a bit like a chemical factory. Different raw materials arrive (such as carbon dioxide and water in plants, and foods in animals) and they are changed or processed in a variety of chemical reactions to make new products. At this very moment there are millions of chemical reactions going on inside you!

These chemical reactions take place inside the cells. In plants, for example, photosynthesis occurs in the chloroplasts which contain chlorophyll, while respiration occurs in very small structures called mitochondria. Not all of the products of these reactions are useful, and some of them may be harmful if they accumulate. Living things therefore get rid of, or **excrete**, these waste materials. Excretion is the removal of waste products made by the activity of living cells. Cars also get rid of exhaust gases, but the fuel has never been part of the car's body.

Living things respond

Living things are affected by changing conditions around them and inside them. These changing conditions are called **stimuli**. Organisms have to **respond** correctly in order to stay alive.

Plants growing in the soil have to grow their roots near to sources of water and mineral salts. They have to grow their leaves so that they can catch the rays of the sun. Some plants, such as the sensitive plant, also respond to touch, and sunflowers twist each day to face the sun.

Animals respond to their surroundings, for example by looking for food, while also avoiding being eaten themselves. They usually have special sense organs to help them pick up stimuli, and muscles with which to respond.

(a) A praying mantis

(b) Fish

(c) Sunflowers facing the sun

(d) A pig with her litter

Living things move
Within plant and animal cells the material or protoplasm moves continually. The main way in which living things respond to stimuli is by the movement of parts of themselves. In contrast, a car moves when we make it move, it cannot move *on its own*.

Movement in living things may be easy to see, for example, you move your hand away from a hot object, or an animal moves around in search of food. The movement may also be inside you, for example, the movement of your jaw or of your stomach as it churns up your food and helps to digest it.

Plants have roots and they do not move around from one place to another. But they can move parts of themselves. A plant's movements are usually very slow and they are brought about by growth. A plant's stem bends towards the light because it grows more on one side of the stem than on the other.

Living things grow
Living things can grow if they make or eat more food than is needed for staying alive. New substances are built up in the body, and the organism grows: it becomes bigger or more complex and increases in mass, length or width.

Cars certainly do not grow, but we can say that crystals grow (page 16). Crystals are non-living and they 'grow' in the sense that extra substances of the same kind are just added to the outside. However, this is not at all like the growth in living things.

Living things reproduce
New organisms are formed by **reproduction**. To *reproduce* means to 'make again'. It means that another organism, similar in many ways to the parent, is made and can live separately. It would be very convenient if cars could also reproduce, but unfortunately this is *only* a characteristic of living things.

Organisms need to grow before they can reproduce. They have to become mature. We have seen that growth is a characteristic of living things. For example, flowering plants grow flowers which produce pollen and eggs before they can reproduce. Animals, such as ourselves, have to reach puberty before they are able to produce the sperm and eggs necessary for reproduction. This is called sexual reproduction (page 38).

Some organisms reproduce asexually. A part of the organism grows and becomes a new, separate organism (pages 39 and 73).

Questions
1. What is meant by (a) a living thing and (b) a non-living thing? Give an example of each.
2. What are the seven characteristics of living things? How is a non-living thing, such as a car, different from a living thing for each characteristic?
3. What characteristics of living things are shown in pictures (a) to (d) above? (*Note*: each picture may show more than one characteristic.)

What are the units of measurement?

A system of measurement

There is a system of measurement which is used by most of the leading nations of the world. The system is called the Système International d'Unités or SI (metric) system and its units are meant to replace all other types of units of measurement. These measurements include measurements of mass, length (distance), time, force, pressure, energy, temperature and electricity. The table below gives the different types of measurements with their units and symbols which you are likely to find in the text. These units have been adopted by all scientists and are essential for accuracy in the recording, transfer and interpretation of data.

SI units

Measurement	Quantity	Unit	Symbol
length, mass and time	length	metre	m
	area	square metre	m^2
	volume	cubic metre	m^3
	mass	kilogram	kg
	density	kilogram per metre cubed	kg/m^3
	time	second	s
	frequency	hertz (= per second)	Hz
force and pressure	force	newton	N
	weight	newton	N
	moment of force	newton metre	Nm
	pressure	pascal (= newton per square metre)	Pa
energy and heat	energy	joule	J
	work	joule (= newton metre)	J
	power	watt	W
	temperature	degree Celsius	°C
	absolute temperature	Kelvin	K
electricity	electric current	ampere	A
	electromotive force	volt	V
	potential difference	volt	V
	resistance	ohm	Ω
	electrical energy	joule	J

There are certain prefixes that can be used to change the standard unit of measurement. For example, the term *kilo* is used with *metre* to derive the term *kilometre*. Since kilo stands for 1 000, a kilometre = one thousand metres.

Often scientists use short forms of mathematical expression to deal with very large or very small numbers. Numbers are expressed as *powers of ten*. For example, one hundred is ten to the power two (or ten squared): $100 = 10 \times 10 = 10^2$.

The table below gives a few examples of the powers of ten of large numbers and shows how the prefixes are used.

Multiple	Prefix	Symbol	Example
10^9	giga	G	gigawatt
10^6	mega	M	megajoule
10^3	kilo	k	kilometre

One metre can be divided into smaller units, for example into one thousand parts, each of which is called a *millimetre*. The prefix *milli* means that the particular unit to which it is attached is divided by one thousand. In the short form of mathematical expression, one hundredth is ten to the power minus two:

$$\frac{1}{100} = \frac{1}{10^2} = 10^{-2}$$

This table gives a few examples of the powers of ten for smaller numbers and shows how the prefixes are used.

Multiple	Prefix	Symbol	Example
10^{-1}	deci	d	decimetre
10^{-2}	centi	c	centimetre
10^{-3}	milli	m	millimetre
10^{-6}	micro	μ	micrometre
10^{-9}	nano	n	nanosecond

Are there other measurements in general use?

A visit to the local grocery store or market place will soon reveal that many things are still sold in *pounds* and *pints*. Tailors still use *inches*, cloth is sold in *yards* and *gallons* of paint can be bought from the hardware store. Racehorses still run *furlongs*, weather reports often give wind speeds in *miles per hour* and ships travel in *knots*. While these units are no longer taught in our schools, they remain part of our everyday experience and we still need to know how they relate to the new units we use more often. Here is how some of the older units relate to the metric system (SI units).

Unit	How used	Metric equivalent
Inch	Length measure	2.54 cm
Foot	Length measure	30.48 cm
Yard	Length measure	0.9144 m
Furlong	Length/distance measure	201.18 m
Mile	Length/distance measure	1.6093 km
Mile per hour	Speed measure	1.61 km/h
Knot	Speed measure	1.85 km/h
Pint (US)	Volume measure	0.473 l
Gallon (US)	Volume measure	3.79 l
Gallon (Imperial)	Volume measure	4.55 l
Ounce	Weight/mass measure	28.38 g
Pound	Weight/mass measure	0.454 kg
Pound/square inch	Pressure measure	7.038 kPa
Ton	Weight/mass measure	1016.05 kg

Astronomical distances

Light year = 9.45×10^{15} km
(The distance light travels in a year.)
Astronomical unit = 149.6×10^6 km
(The average distance of the Earth from the sun.)

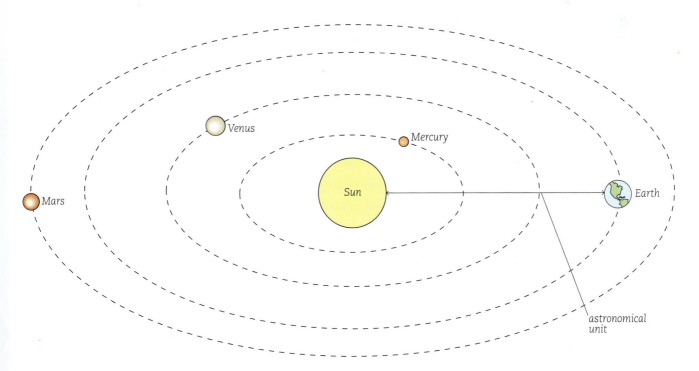

How are living things built up?

What are living things like inside? How do the various parts carry out the processes of life? On pages 8–9 you saw how we measure in science. The **cells** which are the building blocks of living things are very small. We need to use a microscope to see them.

Activity | Looking at plant and animal cells

1. Pull away a little piece of the purple skin from the lower side of the long, green and purple leaf of Moses-in-the-bulrushes (*Rhoeo*).
2. Put it onto a microscope slide with a drop of water and cover it with a cover-slip (see page 34).
3. Examine your slide under the low power of the microscope (see page 35) and see how many of the features shown in the plant cell below, you can see. Notice also how the cells are fitted together to make a **tissue**, in this case the lining skin of the plant.
4. Your teacher will give you a slide with animal cells on it. Examine your slide under the low power of the microscope and see how many of the features shown in the animal cell below, you can see.

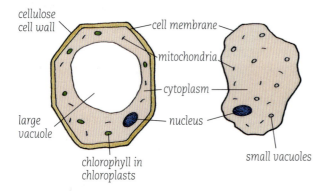

Similarities between plant and animal cells Both have a nucleus, cytoplasm, cell membrane and mitochondria.

Differences Copy the table into your practical notebook and fill in the right-hand side by looking at the pictures above.

Typical plant cells	Typical animal cells
1 Have a cellulose cell wall outside cell membrane	
2 Contain chlorophyll in chloroplasts	
3 Have large vacuoles with cell sap	

Inside the nucleus

Each cell has a dark area – the nucleus. The nucleus controls the activities or processes of the cell. When we magnify the nucleus we see thread-like **chromosomes**. Below is a diagram of human chromosomes.

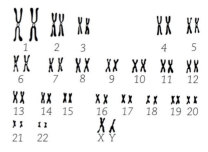

The diagram shows the chromosomes in the cells of a male. There are 22 pairs (the ordinary chromosomes) and two chromosomes left over, one of which is larger than the other. These last two chromosomes are called the **sex chromosomes** which are responsible for the differences between the sexes. The long one is called an **X** chromosome, and the shorter one is called a **Y** chromosome. All males have 22 pairs of chromosomes plus XY. All females have 22 pairs of chromosomes plus two X chromosomes, or XX.

Inside the chromosome

When scientists look at some chromosomes under greater magnification they see dark bands called **genes**. It is the genes which have the instructions for giving us our various characteristics. The picture below shows some genes on the chromosomes of the fruit fly, *Drosophila*.

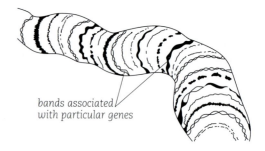

Building up

- Genes are found on chromosomes inside the nucleus of cells.
- Living cells are grouped together to make tissues, such as packing tissue in plants, and muscles in animals.
- Tissues are grouped together to make organs such as leaves in plants, and the stomach in animals.
- Organs working together make up the whole organism.

Organs and systems in a mammal

The photograph below shows a model of an elephant, cut away to show the inside. The **organs** (complicated structures made up of many tissues) carry out one or more jobs or functions. Examples of organs are the heart (pumping blood) and the stomach (digesting food).

- Identify as many parts of the model as you can and talk with a partner about their functions.

Several organs also work together to make **systems,** such as the digestive system, the circulatory system, the nervous system and the reproductive system. In this course you will learn more about organs and systems, and their functions.

- With a partner list the systems you know in a mammal, the organs which make them up, and the functions.

Organs and systems in a flowering plant

When we look at plants we can identify important organs, such as flowers, leaves, stems and roots. The organs working together above ground are sometimes called the shoot system – these are shown below in the photograph of Cassia (also known as the Pride of Barbados). The organs working together underground are called the root system.

- With a partner list the organs in a flowering plant and the functions of each one.

Matter, energy and living things

Matter and energy make up our whole universe and make it possible for things to work.

Matter Matter is the material of which everything is made – both living and non-living things. Matter makes up the *things* around us. We can see matter, even though we need to use special microscopes to look at cells, chromosomes and genes, or to investigate the chemicals of which things are made. Matter has mass – though this mass may be very, very small. We will be introduced to what matter is made of on pages 16–19 and 22–5, and learn more throughout the course. Matter is the chemicals which are the building blocks that are put together in various ways to make living and non-living things.

Energy We cannot *see* energy; we can only see or feel what it does. Energy also does not have mass. We see or feel the effect of different forms of energy, such as light, heat, sound, electricity and the energy of movement. Energy allows things to work. We will be introduced to the forms of energy on page 20, and learn more throughout the course.

Activity | Matter and energy

1. Look around you in the classroom and outside. Look for things made of matter – living and non-living. Search for different forms of energy that you cannot see, but know are there because they are having some effect, such as the wind blowing the trees.
2. Copy this table into your notebook. A few examples of matter and energy have been filled in for you. Find other examples and put them in the correct columns.

Matter (material, things that have mass)		Energy (the power to do work of some kind)
Living things	Non-living things	
Yourself	Your desk	Sound from a CD
Birds	Your clothes	Light from a bulb
Flowers	Clouds	Warmth and light from the sun
Trees	Stones	

Living things Living things have a special ability: they are able to take their own energy from their surroundings – either by photosynthesis, or by taking in ready-made food. They then use this energy for all their activities, and for growth and reproduction. When living things lose this ability, they die. The variety of living things is introduced on pages 12–15, and you will learn more about the life processes throughout the course.

Who's who among living things?

Living things have basic characteristics which make them different from non-living things (pages 6–7). All but the simplest of living things are also built up from cells, tissues and organs (pages 10–11). But living things show a great variety in their appearance and structure. These differences allow us to separate or **classify** them into many different groups.

We will first look at the important similarities and differences between a plant (Pride of Barbados, which is a flowering plant) and an animal (a fish, which is an example of a vertebrate).

A typical plant	A typical animal
Uses simple inorganic substances (page 66) to make its own food	Feeds on complex organic substances (page 66) containing trapped energy
Has chlorophyll and can carry out photosynthesis	Does not have chlorophyll and cannot photosynthesise
Does not digest food	Has structures to digest food
Usually rooted in the ground	Not rooted in the ground
Does not usually move from place to place	Moves around to get food
Has no nerve or muscle cells	Has nerve and muscle cells
Does not have special senses	Has special sense organs

Activity | Comparing a plant and an animal

1. Look back at the characteristics of living things (pages 6–7). Describe in your notebook why you think the plant and animal shown below are living.

Classifying living things

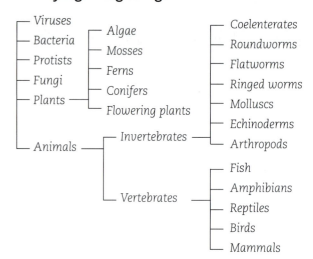

Plants and animals make up a large part of the living things around us. There are also many other organisms which we cannot see unless we use an ordinary microscope or a microscope which uses electrons instead of light rays.

Instrument	Magnification	Can see
Naked eye	Life size (× 1)	Many multicellular organisms
Hand lens	× 10	Cell as a dot
Low power microscope	× 100	Nucleus in a cell
High power microscope	× 400 to × 1 000	Some cell structures
Electron microscope	× 40 000 to × 500 000	Internal structure of mitochondria
100 cm = 1 m 10 mm = 1 cm	1 000 µm (micrometres) = 1 mm 1 000 nm (nanometres) = 1 µm	

2. Write down the different ways in which the plant and animal carry out the characteristics of living things. For example, how does each carry out nutrition? How do their differences in appearance and structure affect how they carry out the characteristics of living things?

Part 1

Viruses about 100 nm, seen only with electron microscope, no cell structure, can only reproduce inside living organisms.

Tobacco mosaic disease virus

Influenza virus

Protists about 10 μm–1 mm, some seen with low power, single-celled, with nucleus. Plant-like with chlorophyll, or animal-like.

Plant-like: diatom

Animal-like: Entamoeba

Bacteria about 0.001 mm, seen under high power, single-celled or joined in chains, no nuclei.

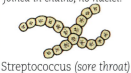
Streptococcus (sore throat)   Bacillus typhosus (typhoid fever)   Vibrio (cholera)

Fungi about 5 μm–20 cm, mostly many-celled with nuclei. No chlorophyll.

Yeast

Pin mould on a banana

Mushroom

PLANTS
Small to very large, visible to the naked eye (i.e. without a microscope), many-celled with nuclei, chlorophyll, mostly stationary and rooted in the ground.

Algae about 5 mm–100 cm, mostly small, no roots, stems or leaves. Threads or divided sheets.

Spirogyra green threads   Sargassum brown seaweed

Flowering plants 5 cm–30 m, small to tree-like, roots, stems and leaves, reproduce by seeds inside flowers.

Monocotyledons	Dicotyledons
Narrow leaves with parallel veins. Bunch of small roots (fibrous roots). One seed leaf (cotyledon) in seed, for example, grass, banana, coconut palm.	Broad leaves with branching veins. Main (tap) root with many branches. Two seed leaves (cotyledons) in seed, for example, balsam, Hibiscus, mango tree.

Mosses about 5 mm–15 cm, simple roots, stems and leaves, reproduce by spores.

Moss 'leaves'

Spore case

Ferns about 5 cm–10 m, small to tree-like, roots, stems and leaves, reproduce by spores.

 Fern   Spore cases   Tree fern

Conifers about 5 m–30 m, mainly tree-like, roots, stems and leaves, reproduce by seeds inside cones.

 Pine tree    Cone

Grass

Banana

Coconut palm

Balsam

Hibiscus

Mango tree

Who's who among living things?

ANIMALS
Small to large, visible to the naked eye, many-celled with nuclei, no chlorophyll, usually move around and feed on other organisms.
INVERTEBRATES   Animals without backbones. Mostly small. Either soft-bodied or with a hard outer case (exoskeleton).

Coelenterates about 1 cm–many metres. Bag-like with tentacles around mouth, no legs. Live singly or in groups (coral). Most live in the sea.

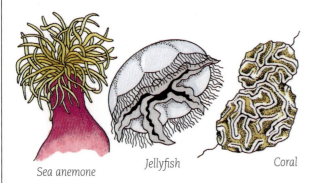
Sea anemone   Jellyfish   Coral

Roundworms (nematodes) about 100 mm–30 cm. Body long and thin, round in cross-section, no legs, no rings. Most are parasites, others live in the soil.

Hookworm   Threadworm

Molluscs about 3 cm–30 cm. Soft body, undivided and often inside a shell, no legs. Some with tentacles. Most live in water, a few on land.

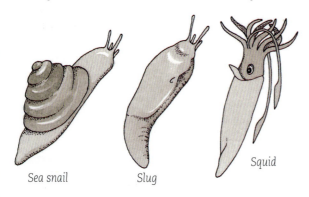
Sea snail   Slug   Squid

Flatworms about 10 mm–5 m. Body long and thin, flat in cross-section, no legs, no rings. Have hooks and suckers. Most are parasites of animals.

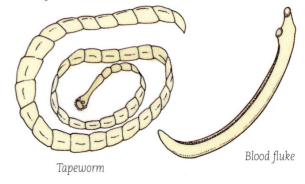
Tapeworm   Blood fluke

Echinoderms about 10 cm–15 cm. Ball-shaped or star-shaped based on a five-part pattern, no legs, tough outer skin with spines. Live in the sea.

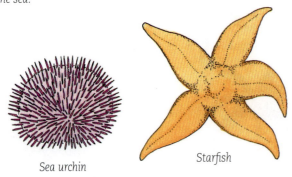
Sea urchin   Starfish

Ringed worms about 5 cm–1 m. Body long and divided up by rings, round in cross-section, no legs. Most aquatic, others live in the soil.

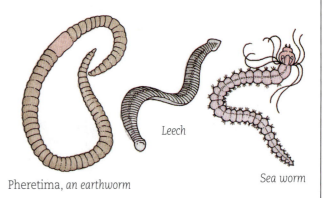
Pheretima, an earthworm   Leech   Sea worm

Part 2

INVERTEBRATES (continued)
Arthropods Mainly small, crawling animals with a hard outer case (exoskeleton). The only invertebrates with jointed legs.

Insects 3 pairs of legs, body divided into head, thorax, abdomen, 2 pairs of wings. Live on land, for example, butterflies, roaches, beetles, termites and lice.	**Arachnids** 4 pairs of legs, head and thorax joined, no wings. Some are parasites. Live on land, for example, spiders, ticks and mites.	**Crustacea** 4–10 pairs of legs, usually with shield over front of body, no wings, breathe with gills. Mostly live in water, for example, crabs, shrimps, woodlice.	**Myriapods** More than 10 pairs of legs, body long and divided into segments, no wings. Live on land, for example, centipedes and millipedes.
 Butterfly	 Spider	 Crab	 Centipede

VERTEBRATES Animals with backbones. Skeleton inside the body (endoskeleton). Mostly large. Some live in water and have gills. Others live on land or in the air and have lungs. Land forms have four limbs.

Fish Covered in scales. Live in water and breathe with gills. Streamlined body and fins for swimming. Eggs laid and develop in water, for example, Nile perch, flying fish, seahorse, shark.

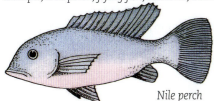
Nile perch

Birds Covered in feathers. Breathe with lungs. Live on land and in the air. Have wings for flying. Eggs laid inside a hard shell in nests, for example, vulture, egret, chicken, owl.

Vulture

Amphibians No skin outgrowths. Young stage (tadpole) lives in water, breathes with gills. Adult lives on land and in water, has 4 limbs and breathes with lungs. Eggs laid and develop in water or moist places, for example, toads and frogs.

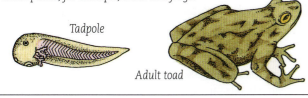
Tadpole Adult toad

Mammals Covered in hair. Breathe with lungs. Most live on land, some live in water. Either walk (dog) or swim (whale). Eggs develop inside female, and the young are born. The young are fed on milk.

 Dog  Human

Reptiles Covered in scales. Breathe with lungs. Most live on land, others in water. Either slide (snakes), swim (turtles), or walk (lizards). Eggs laid inside a leathery shell on land.

 Lizard 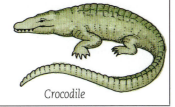 Crocodile

Activity | Classifying organisms
Look at pages 13–15. Read through the descriptions and look at the pictures of the organisms belonging to each of the groups.
1. Your teacher will give you some pictures or diagrams of living things. You should try to find out which of the groups each one belongs to.
2. Collect pictures of living things and prepare some classification charts to stick on your classroom wall.

What is matter made of?

Why do we need to know about matter?

The world around us is made up of a vast number of different substances. Scientists want to know *what* both the similarities and the differences between substances are. They also want to know *why* there are these similarities and differences. They search for patterns in the behaviour of these substances. For example, there may be substances which behave in the same way – this may suggest that these contain matter that behaves in the same way. To find out about this we need to investigate the substances in more detail to find out what they are made of. This may help us to explain the properties we observe.

Quartz crystals which have grown naturally. Note the regular shapes

Activity | Growing crystals

1. Make up some *very* concentrated solutions of copper (II) sulphate and potassium aluminium sulphate (potash alum) in warm water.
2. As the warm solutions cool, use a glass rod to put a few drops of each solution onto a glass slide and observe them using a microscope (page 35).
3. Allow the solutions to cool and observe them every day for 4 to 5 days. You may need to use a seed crystal to encourage crystal growth.
4. Record your observations.

Activity | Diluting a solution

1. Dissolve a small crystal of potassium manganate(VII) (potassium permanganate) in a small volume of water – about 10 cm^3.
2. Pour 1 cm^3 of the diluted solution into a small measuring cylinder. Make this up to 10 cm^3 by adding more water.
3. Pour 1 cm^3 of the diluted solution into the measuring cylinder and dilute that to 10 cm^3 by adding water.
4. Repeat step 3 until the colour of the solution has disappeared.

Questions

1. Do the crystals you have seen all have regular shapes?
2. What happens to the potassium manganate(VII) as the solution becomes more and more diluted?

Activity | A crystal in water

1. Pour about 200 cm^3 of water into a 250 cm^3 beaker. Very carefully add one crystal of blue copper(II) sulphate. Leave the beaker for a few days.
2. Record your observations.

Activity | Looking at bromine

1. Your teacher will place a drop of bromine at the bottom of a gas jar, and place another gas jar upside down on top of the first.
2. Record your observations.

Activity | Hydrogen chloride gas and ammonia gas

1. Your teacher will clamp a long glass tube horizontally and place one piece of cotton wool at each end of the tube. One piece of cotton wool is soaked in concentrated ammonia solution and the other in concentrated hydrochloric acid.
2. The pieces of cotton wool will each give off a vapour – either ammonia gas or hydrogen chloride.
3. Record your observations. Do you think there has been a reaction?

The white substance, ammonium chloride, is formed by the reaction of ammonia and hydrogen chloride.

These different Activities help us to think about matter. We can summarise as follows:

Solid crystals form in regular shapes. When a coloured crystal dissolves, the solution becomes less coloured as it is diluted. The colour of a solid crystal spreads very slowly through water if it is left without stirring. However, gases seem to move very quickly.

One way to explain these observations is to suggest that matter is made up of very small **particles**. We can try to build up a picture of what is happening with these small particles. We shall think further about what these particles might be in Part 2.

Part 1

Solid crystals We can imagine that crystals are built up from particles arranged in regular patterns. As more and more particles are joined together, the crystals get bigger and bigger. The crystals grow until we see the regular shapes which are formed.

We should also include what happens when steam is cooled to form water and water is frozen to form ice.

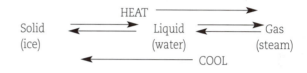

Solid, liquid and gas are called the three states of matter. We know that we have to supply heat energy to change a solid into a liquid and a liquid into a gas. We can use this information to help us think about the way the particles are arranged in each of the three states of matter.

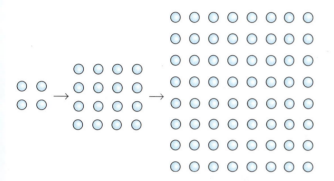

This is how a crystal grows

Diluted solutions Any one crystal of a solid such as potassium manganate(VII) may contain a large number of particles. As the solutions are made more and more dilute, there are fewer and fewer particles in each successive 10 cm³. This may explain why the colour of the solution gets fainter and fainter.

Liquids Particles in a liquid seem to move more slowly than those in a gas. The colour of the blue copper sulphate spreads slowly through water.

Gases The colour of the bromine spreads quickly through the gas jar. This rapid spread suggests that the particles in a gas move quickly.

The white ring of ammonium chloride is formed where the ammonia particles collide and react with the hydrogen chloride particles.

Activity | What are the states of matter?

1. Put a small cube of ice in a small beaker. Heat gently until the solid becomes liquid. Now heat more strongly until the liquid boils. Stop heating and allow the liquid to cool.

When ice (solid water) is heated, liquid water is formed. When water is boiled, steam is formed. Ice is a solid, water is a liquid and steam is a gas (vapour). We can summarise the changes you see in the states of matter in a diagram like the one in the next column.

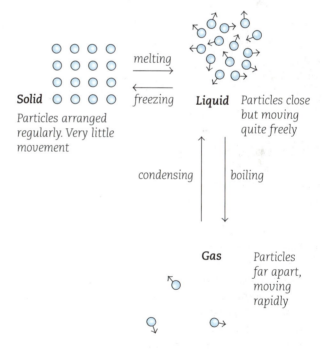

Particle arrangements in solids, liquids and gases

Question

3. How would you define a (a) solid, (b) liquid, (c) gas?
4. What kind of energy is needed to change a solid into a liquid and a liquid into a gas?

What is matter made of?

What are the particles in matter?

In the previous two pages we have looked at some of the evidence for the existence of particles in matter. We have seen that the idea of particles can be used to explain our observations. At this stage, even if we assume that there are particles in matter, we cannot say anything about **what** these particles are. Scientists always search for explanations of observations, and a more detailed examination of matter shows that we need to go beyond the simple idea of particles.

The idea of a basic building block in nature is very strong, and the first complete theory of matter was put forward by John Dalton in the early 19th century. He proposed that the smallest particles found in matter were **atoms**. *An atom may be thought of as the smallest particle of an element which can take part in a chemical reaction.* This definition assumes that you know about **elements**. *An element is a substance in which all of the atoms are the same.* This means that the properties of the atoms of any one element should be the same, as well as different from the properties of the atoms of any other element. Examples of some common elements are hydrogen, copper, oxygen, carbon, iron, nitrogen, aluminium and silicon.

We can consider a very simple reaction of two elements, iron and sulphur. If these two elements are heated together strongly, a new substance, iron sulphide, is formed. We can display this information in the form of a **word equation**. Word equations tell us what we start with and what new substances are formed. Here the **reactants** are on the left-hand side and the **product(s)** are on the right. In this case we have:

iron + sulphur → iron sulphide

The particles which take part in this reaction are **atoms** of iron and atoms of sulphur. The elements carbon and oxygen react together to form the new substance, carbon dioxide:

carbon + oxygen → carbon dioxide

Atoms of carbon react with **molecules** of oxygen to form **molecules** of carbon dioxide. In the Activities you carried out in Part 1, the particles are mostly **molecules**.

A molecule is the smallest part of a substance which can exist by itself and still show all the properties of that substance. Molecules may contain atoms of the same element (an oxygen molecule contains two atoms of oxygen) or atoms of different elements (a carbon dioxide molecule contains one atom of carbon and two atoms of oxygen).

Activity | What happens when elements combine?

We can examine the reaction between iron and sulphur as one example. The elements iron and sulphur contain atoms. When the elements react, as we have seen, a new substance, iron sulphide, is formed. This new substance has properties which are different from those of either iron or sulphur.

1. Make a list of the obvious physical properties of both iron and sulphur.
2. Examine a sample of iron(II) sulphide.
3. How would you try to test whether the properties of iron(II) sulphide are different from those of iron and sulphur?

Iron(II) sulphide contains two elements combined in some way. It is a **compound**. *Compounds are substances in which two or more elements are combined in definite proportions by mass.*

You should notice that this definition does not tell us anything about the nature of the particles when a new substance (compound) is formed by the reaction of two or more elements. In many cases, these may be molecules.

4. Consider the reactions of the following elements:
   (a) Carbon and oxygen
   (b) Hydrogen and oxygen
   (c) Aluminium and iodine
   (d) Iron and chlorine.
5. What would you expect to be the name of the new substance (compound) formed in each case?
6. Write a word equation for each reaction.

Many of the common gases are elements, such as oxygen, nitrogen and chlorine. In these elements, the particles found are molecules, in which two atoms are combined.

Questions

1. How would you define (a) an atom (b) an element and (c) a molecule?
2. Name as many elements as you can that exist as gases at room temperature.
3. Hydrogen chloride and ammonia are gases. Each is a compound containing hydrogen. What would you expect to be the nature of the particles in these gases?
4. When sodium reacts with chlorine, sodium chloride (common salt) is formed. Write a word equation for this reaction.

Part 2

What are atoms made of?

Atoms were thought of as very small particles once Dalton's theory had been accepted. However, towards the end of the 19th century, there was evidence to suggest that there might be smaller particles still. A series of key experiments was carried out early in the 20th century by Rutherford. In one of these, he started with a very, very thin piece of gold foil and directed a stream of radioactive particles at the foil. There were some very interesting observations:
1. Many of the particles passed through the foil without being deflected at all.
2. A very small number of the particles (which were positively charged) were reflected back in the direction from which they had come.

Rutherford needed to find an explanation of these observations and he developed a model of the atom which is still the basis for much of our understanding of basic chemistry. Firstly, he assumed that most of the mass of the atom consists of a small, dense nucleus. (If the nucleus is positively charged, this would account for the reflection of a small number of positively charged particles – like charges repel.)

This small, dense nucleus is surrounded by much smaller particles, mostly at a considerable distance from the nucleus. (This would leave a good deal of empty space, and could account for the fact that most particles passed through the foil.) This model is often called the **solar system model** of the atom, since it is similar to the model of our solar system in which the sun can be thought of as the 'nucleus' of the system of planets.

One of the problems with the early Rutherford model was that no-one was sure what would happen to the very small, negatively charged particles outside the nucleus. Common sense suggested that these should be attracted towards the small positively charged nucleus – unlike charges attract – and then the atom would collapse!

Further development of the model involved the following ideas:
1. The small, positively charged nucleus contained two different particles:
   (a) **Protons**, with a single positive charge and a mass of 1 on a given scale.
   (b) **Neutrons**, with no charge and a mass the same as that of the proton.
2. The particles outside the nucleus (electrons) were very light (mass very much less than that of the protons and neutrons) and had a single negative charge. These electrons were found at fixed distances from the nucleus, such that successive shells of electrons were built up as atoms became heavier. (There are now at least 115 known elements, some of them found only in a laboratory.)

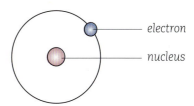
hydrogen atom

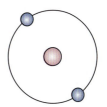
helium atom

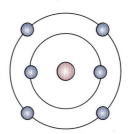
carbon atom

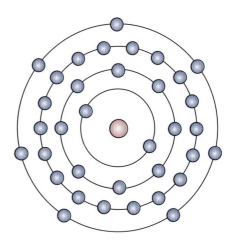
bromine atom

Electron arrangements are often represented by diagrams. These are not meant to be like photographs, showing exactly what real atoms look like. Instead, they are intended to help us understand the structure of the atom

What is energy?

It is very difficult to find a precise yet complete definition of the term **energy**. Energy is described in terms of what it can do, like the purchasing power of money, so energy may be defined as *the capacity for doing work*.

You may know that there are different forms of energy, such as **potential or stored energy**, and **kinetic energy** which is energy in the motion of a body.

Activity | How can energy be stored?

1. Build a toy tank like this, from a cotton reel, a rubber band made from an old bicycle inner tube and some spent matches.

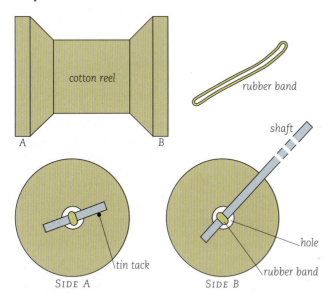

Making a cotton reel tank

  The length of the rubber band should be less than that of the reel. Insert and pull one end of the band through to side B and insert the shaft. The shaft must be about two to three times the diameter of side B.
2. Hammer small tacks into side A to prevent the stick at A from spinning. Lubricate side B with paraffin wax (candle wax).
3. Hold the tank in one hand with side B uppermost. Turn the shaft in a circle several times. This twists the rubber band. Place the tank on the ground. It will move forward as the rubber band slowly untwists.

Energy is stored in the rubber band as it is twisted. The energy is made to do the work of moving the toy tank along as the rubber band untwists. Potential energy is converted into kinetic energy.

Similarly, energy is stored in a spring when a clock is wound up. This energy keeps the mechanism of the clock running.

The head of a match is made of chemicals, which store **chemical energy**. When the chemicals are activated, the chemical energy can be converted into **heat energy**.

Activity | Chemical energy

Heat the tip of a pin until it is red-hot. While it is still red-hot, touch it quickly to a match head. Observe what happens. Do you think the energy released by the match is more than that in the red-hot pin?

Paper, wood, gas and oil can all be considered as stores of chemical energy. Under the right conditions this energy can be released and used. The electrochemical cell is also a store of chemical energy. Here chemical energy is converted mostly into electrical energy (pages 172–3).

In raising a weight up to a height, work is done on the weight. If the weight is allowed to fall it gives up its energy mainly as heat when it falls to the ground. Similarly, water is raised from the sea by evaporation to form clouds. Water vapour condenses and falls as rain, which may produce streams and rivers. In many foreign countries outside the Caribbean, for example Canada, water on high ground is collected into a dam. The large amount of potential energy due to the height of the water is used to power water turbines, which generate electricity.

What are the forms of energy?

Energy appears in different forms, such as in the vibration of particles of matter, causing a temperature increase (heat); in the production of electromagnetic radiation such as light, heat and radio waves; in the movement of electrons through matter (electricity) and in the displacement (movement) of matter (mechanical energy). Other forms of energy include magnetic energy and sound energy.

Sometimes we produce certain forms of energy from others that may be more abundant, for example chemical energy in oil is changed to heat energy by burning, then into electrical energy for us to use as shown at the top of page 21.

How is energy measured?

Energy is measured by the amount of **work** it can do. Work may involve producing motion, overcoming gravitational force, a change of state (for example, water into steam), radiation (for example, production of light), overcoming frictional forces, or countering motion. The unit of work and energy is the **joule**. *The joule is the work done when a weight of one newton is raised to a height of one metre.*

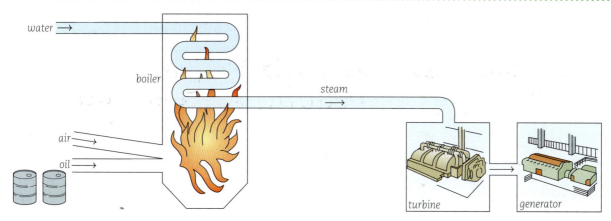

Generating electrical power from oil

What is nuclear energy?

In the late 19th century and the early part of the 20th century, scientists found that some elements emit light and other types of radiation. These elements are called **radioactive substances**. Furthermore, some of these substances change into other new elements because they emit **alpha** and **beta** particles. This process is called **radioactive decay**. Some elements are naturally unstable and these slowly break down, losing energy in the radiation emitted.

One of the denser metals, uranium, exists mainly in two forms, called isotopes. Isotopes are forms of the same element. They have identical chemical properties but their **nuclei** have different numbers of particles in them.

The lighter isotope of uranium has an atomic mass of 235 and the denser isotope has an atomic mass of 238. The lighter isotope can be made to disintegrate by bombarding it with neutrons. When this happens the uranium splits up into lighter elements such as barium and krypton, and releases two neutrons and an enormous amount of energy. This is the energy that binds nuclear particles together and is called **nuclear energy**.

Nuclear energy is the most concentrated form of energy known. In this type of energy matter is directly converted into energy according to Einstein's equation $e = mc^2$ where e = energy, m = mass, c = universal constant (speed of light).

An extremely large amount of energy is produced from a relatively small amount of matter.

The type of nuclear reaction described above is called **nuclear fission**, because matter is disintegrated. The other type of nuclear reaction known is called **nuclear fusion** and this gives a much higher output of energy. The nuclear reactions in the sun (page 194) are of the fusion type. Many of the nuclear weapons stockpiled during the Cold War also used nuclear fusion. For fusion reactions to occur continuously, a temperature of 60 million °C or higher is required. These temperatures exist typically in the centre of stars.

What are the dangers of nuclear energy? One gram of nuclear fuel is over a million times more powerful than one gram of any other ordinary fuel. It therefore has a tremendous power output, which can be used destructively in war, or usefully in nuclear power plants to help solve the world's energy problem.

However, some major problems remain to be overcome. When nuclear energy is produced in nuclear power stations, radioactive waste materials are also produced. These waste materials send out harmful radiation, which is damaging to all living organisms. There is a serious problem in finding adequate means of getting rid of this waste because it will remain harmful for hundreds or even thousands of years.

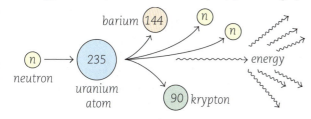

Where nuclear energy comes from

> ### Questions
>
> 1. Trace the energy that lights an electric bulb from its source, namely the sun.
> 2. Are sea waves powered by the sun?
> 3. Would you recommend the introduction of nuclear power plants in the Caribbean?

How can matter change?

How can matter be changed?

In the first few years of your school science course you will almost certainly have seen experiments in which matter has been changed. These changes usually involve one of the forms of energy – see page 20.

The most obvious way of causing a change in matter is to *heat* it. If you heat blue copper sulphate crystals, water is driven off, and a solid white substance (anhydrous copper sulphate) is left. If magnesium is heated, it burns, giving out very intense white light. A great deal of heat is given out too.

If *electricity* is passed through dilute hydrochloric acid, bubbles of gas are given off at each electrode. The electrical energy supplied causes this change.

We can also cause changes by *mixing* solutions together. These must contain chemical substances which react with each other. If iodine solution is added to starch solution a very striking blue-black substance is formed. This is a good test for the presence of iodine. You can also add a solid to a liquid, such as water or an acid. Magnesium metal will react with dilute acid and hydrogen is formed.

We can try to summarise this as follows. *Matter can be changed by supplying energy or by the chemical reaction of one substance with another. In each case, we start with one or more substances and we end up with one or more new substances.*

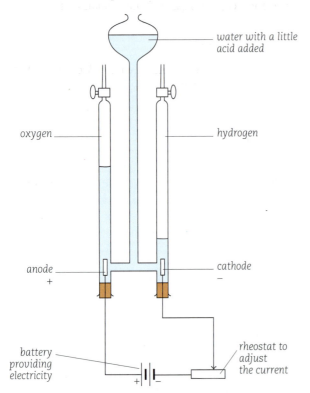

Electricity is used to break water down into its elements: oxygen and hydrogen

How can we show changes in matter?

When crystals of blue copper sulphate are heated, a white solid is formed. The gas given off condenses on the cool part of the tube and a colourless liquid is formed. The original blue crystals are hydrated copper sulphate; the white solid is anhydrous copper sulphate. The liquid formed is water.

When a piece of magnesium ribbon is held in a Bunsen flame, it begins to burn rapidly as soon as it is hot enough. Intense white light can be seen and heat energy is given out. The magnesium combines with the oxygen in the air to form a new substance, magnesium oxide. We can summarise the information we get from our observations by writing a **word equation**, but now we also say something about the energy involved.

(1) blue copper sulphate $\xrightarrow{\text{heat}}$ white copper sulphate + steam

(2) magnesium + oxygen $\xrightarrow{\text{heat}}$ magnesium oxide + energy

Magnesium and oxygen are called the **reactants**, and magnesium oxide is the **product**. In reaction 1, what is the reactant and what are the products?

State symbols tell us about the *physical state* of each reactant and product. Blue copper sulphate is a solid, so we can write this as copper sulphate (s). Steam is water in the gas state so we write this as water (g). Water itself is a liquid, so we write this as water (l). We can show changes in matter by writing more complete word equations, including the state symbols.

(1) blue copper sulphate (s) $\xrightarrow{\text{heat}}$ white copper sulphate (s) + water (g)

(2) magnesium (s) + oxygen (g) $\xrightarrow{\text{heat}}$ magnesium oxide (s)

If you include the energy involved you have added a little more information and the equation may be more useful.

If a substance is dissolved in water, we say that it is **in aqueous solution**. We write the symbol (aq) after the substance to show this. When ammonia gas is dissolved in water we write this as ammonia (aq). When starch solution reacts with iodine solution, we write the word equation as:

starch (aq) + iodine (aq) → blue-black substance (aq)

When you investigate chemical reactions, you should always try to write word equations for the reactions.

How can we classify substances?

On page 18 an element was defined as a substance in which all of the atoms are the same. We can define elements more simply. When substances break down into two or more simpler substances we say that they **decompose**. These simpler substances can sometimes be broken down into even simpler substances. At some stage it will not be possible to break the substances down any further. When this happens we say that the substances are elements: they cannot be broken down any further by chemical means. *An element is the simplest substance which can be obtained by decomposition.*

The substance magnesium oxide contains two elements – magnesium and oxygen. Blue copper sulphate contains four elements – copper, sulphur, oxygen and hydrogen. White copper sulphate does not contain water, so there is no hydrogen present. It contains three elements – copper, sulphur and oxygen. Both magnesium oxide and copper sulphate are examples of substances called **compounds**. Compounds consist of two or more **elements**. *The elements in compounds are combined together in fixed proportions by mass.* Compounds such as blue copper sulphate can be broken down into simpler substances by heat energy.

Elements are joined together chemically in compounds. Water is a compound containing the two elements, hydrogen and oxygen, in fixed proportions. It is possible to mix the two gases, hydrogen and oxygen, together, and obtain a mixture of the two gases. The properties of the mixture are not the same as those of the compound. The mixture can contain the gases in any proportions, and the gases are not combined chemically.

Mixtures can contain elements or compounds or some combination of the two. The substances in a mixture are not chemically combined.

Questions

1. Define the following terms and provide an example:
   (a) decomposition (b) element (c) mixture and (d) compound.
2. Which of the following are compounds: (a) water (b) carbon dioxide (c) rum (d) air and (e) sodium chloride?

How do substances combine?

When hydrogen reacts with oxygen a new substance is formed – water. The properties of this new substance are not the same as those of either hydrogen or oxygen. We can try to explain this by suggesting that there are new particles formed when the particles of hydrogen and oxygen react. These 'compound particles' are called **molecules**. On page 18 we defined a molecule as the smallest part of a substance which can exist and still show the properties of the substance. Water is formed when *two molecules* of hydrogen react with *one molecule* of oxygen to form *two molecules* of water.

The word equation is:
   hydrogen (g) + oxygen (g) → water (l)

We can show this in a diagram like this:

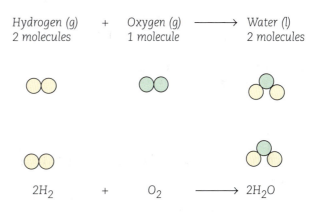

Carbon reacts with oxygen to form carbon dioxide. Carbon is a solid and oxygen is a gas. The word equation is:

   carbon (s) + oxygen (g) → carbon dioxide (g)

One *atom* of carbon reacts with one *molecule* of oxygen to form one *molecule* of carbon dioxide gas.

Carbon (s)	+	Oxygen (g)	→	Carbon dioxide (g)
1 atom		1 molecule		1 molecule

C + O_2 → CO_2

Most of the common gases are found as molecules. Each molecule of oxygen contains two atoms. Ammonia gas contains nitrogen and hydrogen. Each molecule of ammonia contains one atom of nitrogen and three atoms of hydrogen. When new substances are formed, new particles are formed. The new particles may be molecules but this is not always so.

Why is carbon so important?

What do you think the following things have in common: sugar, cheese, kerosene, gasoline, a compact disc and a plastic cup? One answer is that they all contain carbon. Carbon is a very important element in all living things, as well as in foods, fuels and plastics. Carbon is an element and it may be found in forms as different as diamonds (which are very hard and shiny) and graphite (which is soft and used as a lubricant). Charcoal is an impure form of carbon.

Activity | Where can we find carbon?
1. Heat a small piece of wood in a test-tube until there is no further change. What do you see remaining in the test-tube?
2. Hold a clean test-tube or tin lid in the yellow part of a Bunsen flame or a candle flame. What is formed on the outside of the test-tube or on the lid?
3. Your teacher will put a small quantity of white sugar in a small beaker and *carefully* add a few cm^3 of concentrated sulphuric acid. What happens to the sugar? What is left?
4. Put some kerosene in a bottle top, and then light it. Hold a test-tube or lid over the flame. What is formed on the outside of the test-tube or on the lid?

In each case, a black substance is formed, and it is reasonable to guess that this may be carbon. Most of the substances containing carbon are called **organic compounds**. You have looked at some of the properties of a few compounds – but there are many others containing carbon.

How can carbon form so many compounds?
Methane is an example of a simple compound of carbon. It contains carbon and hydrogen. One molecule of methane contains one atom of carbon and four atoms of hydrogen. Turn back to page 18 to remind yourself about the meaning of the terms atom and molecule. The diagram shows a model of a **methane** molecule.

Now look at the diagram of an **ethane** molecule.

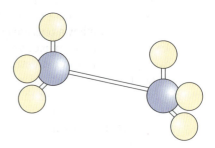

Ethane

Questions
1. How many carbon atoms are there in an ethane molecule?
2. How many hydrogen atoms are there in an ethane molecule?
3. How many hydrogen atoms are joined to each carbon atom?
4. Are the carbon atoms joined together?

Methane and ethane are examples of compounds which contain carbon and hydrogen only. These are called **hydrocarbons** (see pages 188–9). There are many carbon compounds in which the molecules contain carbon atoms joined to *four* other atoms. This is one reason why there are so many carbon compounds. The molecule of a carbon compound may be quite small – methane, for example – or larger, for example, the sugars. There are also very large molecules called **polymers** (see pages 204–5).

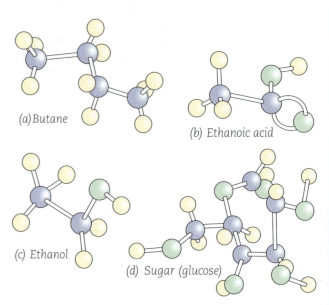

(a) Butane
(b) Ethanoic acid
(c) Ethanol
(d) Sugar (glucose)

Key
- Carbon atom
- Hydrogen atom
- Oxygen atom

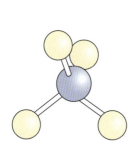

Methane

Look at the diagrams at the bottom of the previous page again. Each diagram represents a molecule of a different carbon compound. For each molecule, count the number of carbon atoms. Then work out how many atoms are joined to *each* carbon atom in the molecule of the compound.

The molecule shown in (a) is a hydrocarbon – it contains hydrogen and carbon only. There are three other compounds shown in (b) to (d). Each of these compounds contains *oxygen* as well as carbon and hydrogen. You can check this by looking at the key at the bottom of the previous page. Many of the compounds containing carbon, hydrogen and oxygen only are called **carbohydrates** or **fats**. Another important element found in many compounds of carbon is *nitrogen*. This is found in **proteins**, for example. You need to know about proteins, fats and carbohydrates when you are thinking about eating a healthy diet (see pages 74 and 76).

What are some of the important groups of carbon compounds?

We have used a number of different names for groups of carbon compounds. Although there are many groups, we shall concentrate on three in particular.

Fuels Fuels are compounds which burn in air. When fuel burns, energy is given out. Gasoline and charcoal are both examples of common fuels. Can you think of a few other examples of fuels?

When a fuel is burned new substances are formed and energy is given out. We have to think about **where** this energy comes from. It obviously has something to do with the chemical composition of the fuel, since different fuels burn in different ways – some give out a great deal of energy, and others less. You should remember that when new substances are formed, energy is either given out or taken in. When a fuel such as kerosene is burned, the new substances formed include compounds of carbon and hydrogen – as you might expect. Look at pages 188–91 for more information on fuels.

Foods There are three classes of foods: carbohydrate, protein and fat. Each of these contains carbon. Sugar is a carbohydrate, and therefore contains carbon, hydrogen and oxygen only. Proteins contain the same elements but also contain nitrogen. Some proteins contain sulphur as well. Plants can make food during **photosynthesis** (see pages 66–7). Energy is taken in as the new compounds are formed. This energy can be released during respiration in plants and animals – see pages 104–7. Foods act like fuels for living things.

Polymers These are compounds which contain very large numbers of atoms in their molecules. For example, the substance known as polythene – or more correctly, polyethene – contains very large numbers of carbon and hydrogen atoms. There are two hydrogen atoms for every carbon atom.

The diagram below shows a very small *part* of a model of a polyethene molecule. It shows how the atoms are joined together (see also page 204).

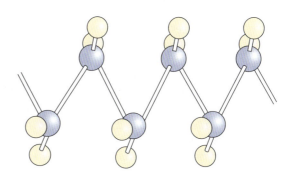

There are many polymers in nature, as well as polymers prepared industrially (synthetic polymers). Examples of natural polymers are cotton, silk and wool. There are now many synthetic polymers, such as plastics (see pages 204–5).

Objects made from a common polymer: plastic

Questions

5 What is meant by (a) a hydrocarbon and (b) a carbohydrate?
6 Which fuel is used for (a) aeroplanes (b) trucks and (c) heating?
7 Name three common plastics.

Why study science?

Imagine being a young person anywhere in the world five hundred years ago and being asked, 'Do you want to learn to write?' You might think or say, 'Who, me? Do I want to be a scribe?' Or if you were living one hundred years ago and you were asked, 'Do you want to learn to drive?' Again you might say, 'Me? Do I want to be a driver?'

In today's world, it is essential to be literate and in developed countries nearly everyone expects to have a driver's licence.

Now ask yourself, what good reason could there be to study science if you are sure that you do *not* want to be a scientist?

The kind of science presented in this text is called **integrated** science because it tends to involve many branches of science in the discussion of the various topics. For example, a study of movement in animals would involve **biology** for the muscles and nerves, **physics** for the lever action and forces and **chemistry** for the breakdown of substances to supply energy.

Can you guess what branches of science are involved in plant growth?

What mental skills are needed to study science?

The 'facts' of science can be read from books and memorised, but these can be of little value if the facts are not properly understood. There are higher-order skills beyond the knowledge of facts that will enable the student to examine information critically, to reorganise information and come up with new ideas. These higher-order skills are called **comprehension, analysis, synthesis** and **application**.

Science is very practical. In some cases the only way science can be understood is through practical work and demonstrations. When you do something practical, the experience becomes your own and you will understand and learn much quicker.

Practical approaches to the study of science involve skills such as **drawing, manipulating, observing** and **recording** (see pages 30–1 and pages 34–5). You will develop these skills while working through this book.

When important scientific principles are taught the learning process should involve many areas of the brain so that you use as many higher-order learning skills as possible.

While you study science, keep in mind that you have many different abilities or **aptitudes** which influence the way you learn. These aptitudes can also be called **intelligences**. Learning situations should involve as many intelligences as possible. These may include: interpersonal, intrapersonal, bodily-kinaesthetic, linguistic, logical-mathematical, musical, naturalistic and spatial intelligence.

Should the artist study science?

Artistic people are usually good at visualising and expressing their ideas through drawing, painting and designs. They have a high spatial intelligence. They can benefit from the study of science, if only to get a better understanding of the properties of the materials they have to work with. A good grounding in colour mixing, staining, fabrics, lighting, angles, the principles of photography, etc. can all be very beneficial to the artist.

Should the sportsperson study science?

The sportsperson is usually very aware of how the body works. They have a high bodily-kinaesthetic intelligence. They can benefit from the scientific knowledge of the body's functioning and of the corrosion of materials.

Should the musician study science?

What the musician can gain from the study of science is a good understanding of the principles of sound production, how the various instruments generate different sounds and how sounds are recorded and reproduced.

Are there mental skills specific to science?

The study of science will help you improve your linguistic and logical-mathematical skills. These skills benefit people in all sorts of fields. For example, lawyers who argue their cases must do so logically and the jury, made up of ordinary people, must use logic and reason to come to a conclusion. The restaurant chef too must realise the logical outcome of the mixing of the various ingredients to produce the desired outcome.

What science does a chef need?

Can the study of science help to develop new skills?

The best approach to the study of science is an experimental approach. You can do experiments to test and verify logical outcomes of reactions. The results can be presented in tangible or concrete forms rather than abstract form. Hypotheses can be put forward and tested and the results can be displayed (pages 32–3). In this way you can make connections between cause and effect and you will be able to distinguish between the reasonable and the unreasonable. Disguised, logical-sounding arguments may also be unmasked.

What other skills can be used in the study of science?

Students should use all their skills when studying. If you are musical, you could prepare amusing jingles to learn formulae or to link concepts. If you love nature, you could accept responsibility for the animals and plants in the classroom. If you have high bodily-kinaesthetic intelligence, you will probably easily remember the rules governing electric current, magnetic and electric field from the finger positions on your hand.

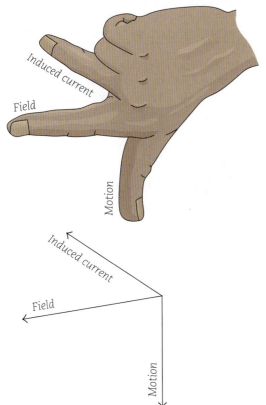

Induced current

If you are creative and have a high spatial intelligence, you will gain much from the preparation of diagrams and maps illustrating and linking concepts.

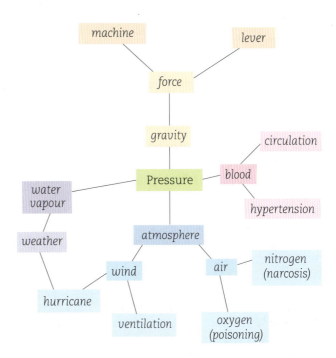

Can science help you make decisions?

In a democracy, the voting public choose the government. In the Caribbean, as the differences between the political parties become more and more vague, the election of governments tends to be based more on current issues than political ideologies. These issues could include: Is there adequate waste disposal? Is there enough water? Shall we clear more forests? Are we protecting or destroying the environment? Are we prepared for the next hurricane season? Shall we drain another mangrove swamp? What about the wildlife in the swamp? Are we controlling the spread of disease?

If the voting public had a good understanding of basic science and scientific principles, they could distinguish facts from fallacies more easily. They could also use their scientific knowledge as a basis for sound arguments.

Questions

1. Discuss how science can help one to become a better (a) taxi driver (b) nursery nurse (c) security guard and (d) dental assistant.
2. Can science help one develop better argumentative skills? Discuss.

How do we use information?

Why is information needed?
Living organisms, even single cells, are aware of their surroundings. Being sensitive to stimuli, and making the appropriate response, are important characteristics of life (see pages 6–7). Complex organisms have highly developed sense organs. The structure and functions of some of these organs will be described later (pages 262–85).

Humans attach so much importance to gathering information that a considerable amount of time and effort have been devoted to devising and developing apparatus and instruments which can extend human senses.

How is information stored by living organisms?
If you touch a hot stove, your sense of touch makes you painfully aware of the fact and an immediate response is called for. However, if you see that it is raining and have no intention of going out, no action is taken, but that information may be stored in your brain. Throughout your life, you receive information through your senses and add this to your information store or **memory**.

All living organisms have an information store. In single cell organisms, the information store is **DNA**. This ensures that the basic functions of the cell are carried out. Organisms with a nervous system also have an information store in their **nervous tissue**. Organisms with **brains** have a very great capacity for storing information. The human brain is large and particularly good at storing information.

Long ago, in some human societies, human memory was the most important and perhaps the only store of information. This information was passed on from parent to offspring, through countless generations. In some tribes, there were a few people with the entire history of the tribe recorded in their memories. Think what would happen if these few individuals suddenly died.

Anthropologists think that one of the most significant leaps forward in the development of human society was achieved when writing was invented. Once humans could write, information could at last be preserved outside the human brain, it could be shared and studied by a large number of people and could more easily influence the development of society.

How is information stored today?
The **written word** in newspapers, magazines and books is still the most important means of recording and storing information. The written word forms the basis of many kinds of information systems.

Other information systems include **photographic film**, **gramophone** records (old and new optical types) and **magnetic (video) tapes**. More recent devices include the **computer chip**, **hard and floppy (magnetic) discs**, the **(optical) compact disc (CD)**, **CD-ROM** and **DVD** (digital video disc).

How is information retrieved from different stores?
You must be able to read to get information from books. In some cases, however, specialised equipment is needed. For example, a video player and viewer (television) are needed for videotape, and for photographic reel films, projectors are required. The computer chip is part of a computer and special procedures are required to get the information from the computer chip (memory), floppy disc, hard disc or CD-ROM.

Activity | How can you find a particular book?
Visit a library or a bookshop and see how the books are arranged. Try to find a book by a particular author. If you do not find it, can you be sure it is not there?

The arrangement in a library or bookshop enables you to find what you are looking for very quickly, once you understand the system. It also helps you to see at a glance how many other books by the same author or on the same subject are available. You can also make comparisons between the numbers of books there are on, for example, football compared with cricket.

Would you be able to do this if the books were in no particular order? By arranging objects or information in an orderly and systematic way, comparisons can easily be made. New relationships can be established and new and useful conclusions can be drawn.

The ways of comparing information, and searching for and selecting relationships out of given sets of information (data), are called **data processing**. The human brain has created the computer to help deal with these problems.

What do computers do?
Computers are particularly useful because of the speed at which they can select and display information. Of course, it is essential that the information is stored in a special way. **Magnetic discs** store computer information in a special **binary code**, which is understood by the computer. Information may be selected for display or modified according to the instructions given in a **computer program**, which controls the computer's operation.

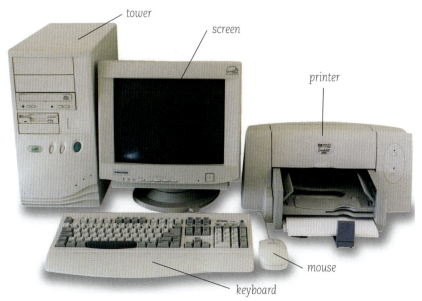

A personal computer (PC), also called a micro-computer, with accessories

Screen – (sometimes called a visual display unit, VDU) displays information in the form of letters, numbers or graphics

Tower – containing hard drive, floppy disc drive, CD-ROM drive and a central processing unit (CPU)

Keyboard – usually like a typewriter, for communicating with the CPU

Mouse – input device for rapid communication with the CPU

Printer – prints out data/information on command from the CPU

What are computers used for?

Electronic games machines Electronic games machines may be computers built specially for the purpose. In some cases, special screens are constructed for displays. Games programs allow different amounts of user interaction (active participation). They are also able to keep records of responses and scores.

Educational software A large number of learning programs is available. Some of these programs are simulations of physical systems which are too difficult to explore in reality. A student can learn a lot about the system by experimenting with the simulations.

Other types of programs are used for data analysis and data processing. These consist of exercises in the use of language or mathematics, or simplified versions of the programs used in business.

In business In offices, manual and electronic typewriters have almost all been replaced by computers with **word-processing** features.

A modern supermarket has hundreds or even thousands of different items on display. Records have to be kept of items sold, replacements have to be made from stock, new stock has to be ordered, prices have to be controlled and the shelf-life of perishable food has to be monitored. This kind of control requires many trained people unless a computerised system is used.

There is a computerised system for coding every item in a supermarket using **bar codes**. The bar code is read by the 'magic eye' at the checkout.

A bar code

The information is fed directly into a central processing unit (main computer) and a stock inventory is continually updated.

The internet One of the most spectacular and interesting uses of the personal computer (PC) is to access the international information network, called the **internet**. The internet is a network of linked computers. You can access the internet by using a modem which is linked to your PC. Text, sound and pictures can be sent to or received from anywhere in the world in an instant.

Questions

1. Discuss ways in which information can be recorded.
2. What special instruments and skills are required to retrieve recorded information?
3. Find out how the computer may be useful in an industry or business near you.
4. If you had a personal computer, how could you use it at home or at school?

Practical work

School-based assessment
Throughout your course your teacher will be assessing the skills related to practical work. School-based assessment (SBA) provides a framework for this process, and at the same time enables students to gain credit towards their final examination grades.

Keeping a practical notebook
Each student has to keep a practical notebook. The notebook should include a list of contents showing all the practical activities undertaken, the date on which each activity is carried out and the page on which it is recorded. The teacher will add the skills which have been assessed and the marks given.

The Caribbean Examination Council (CXC) will require a sample of the students' practical notebooks to be sent in, and will indicate the names of the students whose books should be submitted. All other notebooks must be kept by the school for at least three months after publication of examination results.

Throughout this book you will find instructions for activities. Your teacher will select those that you should do. The activities and your reports should then be recorded in your practical notebook.

Skills to be assessed
The SBA skills are described below and there is more information on pages 32–7. Also refer to the CXC Integrated Science (single award) syllabus.

Observation (O) The ability to:
- select observations relevant to the particular activity
- use the senses, in a safe way, and perceive objects and events accurately with the necessary attention to detail
- make accurate observations and record and report them in a suitable way.

For more on observation see page 34.

Make a list of at least 20 observations relating to this scene

Recording and Reporting (RR) The ability to:
- accurately record observations of living things and changes in experimental set-ups
- report and recheck unexpected results
- present a written and oral report, drawing or graphical representation which is clear, precise and pertinent to the investigation
- use appropriate headings for the sections of a report and arrange them in logical sequence
- use appropriate forms of reporting – diagrams, tables, line graphs, pie charts, bar charts and histograms
- prepare graphs with appropriate scales, adequate labels, accurate plotting of points and drawing of the best straight line or smooth curve.

For more on recording and reporting see pages 36–7. Along with observation, these are assessed together as **observation/recording/reporting (ORR)**.

Drawing (D) The ability to:
- make large, clear, accurate, drawings/diagrams of specimens, apparatus or models
- use pencil and add label lines and appropriate labels
- include magnification, view/section where appropriate.

For more on drawing see page 34.

Manipulation (M) The ability to:
- correctly and carefully handle equipment and materials
- set up and use carefully and competently laboratory apparatus such as Bunsen burners and reagent bottles
- prepare specimens and material for observation/investigation, for example, make slides, and construct simple series and parallel circuits
- show mastery of laboratory techniques, for example, simple distillation, heating of solids and liquids in test tubes, detection of gases, and filtration.

For more on manipulation see pages 34–5.

Measurement (M) The ability to:
- use equipment correctly and read scales accurately

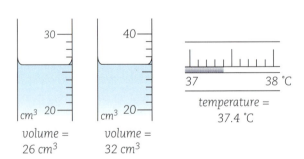

Scales on measuring cylinders and clinical thermometer

- use basic measuring equipment with competence and skill and make accurate readings, for example, ruler, balance, thermometer, measuring cylinder, burette, syringe, watch/clock or any timing device, voltmeter, and ammeter
- give readings with the correct units.

For more on measurement see page 35. It is assessed under **manipulation/measurement (M/M)**.

Planning/Designing (P/D) The ability to:
- plan an investigation
- suggest appropriate hypotheses
- outline a plan to test your own or other hypotheses, in an appropriate sequence of operations, within time allotted
- identify important variables to be controlled or measured
- suggests controls where appropriate
- state possible outcomes and/or the way in which data will be analysed and presented
- describe what is to be done with appropriate practical detail
- modify your original plan in the light of difficulties experienced in carrying out experiments or obtaining unexpected results.

For more on planning and designing see pages 32–3.

Analysis and Interpretation (A/I) The ability to:
- identify parts of a whole, recognise patterns and relationships, and extract information
- make necessary and accurate calculations
- make logical inferences from observations and data
- predict from data
- draw conclusions based on the data
- recognise the limitations and assumptions of the data
- evaluate data (including sources of error).

For more on analysis and interpretation see pages 36–7.

Notes:
1. Your teacher will carry out the practical assessment for SBA during terms 2, 3, 4 and 5.
2. Plotting and drawing of graphs are assessed under **observation/recording/reporting**, but inferences from graphs are assessed under **analysis and interpretation**.
3. The following mathematical skills are needed throughout your CXC Integrated Science course:
    - the four basic operations ($+$, $-$, $\times$, $\div$)
    - decimals
    - change of subject of simple formulae
    - substituting values into simple formulae
    - means and modes
    - graphs, histograms, charts and tables.
4. SI units are used on all examination papers. These units include kilo, centi, milli, etc.

Profiles of the CXC examination

For Integrated Science you have to show various skills and abilities which are called profiles. How well you do on the **profiles** is recorded as part of your examination report. Refer to the CXC Integrated Science (single award) syllabus for the details. A simplified summary is given below. The profiles are:

Knowledge and Comprehension (KC) The ability to:
- recall, state and identify basic facts and ideas
- show understanding by selecting appropriate facts and ideas and giving examples for familiar situations.

Use of knowledge (UK) The ability to:
- change data for new situations, classify living and non-living things, and use mathematical formulae
- identify the parts of a whole and the relationships between them – for example, studying a local habitat, and identifying factors causing change and how they interact
- combine parts to form a whole – for example, studying parts of the systems in a mammal, making predictions and solving problems
- evaluate (give reasons for the value of) ideas and actions, and make recommendations based on the likely outcomes of different actions.

Practical skills (PS) These will be assessed by the teacher as part of the SBA which counts up to a quarter of your final CXC grade. The headings are:
- Observation/Recording/Reporting (ORR)
- Drawing (D)
- Manipulation/Measurement (M/M)
- Planning/Designing (P/D)
- Analysis and Interpretation (A/I).

The CXC examination

Revise in plenty of time and do the practice questions starting on page 321.

CXC Basic proficiency

Paper I	60 multiple choice items	$1\frac{1}{4}$ hours
Paper II	Part A: 4 structured questions	
	Part B: 2 extended essay-type questions or handling of data	2 hours

CXC General proficiency

Paper I	60 multiple choice items	$1\frac{1}{4}$ hours
Paper II	Part A: 3 structured questions	
	Part B: 3 extended essay-type questions or handling of data	2 hours

School-based assessment Practical work is assessed by your teacher. The score makes up to a quarter of your total mark.

Setting up experiments

Why do we do practical work?
Practical work is an important part of science. If you only learn theory you may learn something about science, but you will not begin to understand what science is really about. We use our practical skills to find out how things behave.

Hypotheses Scientists often find problems or make observations which they cannot explain. In order to try to solve the problems or explain the observations, scientists put forward **hypotheses**. A hypothesis is a statement which makes an intelligent guess to answer a question. You cannot just make any guess – it has to be likely to be correct, and based on information you already have.

The question may be one which is important for an industry. Imagine that you want to sell fruit juices in cans. These fruit juices contain acids. The problem is: Which materials could I use for these containers? You may then make the hypothesis that certain metals could be used. Then you would have to find out how different metals react when they come into contact with the dilute acids before you can make a choice. In other words, you would experiment to *test* the properties of the metals. Experiments are important because they allow us to test our hypotheses. The *results* of the experiments may lead us to other problems, and therefore more hypotheses. For example, you would have to find out if the chosen metal was available, and what thickness you should use in relation to safety and price, and so on.

Laws Scientific laws are general statements that have so far been shown to be true in the situations under which they have been tested. But we do sometimes modify laws.

Experiments Many of the experiments which you do in school have been done many times before. This does not mean they are a waste of time. They are important because *you* are carrying them out. You behave like a scientist. You have to follow instructions, plan your experiment and handle pieces of equipment (pages 34–5).

There are *skills* which you have to develop. One of the most important of these is **observation**. Did you know that the same event is often described very differently by two different people? This is important not only in science but also for the police. If a suspect cannot be identified by witnesses, there will be no chance of showing that the suspect committed the crime!

When you have observed what happens during an experiment, you need to **record** these observations. Then you need to think very carefully about your observations to see if you can reach a **conclusion**.

There are two other very important points to remember.
- If something does *not* happen, this may be just as important as when something does happen. In the fruit juice problem it was important for us to find a metal that does *not* react with an acid and this tells us something useful about the properties of that metal. All observations may be important.
- Also, when you carry out experiments and record results there may be some experimental error. This is the reason why we repeat experiments several times. This is especially important when using living things. So if your results are different from other people's, record them, but then check them again. Try to explain why they are different, and then discuss them with your teacher.

Why is our method so important?
Science involves a way of looking at the world around us. Scientists look for patterns. The patterns may be in the properties of metals, the classification of living things, or the conduction of heat by solids. Some scientists investigate the behaviour of the solar system, while others investigate the particles which are found in atoms. They all use the same method to try to solve problems. They make hypotheses which they can try to test by experiment.

How can we try to solve scientific problems?
Many scientific problems arise in the world outside the laboratory. We have already discussed the problem of canning fruit juices. Another problem might be: What happens to the air around us when we burn fuel in cars, buses and trucks? We could try to answer this for a particular area of a town, and it would be possible to test this experimentally. But, we might try to answer this for the atmosphere in general, and then the problem is much more difficult. You may have read about lead being removed from gasoline. That was the result of answers to problems such as the one above.

In every case, it is important to decide exactly what the problem is, and then how to try to solve it. We need a **method** for solving problems in science. We can try to summarise a useful method in a simple diagram.

1. **The problem**
   This may lead to
2. **A hypothesis**
   which can be tested by
3. **Experiment(s)**
   from which we make
4. **Observations**
   which, when we have recorded them properly, and thought about them, should lead to
5. **Conclusions**
   which may help to solve the problem, or, may lead to further problems

A scientific method

What happens in controlled experiments?

It is not enough just to do experiments. We have to make sure the experiments provide as much useful information as possible. At the end of the experiment we want to be able to say what particular condition or **variable** caused the change we have observed. For example, to investigate what causes the rusting of iron you would start with a number of iron nails, and make sure they are investigated under conditions which are *each* different in only one important way.

You may have guessed (made a hypothesis) that both air and water have to be present for rusting to occur. You can then set up experiments to check this hypothesis. One nail can be exposed to the atmosphere which contains air and water vapour. Then other nails should be exposed to air only, water only, and neither air nor water. In this way, you can find out whether air or water alone is enough to cause iron to rust, or whether both are needed (pages 212–13).

What this means, in scientific language, is that we try to make sure that *only one variable* is changed at a time in any experiment. We can therefore ensure that it is the effect of that variable which is investigated in the particular experiment and that variable that causes any change we observe. We call this a **fair test**.

Why do we need to plan experiments?

You may have to plan your own experiment, or you may be given instructions. Read the instructions before you start the experiment. Then you will need a period of discussion, which should include any safety precautions. Decide which pieces of apparatus and other items are required.

Before starting make sure that the equipment is properly set up and that all safety precautions are being observed. This may include wearing laboratory coats and safety goggles. Do not plug anything into the mains electricity until your teacher has checked your work. Make sure you have the necessary skills to use the equipment correctly and safely (pages 34–5).

Record your observations *as you make them* in your **practical notebook** (page 30). Remember that CXC can ask to see your notebook. Make sure that the discussion of your recorded observations is sensible and to the point.

An example of a controlled experiment

We set up a controlled experiment so that we can test only *one* variable at a time; that is, we carry out a fair test.

Planning and Designing

Problem   Do seeds need water to germinate?
Hypotheses   Seeds left without water will not germinate.
Seeds kept moist will germinate.
(You may be given the hypotheses, or have to suggest your own based on a problem.)

Identify the variables

- **A manipulated variable** is the variable we change. In this example the manipulated variable is moisture. We leave seeds with or without moisture. Use large numbers of seeds to reduce the effect of any chance differences. (*Note*: you change *one* variable only.) This manipulated variable is also called the **independent variable** – it has its values determined by the experimenter.
- **Constant variables** are the variables we will keep the same in each part of the experiment. We will use the same number of just one kind of seed, and the same temperature and amount of light for each set-up. Then, at the end of the experiment, we will know that it was the presence or absence of moisture that accounted for any differences we observe. (Be very careful to consider all the important variables which *might* affect the results.)
- **Responding variables** are the result(s) we look for and record. They are also called the **dependent variable** – as they depend on the other changes we have made. In this case it will be the number of seeds which germinate. You need to say what you expect to happen. You need to state possible outcomes and the way in which you will record them and analyse them.

Set up the experiment. Work out what you will need and the order of the steps for carrying it out in the time you have been given. Describe what is to be done with appropriate practical detail. Modify your plan if you have difficulties or if you obtain results different from those of other class members.

Carrying out the experiment

Observations   These may be the number of seeds that germinate, or changes in texture (such as the seed coat's appearance). Colour changes are often important.

Conclusion   What have you learned about the original hypotheses? Which one(s) is/are correct? Do you need to do more experiments?

Questions

1. In a simple electrical circuit (see page 174), what are the variables that determine how much current flows?
2. What do you understand by each of these terms?
   (a) Problem (b) Hypothesis (c) Variable (d) Fair test
3. You have noticed that a new kind of bread, advertised as having 'no preservatives', develops mould quickly. Set up a controlled experiment to compare the rate at which mould develops on the new bread and your usual brand.

How do we use equipment and materials?

When we carry out practical work we need certain skills. These skills are assessed under the SBA headings of **observation**, **drawing**, **manipulation** and **measurement**.

Observation
Observation means using all your senses safely. Make your observations as accurate and detailed as possible.
1. Sight: colours and shades (and how they change), shapes, sizes (either compare or measure them). Measure or count when appropriate.
2. Touch: texture, such as smooth, hairy, tough, soft. Check with your teacher for safety.
3. Hearing: compare to known sounds, for example, a metal that rings like a bell when struck.
4. Taste and smell: record for foods and items only as given by the teacher, such as sweet, sour, salty or bitter. Waft a little of an unknown gas towards the nose. Never taste or smell chemicals without the teacher's instruction.

Observations are written up (see **recording and reporting**, pages 36–7), or made into **drawings**.

Drawing
1. Use a sharpened HB pencil, not a pen or crayon.
2. Make the drawing big. Leave space for labels on both sides.
3. Make a faint outline first, about twice the length of the object. This drawing will have a magnification of × 2.
4. Make the parts of your drawing in the same proportion as the parts of the object.
5. Make your final lines continuous. If they are straight lines on apparatus, use a ruler.
6. Do not use shading, it often hides bad drawing.
7. Use a ruler to draw label lines. They should be straight, to the right or left of the drawing and should not cross each other. They should stop immediately next to the part being labelled. They do not need arrow heads.
8. If you are drawing apparatus, draw a single cut edge of a longitudinal section.
9. Label in pencil or ink.

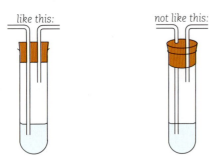

10. Give your drawing a title which describes the apparatus or the specimen you have drawn.

Manipulation
Correct handling of glassware and chemicals
1. Make sure all glassware is clean before use, and keep it a safe distance from the edge of the bench at all times.
2. When carrying reagent bottles or measuring cylinders keep them upright and use two hands.
3. Use only small amounts of chemicals. Do not return unused chemicals to the bottle.
4. When heating test-tubes, have the neck pointing away from people. Move the test-tube in the flame.
5. Take care how you hold and put down hot glassware.

Correct handling of electrical appliances
6. Set up circuits and check them with your teacher before making the final connections.
7. Use dry hands to plug in electrical appliances.
8. Do not plug in more than two appliances per socket to prevent overloading. See safety rules (pages 180–3).

Careful handling of living things
9. Some organisms sting or bite; handle everything carefully.
10. Treat living things with respect. Lift soft organisms carefully, so that you do not squash them.

Cutting sections and making slides We may want to look at the structures inside a stem. To do this use a safety razor and either cut *across* the stem (**transverse section**), or lengthways *down* the stem (**longitudinal section**).

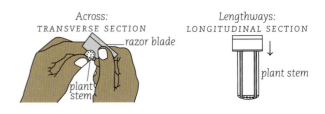

1. You should cut *very thin* sections, or parts of sections and put them in water in a dish.
2. You may want to stain your section or other material by putting it in a drop of stain. Then transfer the section to a drop of water on a microscope slide.
3. Using a pointed needle or a large pin carefully lower a cover-slip over the material.

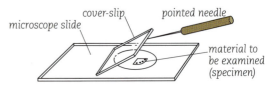

Using a microscope A microscope magnifies materials so that we can see their structures more clearly. The total magnification is the magnification of the objective lens (near the slide) × the magnification of the eyepiece lens (near the eye).

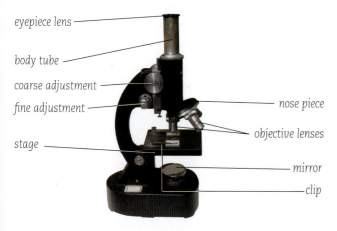

1. Carry the microscope with two hands and put it down carefully onto a flat surface away from the edge of the bench.
2. Clean the lenses with soft tissue paper.
3. Put the × 10 eyepiece lens into the top of the body tube.
4. Screw the × 10 objective lens into the nosepiece, and place it above the hole in the stage.
5. Adjust the condenser so that its top is level with the top of the stage.
6. Look through the eyepiece, open the iris diaphragm and adjust the mirror so that it collects light from a light bulb or patch of sky.
7. Place the slide on the stage and secure it with the clips so the specimen is centred over the opening in the stage.
8. *Looking from the side* turn the coarse adjustment *down* so the objective is close to the slide.
9. *Looking through the eyepiece*, turn the coarse adjustment *up* until the object comes into focus.
10. Use the fine adjustment, if necessary, to bring the specimen into clear focus.
11. When you make your drawing the magnification is worked out in this way:

$$\text{Magnification} = \frac{\text{length of drawing}}{\text{length of object}}$$

If the object is 2 cm long and the drawing is 4 cm long, the magnification = $\frac{4}{2}$ = × 2

If the high power of the microscope is needed, *first* focus under low power as described. Then swing the × 40 objective lens so that it lies above the specimen. Use the fine adjustment *only* to bring the object into focus.

Using a hand lens Keep the hand lens close to your eye. Move towards the object until it comes into focus.

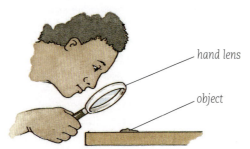

Using a Bunsen burner
1. Twist the sleeve on the barrel so that the air hole is open.
2. Hold a lighted match so that the bottom of the flame is just below the top of the barrel.
3. Turn on the gas and the burner will light.
4. Adjust the air by twisting the sleeve so the flame is the desired shape and colour, for example, an inner blue cone and an outer almost invisible cone. If the flame is large and luminous let in more air. (See also page 191.)

Measurement

You should use the following with competence and care to make accurate readings: ruler, metre rule, balance, thermometer, measuring cylinder, burette, syringe, watch/stop clock, newton-meter, voltmeter and ammeter.

1. Handle all glass measuring apparatus carefully and clean them before use. Rinse them with a little of the liquid to be measured.
2. Observe the scale carefully and work out the value of each small division by using the relationship:

$$\text{Value of small division} = \frac{\text{value of large division}}{\text{number of small divisions}}$$

3. Read the instrument straight-on at right angles. For liquids, for example in a measuring cylinder or burette, read the lower level of the curve, the meniscus.
4. Carefully give a thermometer a quick flick before use to shake the mercury down. Leave the thermometer in place for two minutes before taking a reading. Read the top of the curve in a mercury-containing thermometer.
5. Observe the scale and record the readings to the nearest small mark, together with the correct units, like this:
Thermometer: Temperature to nearest 0.1, 0.5 or 1.0 °C
Ruler/metre stick: Length to nearest 1 mm (0.1 cm)
Measuring cylinder: Volume to nearest 1 cm^3
Stop clock: Time to nearest second (s)
Lever-arm balance: Mass to nearest 1 gram (g)
Spring balance/newton-meter: Weight to nearest 0.1 or 1 newton (N)

How do we use our results?

How do we get the most out of experiments?

When you are carrying out experiments, you need a number of **practical skills** (pages 34–5). You need other skills to make the most of your practical work. These skills are assessed under the SBA headings **ORR (observation/recording/reporting)**, and **A/I (analysis and interpretation)**.

The main skill you will use during and after experiments is **observation**. You need to be able to describe *precisely* and *accurately* what you observe. For example, when dilute acid is added to a carbonate, a colourless gas without a smell is produced. This gas extinguishes a burning splint, turns moist blue litmus faintly red, and turns lime water milky.

When writing up practical work, you should have a **method** of approach.

- Firstly, write about *why* you did the experiment. This is usually because you wish to solve a particular **problem**, or want to investigate a **hypothesis** (pages 32–3).
- Secondly, describe *how* you did the experiment. This should include the apparatus and/or chemicals which were used, or the living organism studied. If necessary, it should include diagrams of the apparatus.
- The third part should be a **record** of your observations. This is a written summary of what you have seen, such as the temperature changes in a cooling liquid – we call these observations the results. A table of **results** for the cooling of naphthalene might look like this.

Time/min	Temp/°C	Time/min	Temp/°C
0	88.0	10	80.0
1	85.5	11	79.9
2	83.2	12	79.9
3	82.0	13	78.5
4	80.3	14	77.0
5	80.1	15	76.8
6	80.0	16	76.0
7	80.0	17	74.8
8	79.9	18	74.6
9	80.0		

You then need to show how you have thought about your results. This may mean looking for **patterns** in the results. This can often be easier if the results are presented in a way that reveals patterns, such as **graphs**. This discussion of the results is the most important since you are being asked to think about what the results *mean*.

The final stage really involves going back to the first. Have you answered the question or solved the problem? If so, you can write a **conclusion** to your experiment. If not, your conclusion may suggest other hypotheses to be tested. Many more experiments may be needed; one experiment cannot prove or disprove a law.

Questions

1. Write a brief summary of the method to be used when writing up experiments.
2. Choose one experiment you have done. Explain clearly how the *conclusions* were obtained from the *results*.

How can we tabulate results?

You need *method* in your approach to scientific problems. Part of this method involves *planning* experiments carefully (pages 32–3). One important part of the plan is how you intend laying out your results. This means knowing the type of observation or measurement you will take.

A **table of results** is often a good way of summarising observations or measurements. For example, you may be investigating the temperature change in a liquid as it cools down. If you heat a liquid such as naphthalene in a boiling tube, and then allow it to cool with a thermometer in the liquid, you can measure the temperature of the liquid at regular intervals, such as every minute. The table of results might look like this.

Time/minutes	Temperature/°C

When you have completed the experiment, you will have a table with times and temperatures. You can look at the figures and see if there are any obvious patterns. But looking at the figures is not enough. You will have to decide if the results can be presented in such a way that you can obtain conclusions from them.

How can graphs be drawn and used?

Let us assume that you have obtained a table of results like that on the left. You will have to decide about the **axes** of a graph from these results. You are interested in how the *temperature* of the liquid changed with *time*.

- Time is the independent variable. You have decided its values by taking readings every minute. It is put along the **horizontal x–axis**.
- Temperature is the dependent variable. Its values depend on the time at which you take the readings. It is plotted on the **vertical y–axis**.

You can decide on the **scales** for each axis. Try to make use of as much of the graph paper as possible; do not cram the graph into a corner. Scales do not have to begin at zero.

Once you have decided on the scales, and recorded these on the axes, you plot the **points** on the graph. Record each point as a dot, and surround each dot with a small circle. The graph from the table of results would look like this.

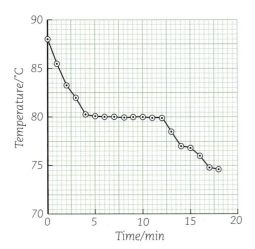

You should draw the best line or curve possible through the points. This does not mean that you join up each point to the next. Now try to **interpret** the graph. You should be able to read off the *freezing point* of naphthalene. As you can see the temperature stays about the same over the period minutes 4 to 12. During this time liquid naphthalene changes to solid. You can say that: Heat energy is needed to change liquid naphthalene to solid *without* a change in temperature.

Pie charts

Graphs are not the only methods which we have for trying to interpret tabulated results. A **pie chart** is a circular diagram that looks like a pie which has been cut into portions ready for eating. The size of each portion depends on the angle at the centre of the circle and is proportional to the amount of the item shown. The proportions of carbohydrates, proteins, fats and oils, and water in lean beef and rice are shown.

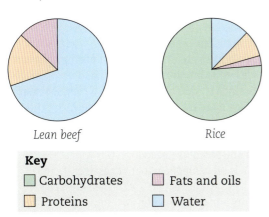

Questions

3   Measure the angle at the centre of the pie for each substance contained in (a) lean beef and (b) rice. Calculate this angle as a fraction of 360°. Then work out the *percentage* of each substance contained in lean beef and rice.

4   Refer to page 79. Convert the figures given for groundnut, maize and rice into suitable angles, and plot the corresponding pie charts. Note that the remaining angle in each pie chart represents water.

Bar charts and histograms

Both of these are convenient ways of presenting results. The vertical axis shows the scale on which things are being compared. The horizontal axis shows the bars for things being compared. We could plot a histogram of these results.

Height range/cm	Number of students in each group
120–124	2
125–129	7
130–134	10
135–139	11
140–144	6
145–149	4

The histogram shows the measurements of height of the students, but not the heights of individual students. The heights are *grouped* in **ranges**.

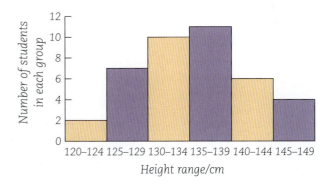

It is important to remember that the purpose of presenting results is to allow us to think about and interpret the results which we have obtained.

Question

5   Look at page 46 where you will find a table of the average heights of boys and girls at yearly intervals. Make two histograms of the information.

How do organisms reproduce?

Activity | **How does bread mould reproduce?**
1. Take a piece of bread, dampen it with water and lay it on a plate. Put an empty jar over it. This will keep the cockroaches out, and keep the bread moist.
2. Look at the bread every day. Soon you will see white threads growing on it. This is mould. After two or three days you will see some upright threads with swollen heads, looking like pins. The mould is reproducing. What you see are the cases which contain **spores** (below). When the cases break open the spores drop onto the bread and can start to grow a new thread. This kind of reproduction is called **asexual reproduction**.

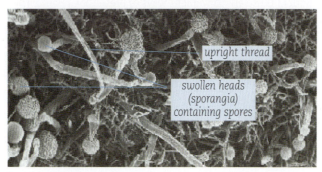

Bread mould showing spore cases

3. After a few more days you may see some small dark balls on the surface of the mould (below). These contain **zygotes** and have been made by the joining together of special cells. This is the simplest kind of **sexual reproduction**.

Bread mould showing zygotes

4. Look at the spore cases and the zygotes with a hand lens and with the low power of a microscope (page 35).

Sexual reproduction
1. Two parents are usually needed.
2. Two special cells, **gametes**, have to join together.
3. The new organisms that are formed are *not identical* to the parents nor to each other.

Sexual reproduction occurs when **male** and **female** gametes join together. These gametes may be made in one organism, for example, in many flowering plants the flower produces the pollen (male gamete) and the egg (female gamete) (page 40).

In animals it is common to have male and female organisms that look different. In this case the male produces sperm (male gamete) and the female produces the egg (female gamete) (page 51).

In all organisms that reproduce sexually the male gamete joins with the female gamete. This is called **fertilisation**. The result is a fertilised egg called the **zygote**, which can then grow into a new organism. This is illustrated in the diagram below.

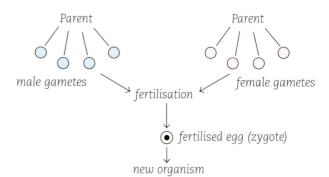

The gametes that are produced in sexual reproduction are all different from each other (page 60). So when they join together to make zygotes, the zygotes are also different and grow into organisms which are different from each other and from the parents.

Advantages of sexual reproduction
1. New organisms (**offspring**) are not identical to each other nor to their parents. This **variety** allows the plants or animals to spread into new areas and to withstand changing conditions.
2. Seeds that are produced may be **resistant** to poor conditions such as drought. This is one way in which plants can survive the dry season.
3. New plants might arise which, for example, have bigger seeds and so produce more food.
4. New kinds of animals might arise which, for example, produce more milk or have better meat.

Asexual reproduction
1. Only one parent is needed.
2. There is no joining together of gametes.
3. The new organisms that are formed are *identical* to the parents and to each other.

Asexual reproduction occurs when the whole or a part of an organism can produce a new organism. The simplest example is a protist called *Entamoeba* (page 13). These can reproduce by just splitting into two halves (**binary fission**). Each half grows into a new adult, which again can split into two.

Another protist that can reproduce asexually is *Plasmodium*, the organism which causes malaria. This malaria parasite burrows into a red blood cell and divides up to form very many small organisms just like itself. The blood cell then breaks to release the new parasites which repeat the cycle.

Some simple animals, and plants such as mould, mosses and ferns, reproduce by making **spores**. Each of these spores can then grow into a new organism. In this way many more organisms are produced.

In flowering plants, new plants can be formed asexually from the stems, leaves and roots. This is called **vegetative reproduction**. An example is *Bryophyllum* (leaf of life). Small **buds** grow out of the leaf and each can become a new plant.

In other cases, a gardener only needs to cut off a part of the stem, called a **cutting**, and put it in water, or in the soil. It soon grows new roots and makes a whole new plant. Cuttings are used, for example, in the growing of *Coleus*, balsam, sugar cane and cassava.

Another example is a **runner** such as water grass. This runs along the ground and produces new sets of roots. The runner can be broken between the sets of roots and so new plants are formed.

Part of the plant may become swollen with food (page 73), for example onion, ginger, cocoyam, Irish potato, and ornamental coco (caladium). On the plant there are young buds which make use of the stored food and grow out to make new plants. These can become separated from the parent plant as new individuals.

Advantages of asexual reproduction
1 Asexual reproduction usually produces large numbers of offspring.
2 Offspring are identical to each other and to the parent and so a farmer can produce more copies of a particularly 'good' plant.
3 As only one organism is involved it is easier for reproduction to occur without problems.
4 Asexual reproduction produces a new independent organism more quickly than sexual reproduction.
5 Asexual reproduction is essential in plants such as banana where seeds are not produced.

How is asexual reproduction useful to us?
We make use of the fact that flowering plants can produce new plants by asexual reproduction. We call this **artificial asexual reproduction**.

1 We grow plants which reproduce asexually and then separate the parts to make new plants (also page 73). Here are some examples:
(a) Corms (food stored in a short stem), for example, dasheen, eddo, and ornamental coco (caladium) (below).

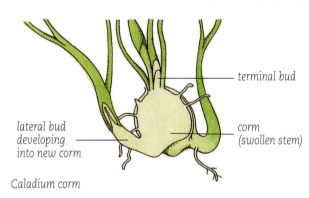

Caladium corm

(b) Bulbs (food stored in the leaves), for example onion, garlic and scallion.
(c) Rhizomes (food stored in horizontal stems underground), for example ginger and canna lily.
2 We cut pieces of the stem (cuttings) and plant them to make new plants. This takes less time than growing the plant from seeds. It is less expensive and there is less likelihood of the plants dying. Some examples are:
(a) sugar cane 'setts' are used to re-plant the field
(b) cassava stems are planted, and new swollen roots provide food
(c) sweet potato stems are planted, and new swollen roots provide food.
3 We cut off part of the plant which then grows into a complete new plant, for example:
(a) yam: buds from the top of the yam, or slices with skin around them can grow into new yams
(b) pineapple: the 'head' can grow into a new plant
(c) Irish (English) potato: the 'eyes' (buds) can grow and make new plants.
4 We can take a small piece from a plant in order to grow a new plant. This is called **tissue culture** and is used, for example, for growing orchids.

Questions
1 List the different methods of asexual reproduction and name an animal or plant which uses each method.
2 In what ways are asexual and sexual reproduction different?
3 How does a farmer make use of asexual reproduction in plants? Give three examples.

What are flowers?

Flowers are the parts of the plant which are involved in sexual reproduction. Let's find out about their structure.

Activity | What are the parts of a flower?
Collect a flowering shoot of Pride of Barbados.

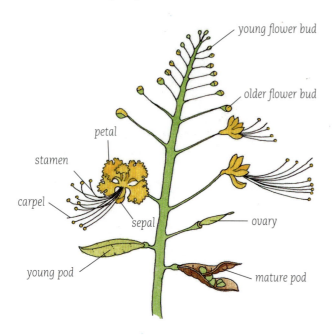

Flower buds at the top of the shoot open to become flowers, which later form pods (fruits) containing the seeds

Use a hand lens to examine your shoot. Record your drawings and observations in your practical notebook.
1. Look at a flower bud. See how the **sepals** form the outside protection to the bud.
   - What colour are the sepals?
   - Take off one sepal and draw it.
2. Take the bud and remove the sepals. Inside the sepals you will find the **petals**.
   - What colour are the petals?
   - Take off one petal and draw it.
3. Look at an open flower of Pride of Barbados. Find the sepals and the petals. Carefully remove them.
4. Inside the petals you will find the **stamens**. These look like long pins. The head part is called the **anther**, and the stalk part is called the **filament**.
   - Take off one stamen, draw it and label the parts. Look for yellow powder on the stamens. This is called **pollen** and it contains the male gamete of the plant.
5. At the very centre of the plant is the **carpel**. It has three parts. The swollen bottom part is the **ovary**. The stalk part is the **style**. The small bump at the top of the style is the **stigma**.
   - Draw the carpel.
   - Label the ovary, style and stigma.
   - Cut the ovary in half along its length. The small round structures inside are called ovules. Each ovule contains an egg (the female gamete of the plant).
6. Look for a pod on your plant. Feel the outside of the pod.
   - Can you tell what is inside? Open up the pod and look inside. Describe in your notebook what you see. The pod is the fruit of the plant and it contains the seeds. The seeds are made when the male gamete (from the pollen) combines with the female gamete (the egg inside the ovule) (see page 43).

Questions
1. What use are the sepals?
2. What use are the petals?
3. What use are the stamens?
4. What use is the ovary?

Activity | Do all flowers have male and female parts?
1. Collect a flower of sunflower or railway weed (*Tridax*). The 'flower' is really made up of lots of very small flowers called **florets**.
   - Take off some of the florets. Pull them apart and look at them with a hand lens.
   - Look at a floret from the outer part of the flower. Does it have a stigma? Does it have anthers?
   - Look at a floret from the inner part of the flower. Does it have a stigma? Does it have anthers?
   - Compare your florets with the drawings below.

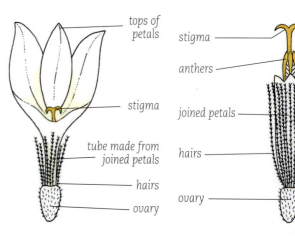

Tridax – outer floret   Tridax – inner floret

The inner florets are *complete*: they have male and female parts. The outer florets are *incomplete*. They do not have the male parts, the anthers.

2   Look at the drawings of the pawpaw flowers below. They are also incomplete. In some of them, namely in the **male flowers**, there are anthers but the ovary is small and undeveloped. In others, namely the **female flowers**, there is a large ovary which forms the fruit, but there are no anthers.

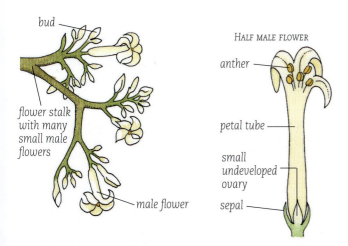

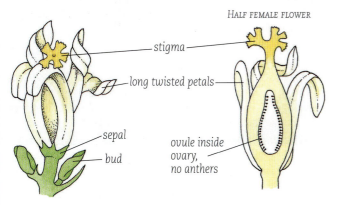

Pawpaw – male (above) and female flowers (below)

Usually one pawpaw tree bears either all male or all female flowers, but sometimes both kinds are carried in a bunch which hangs out of the tree. The female flowers form in small groups close to the tree trunk.

Question

5   Look at the photograph of a lily flower. Is the flower complete or incomplete? Give your reasons.

Activity | *How do we draw flowers?*

Collect a flower of *Delonix regia*, known as flamboyant, or another similar large flower.

1   Make a drawing of the whole flower. Label your drawing.

Whole flower of flamboyant (Delonix regia)

2   Cut the flower lengthways down the middle. Look at one half and make a drawing like this. Label your drawing.

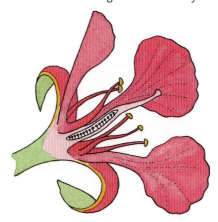

Half flower of flamboyant

How are fruits and seeds formed?

Fruits and seeds are formed after the female gamete (egg) has been joined or **fertilised** by the male gamete (pollen). The gametes are brought together after **pollination**.

What is pollination?
Pollination is the transfer of pollen from the anther to the stigma. There are two main ways in which this can happen.
(a) By the wind blowing the pollen (wind pollination).
(b) By animals such as insects (for example, butterflies, moths and flies), bats and humans carrying the pollen (animal pollination).

Questions
1. Would small, green maize or grass flowers be attractive to insects?
2. Would maize or grass flowers be big enough for an insect to get into them?
3. Would the small, dry pollen grains from the maize tassel blow easily in the wind?
4. Would the long, feathery, sticky stigmas of the maize catch pollen easily?
5. Would large, brightly coloured scented flowers with nectar attract insects?
6. (a) If the anthers were hidden inside a flower could they shed their pollen into the wind?
   (b) Would this pollen be more likely to be shed onto a visiting insect?

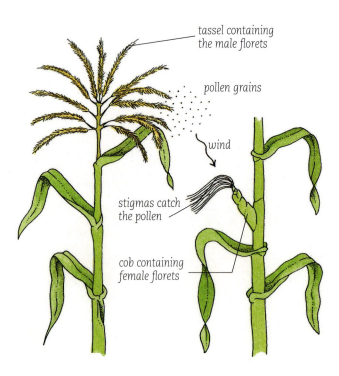

Wind pollination in maize

Insect pollination in Mallow. While a bee feeds it picks up pollen from the stamens. This may then be taken to an older flower where the stamens have withered and the stigma picks up the pollen

What are wind-pollinated flowers like?
1. The flowers are usually small and green.
2. The petals are small or absent and do not get in the way of the pollen.
3. There is no scent or nectar because there is no need to attract insects.
4. The anthers hang out of the flower and they can be shaken by the wind.
5. There are a lot of very small, dry pollen grains which will fly easily in the wind.
6. The stigmas hang out of the flowers; they are feathery and sticky and can catch pollen.

What are insect-pollinated flowers like?
1. The flowers are usually large and attractive.
2. The petals are large and brightly coloured with lines to guide the insects to the nectar.
3. There is scent and nectar to attract insects. The insects use the nectar for food.
4. The anthers are inside the flower where an insect will touch them.
5. The pollen grains are larger and sticky, so they will become attached to an insect.
6. The stigmas are shorter and sticky, so pollen grains attached to an insect will stick to them.

What is self-pollination and cross-pollination?

Self-pollination occurs when pollen is transferred from the anther of a flower to the stigma of the *same* flower.

Cross-pollination occurs when pollen is transferred from the anther of one flower to the stigma of a *different* flower of the same species.

The advantage of cross-pollination is that the gametes come from different flowers and so there will be more variety in the offspring (page 60). This variety gives rise to plants with different characteristics and these are more likely to be able to spread into new areas. At least some of the new plants will also be healthier than either of the original plants.

What happens after pollination?

Pollen grains land on the stigma. Small tubes grow out of the pollen grains and grow down the style and into the ovary. (See (a) below.)

Inside each pollen tube is a male gamete, and inside each ovule is an egg (female gamete). These fuse together (**fertilisation**) (see (b) below). The fertilised ovule grows into a seed. The ovary grows into a fruit. (See (c) and (d) below.)

The fruit may become hard and dry, for example the pod of flamboyant, or it may become soft and fleshy as in the tomato and pawpaw. The seeds are inside the fruit, for example peas and beans inside a pod.

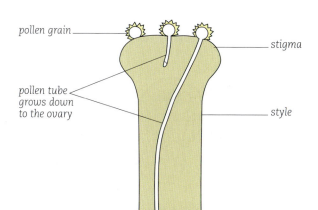

(a) Pollen grains land on the stigma

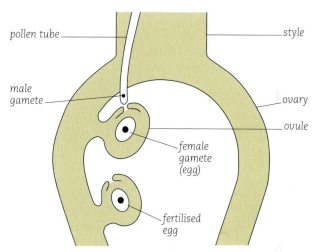

(b) Pollen tubes grow into the ovary

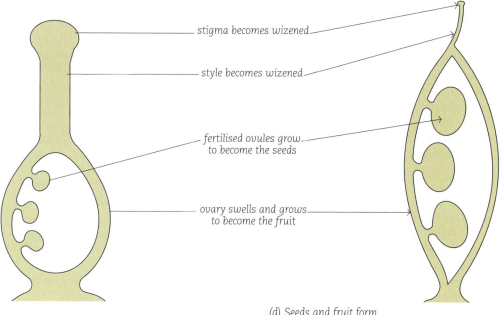

(c) Fertilisation

(d) Seeds and fruit form

How do plants grow?

We have seen how seeds and fruits are formed (page 43). The seeds usually stay inside the fruit for some time, and are then distributed or **dispersed** in some way. When the seeds are mature they can begin to develop or **germinate**.

Activity | What conditions are necessary for germination?

You might think that water is necessary for germination. How could you *prove* this? By carrying out a **controlled experiment** (page 33). You would leave some of the seeds damp, and others dry. All other conditions for the seeds would be the same. The seeds that are left dry would be called the experiment. These dry seeds would be compared to the damp seeds (the **control**) to find out the effect of water on germination.

- The **control** has all the conditions (variables) you think will allow germination to take place.
- The **experiment** changes just one of these variables.

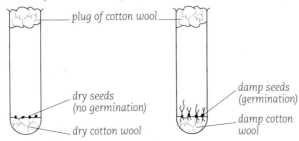

Finding the effect of one variable, water, on the process of germination

In a similar way if we wanted to test if warmth, for example room temperature, was needed for germination we could leave some wet seeds in a cupboard, and other wet seeds in the refrigerator. Notice that in both cases the seeds are wet and in the dark: remember that all conditions, *except* for the one that we are testing, have to be the same.

If we wanted to test if air was important we could cover the seeds with cool, boiled water that had a layer of oil on top. The oil would stop air from getting to the seeds. These seeds would be compared with seeds that were not covered. To test for light we could leave a test-tube that contained damp seeds in a dark cupboard and compare the seeds to those left in the classroom.

Now set up your experiment.

1. Using small bean seeds, cotton wool, boiled water and oil, set up a series of test-tubes to find out what conditions are necessary for germination.
2. Leave the seeds for three days.
3. Write down your results. Try to explain the importance of each of your results to seed germination.

Activity | What is a seed made of?

Use large seeds such as kidney beans or red peas. Soak some seeds in water for a few hours.

1. Examine a dry seed and a soaked seed.
   - What differences do you see in the skin and in the size of the seeds?
   - What might account for these differences?
2. Find a **scar** on the seed. The scar is where the seed was attached inside the fruit.
3. Carefully remove the skin of the seed. It is called the **testa**. Inside you will find two whitish structures which are the seed leaves or **cotyledons**. Separate these seed leaves and find the parts of the new young plant (the **embryo**).
   - Make a labelled diagram of your seed to show these structures.

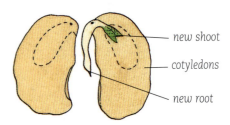

Activity | What happens to the parts of a seed when it germinates?

Take some soaked kidney beans, red peas, balsam seeds or maize grains.

1. Put them into a test-tube or jam jar as shown below. Keep them moist.

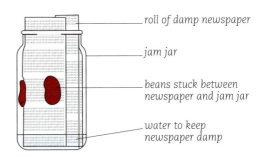

2. Watch what happens in the next week.
   - Each day write down what happens:
     (a) to the testa
     (b) to the new root of the embryo
     (c) to the new shoot of the embryo
     (d) to the seed leaves.
3. Which set of drawings in the next column is most like the germination of *your* seed?

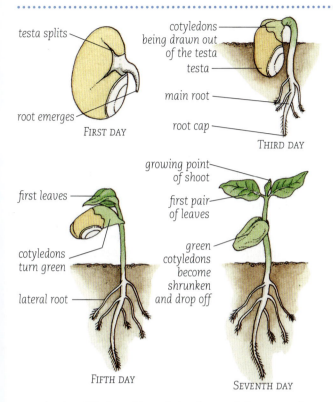

Germination of black-eyed bean (epigeal) – cotyledons grow above the ground

Activity | How does a bean plant grow?

At home, set up a seed such as a kidney bean or red pea in a container similar to that in the last Activity. Once the seed germinates and grows it is called a **seedling**. Watch the growth of your seedling.

1. At the same time each day measure the length of the root and shoot in centimetres.
2. Record your results in a table.
3. Make a graph to show the growth of the shoot. Put the length of the shoot (in centimetres) on the vertical axis and the time (in days) on the horizontal axis. (There is some help with making graphs on pages 36–7.)
4. On day 7 guess/**predict** what the length will be on day 14.
5. On day 14 measure the length of the shoot and see how good your prediction was.
   - Who made the best prediction in your class?
6. On day 14 plant your seedling in the ground in a sheltered place either at school or at home. Over the next few months water it regularly and watch it grow.
7. When the plant is ready to reproduce it is **mature**. Maturity is shown by the formation of flower buds.
   - How long does it take from germination until the flower buds form?
8. When do the flowers open? What happens next?

Sexual reproduction will occur in the flowers, and seeds and fruits will be formed. Watch your plant for the production of pods (the fruit) with the beans or peas (seeds) inside. The germination, growth, sexual reproduction and formation of seeds and fruits make up the parts of the plant's **life cycle** (see below). The seeds then germinate to make new plants and the cycle is repeated.

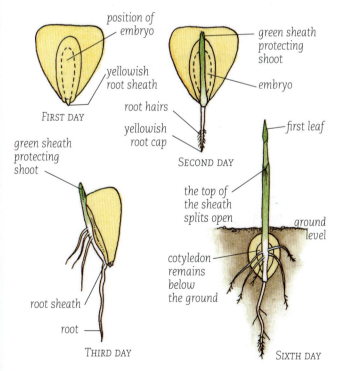

Germination of maize (hypogeal) – cotyledons remain below ground

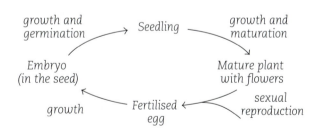

Questions

1. (a) What are controlled experiments?
   (b) Why is it important to use a control?
2. Describe the structure of a bean seed. Draw a labelled diagram as part of your description.
3. What are the main stages in a plant's life cycle?

How do we grow?

We have similar stages of growth as flowering plants. We can represent our life cycle like this.

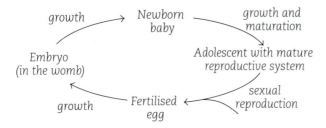

How quickly do we grow?
Information was collected on the average height of males and females from birth to 18 years of age. This is given in the table and graph below.

Age	Male heights (cm)	Female heights (cm)
Birth	50.6	50.2
1 year	75.2	74.2
2 years	92.2	86.6
3 years	98.2	95.7
4 years	103.4	103.2
5 years	110.0	109.4
6 years	117.5	115.9
7 years	125.1	122.3
8 years	130.0	128.0
9 years	135.5	132.9
10 years	140.3	138.6
11 years	144.2	144.7
12 years	149.6	151.9
13 years	155.0	157.1
14 years	162.7	159.6
15 years	167.8	161.1
16 years	171.6	162.2
17 years	173.7	162.5
18 years	174.5	162.5

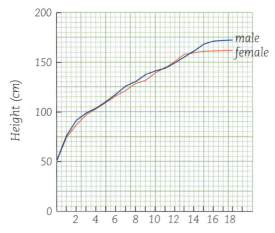

Graph of height against age

What stages of growth do we go through?
Before birth From the moment of fertilisation (conception), growth takes place. The fertilised egg divides into a ball of cells which becomes embedded in the wall of the **uterus** (womb). Here the embryo grows (pages 62–3).

Childhood After birth, the baby grows rapidly for the first years of life. After this the rate of growth begins to slow down (you can see that the line on the graph becomes less steep).

Adolescence and puberty Adolescence is the change from childhood to adulthood. It begins with **puberty** when girls and boys begin to mature and develop the characteristics of women and men (pages 48–9).

Puberty begins at about 10 or 11 years in girls and 12 or 13 years in boys. One feature of puberty is the increase in rate of growth. The increase occurs earlier in girls than in boys.

Another feature of puberty is the beginning of the 'periods' or monthly bleeding in girls. This is called **menstruation**. It usually begins between about 11 and 14 years, but can begin earlier or later.

Adulthood By about 16 or 18 years of age the rapid rate of growth has begun to slow down (see the graph). This marks the end of adolescence and the beginning of adulthood.

Menopause This is when a woman stops having periods. It usually occurs between about 40 and 55 years of age. It is preceded by irregular periods.

Old age By the age of 75 a person will be about 5 cm shorter than his or her tallest height. This is because the discs between the bones in the spine become flattened.

Activity | Growth in height
Find a convenient time for your class to visit other classes in your school.
1 Measure each student in centimetres (without their shoes on) and ask for their ages in months and years.
2 Make a graph (see pages 36–7) of the heights (vertical axis) against the ages (horizontal axis). Plot the points, then draw separate lines for boys and girls.
3 Compare your results with the graph in the first column. What did you find? How can you explain any differences in your results?

Questions
1 Look at the graph in the first column. At what ages are the rates of growth:
   (a) of boys the greatest (i) below 10 (ii) above 10?
   (b) of girls the greatest (i) below 10 (ii) above 10?
2 What do you understand by the terms:
   (a) adolescence and adulthood?
   (b) menstruation and menopause?

What about height and mass?

Instead of growing upwards children and adults may grow outwards! Tables have been prepared for the average mass of adults of different heights. The tables take into account the differences between males and females, and the different kinds of body build (small, medium or large) that people have.

Height (cm)	Average mass (kg)		
	Small build	Medium build	Large build
(a) Males			
157	59	62	65
160	60	63	66
163	61	64	68
165	62	65	69
168	63	66	70
170	64	67	72
173	65	68	73
175	66	70	75
178	68	71	77
180	69	73	78
183	70	74	80
(b) Females			
147	48	52	56
150	49	53	58
152	50	54	59
155	51	55	60
157	52	57	61
160	53	58	63
163	55	60	65
165	56	61	66
168	57	62	68
170	59	64	69
173	60	65	71

To use the tables Find your height on the left-hand side (with shoes with 2.5 cm heels). Then decide if you have a small, medium or large build and read off the mass in the correct column. Note that this reading includes indoor clothing. The tables apply to people over 25 years of age. You should subtract half a kilogram for each year you are under 25.
- If your actual mass is less than the amount in the table then you may be too light.
- If your actual mass is more than the amount in the table then you may be too heavy.

Obesity When a person exceeds the average mass (in the tables) by more than 20% he or she is called **obese**. Obesity is associated with dangerous conditions such as diabetes, stroke, high blood pressure (pages 96–7), and kidney trouble. For example, someone who is more than 40% heavier than average runs twice the risk of dying from a heart attack than a person of average mass.

Abnormal growth in our bodies: cancer

Cancer is not one disease. It is the name given to many diseases in which the growth of certain body cells gets out of control. The cells divide to form a mass of tissue called a **tumour**. With a benign tumour the cells do not spread, and the tumour can be removed. With a malignant tumour the cells keep on dividing and may spread to other organs of the body, and cause secondary cancers.

What causes cancer? Scientists believe about 80% of all cancers are due to cancer-causing substances called **carcinogens**. Some examples are cigarette smoke, asbestos, tar and some food contaminants (page 77). Nuclear radiation and certain viruses can also cause cancer.

What is the treatment? Many cancers, for example those of the cervix, breast, testes, rectum and skin can be detected early. The cancerous cells can then be removed in an operation, and the person will have complete recovery.

But if the cancer has spread it is much more difficult to treat. X-rays may be used (**radiotherapy**) to slow down the growth of the cancer. Drugs can also be used (**chemotherapy**) to kill the cancer cells. Unfortunately the drugs also affect the rest of the body and have unpleasant side-effects. For these reasons early detection of cancer is important (see below and pages 51 and 102).

> **Cancer: early warning signals**
> - A sore in the mouth or on the body which does not heal in three weeks
> - Any unusual bleeding, for example from the vagina or in the urine or faeces
> - A change in the kind of faeces produced which continues for more than three weeks
> - Any slow-growing lump, for example in the breast or testes
> - Rapid loss of mass (more than about 2 kg in five weeks) without a known reason
> - Persistent cough that suddenly gets worse, especially in smokers, and if blood is coughed up

Female cancers (commonest first):
breast, lung, cervix, ovary, lymph glands, uterus, bladder, kidney and stomach.

Male cancers (commonest first):
prostate, lung, colon, lymph glands, bladder, kidney, stomach and pancreas.

> ## Questions
> 3 What is obesity and what are its dangers?
> 4 Why is early detection of cancer important?

living things | reproduction and growth | how do we grow?

What changes occur in adolescence?

Puberty marks the beginning of adolescence. Puberty begins in girls at around 10 or 11 years and in boys at around 12 or 13 years. The pituitary gland (pages 126–7) begins to produce hormones which activate the ovaries (in girls) and the testes (in boys). The changes may occur earlier or later, but this variation is normal.

As a result of this the testes start to produce the male hormone called **testosterone** which in turn causes the development of the male **secondary sexual characteristics**.

The ovaries start to produce female hormones called **oestrogens** and **progesterone** which in turn cause the development of the female secondary sexual characteristics. The girl also starts her monthly bleeding (periods or **menstruation**). If a woman uses contraceptive pills, she still has periods although she does not release eggs. Past the menopause, when egg production ceases, an older woman using hormone replacement therapy (HRT) might also have monthly bleeding.

Menstruation

The **menstrual cycle** (see below) starts with the period. The period lasts for about 5 days during which time blood is shed from the uterus and out of the vagina. On about day 14 or 15 of the cycle, an egg is shed from the ovary and passes into the uterus.

If the egg is fertilised it begins to develop in the uterus wall which has become thickened and filled with blood.

If the egg is not fertilised, the blood-filled lining of the uterus starts to break down and the blood is shed as the next period (day 28, and day 1 of the next period).

Each month an egg in the ovary ripens and is released. This is controlled by hormones from the pituitary gland. Other hormones (oestrogens and progesterone) from the ovary control the repair, building up and breaking down of the lining of the uterus wall (see also page 127). The average cycle is 28 days, but many women have cycles which are shorter or longer than this.

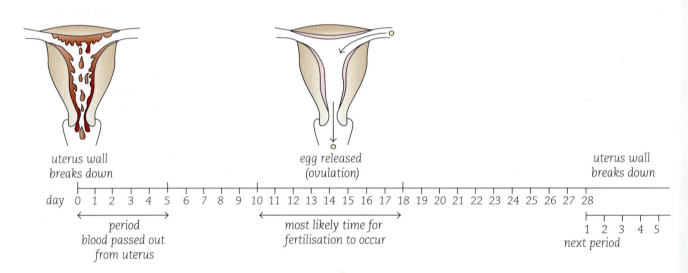

Menstrual cycle

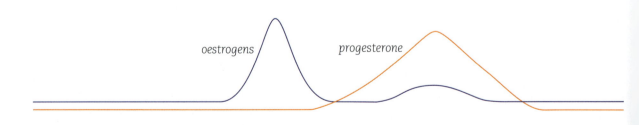

Hormone changes

Secondary sexual characteristics

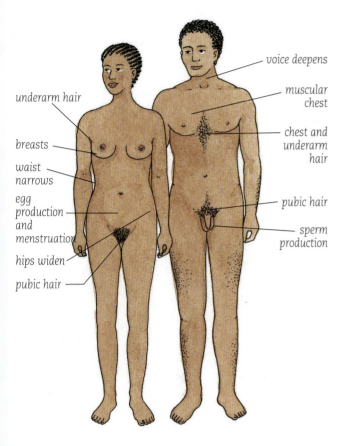

In girls, changes at puberty are caused by the female hormones. The breasts begin to enlarge and hair starts to grow under the arms and around the pubic area between the legs. Fat is laid down beneath the skin and a typically female shape develops with a narrower waist and wider hips (see above).

In boys the development of secondary sexual characteristics is caused by the male hormone. The testes and penis increase in size and sperm production begins. Hair appears on the face, chest, under the arms and around the pubic area. The Adam's apple increases in size and the voice 'breaks' and becomes deeper. The shoulders widen and the chest becomes more muscular giving the typical male shape (see above).

During adolescence many changes occur in the body due to the production of hormones. The exact age at which these changes happen varies from person to person, but with girls' development tending to be earlier than that of boys. Boys and girls also become more attracted to members of the opposite sex and thus more concerned about their physical appearance. As a member of a group they may also want to share in the experiences and activities of the group, some of which may be harmful.

Special problems of adolescence

Emotional changes The hormonal changes which occur during adolescence may lead to emotional problems. Adolescents need to assert their independence while at the same time they have to rely on adults for most of their daily needs. This may cause a certain rebelliousness. Adolescents may also be moody with swings between happiness and sadness. These changes are a normal part of growing up. But loss of sleep, loss of appetite and deep depression may indicate a more serious problem.

Homosexual feelings This is the attraction between members of the same sex. Homosexual feelings are a natural part of life, there are times when we prefer the company of members of our own sex. If, however, the homosexual feelings develop into a deep physical attraction towards a person of the same sex then this is called **homosexuality**. During adolescence there can also be an increase in heterosexual feelings: the attraction between members of the opposite sex. This happens at different times for different people.

Acne This is the development of large numbers of spots and pimples. Tiny pits (follicles) at the base of the hairs become blocked by dirt or old skin. An oily substance, sebum, produced in large amounts during adolescence, accumulates in the pits and so produces the acne.

Acne is most common in adolescent boys. In girls it may be more marked at the time of their periods. Thorough cleaning of the skin, and the use of special soap and cream may help to reduce acne.

Teenage pregnancies A girl can become pregnant before she sees her first period. An egg will be shed from her ovary and if she has sexual intercourse the egg may become fertilised. A young teenager is not yet fully developed. If she becomes pregnant, the strain on her system would be very great. For one thing her bone structure may not be wide enough to allow for the easy birth of her baby. Having a baby would also disrupt her schooling and cause many problems both for her and the child. Serious thought must be given to this by both teenage girls *and* boys.

Questions

1. What is the menstrual cycle? Why is it important?
2. What effects do hormones have on (a) girls and (b) boys as they enter adolescence?
3. What special help might an adolescent need?
4. What are the disadvantages of teenage pregnancy?

How do we reproduce?

The human reproductive systems are shown below. The female reproductive system is shown on the left, and the male reproductive system is shown on the right. The top diagram shows a front view, while the bottom ones show a side view.

Copy the unlabelled side view diagrams into your practical notebook and label them.

Activity | Looking at other reproductive systems
1. Your teacher will dissect male and female rats or rabbits to show the reproductive systems.
2. Examine the dissections. Use the diagrams below to identify the parts of (a) the female and (b) the male reproductive systems. Describe how they are similar to and different from those in humans.

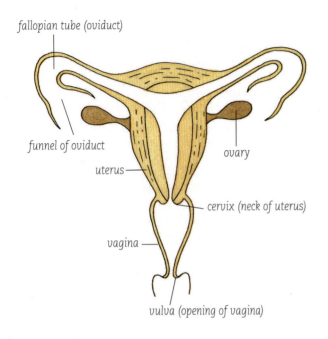

Female reproductive system – front view

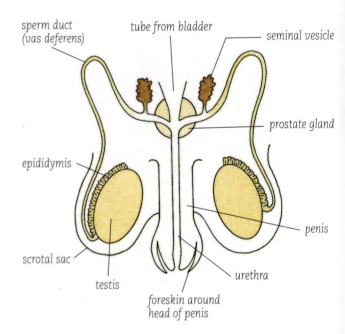

Male reproductive system – front view

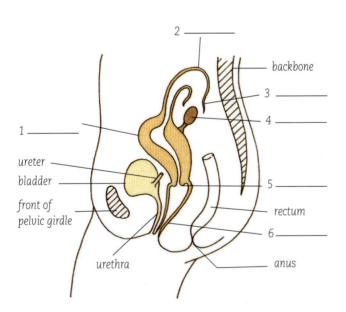

Female reproductive system – side view

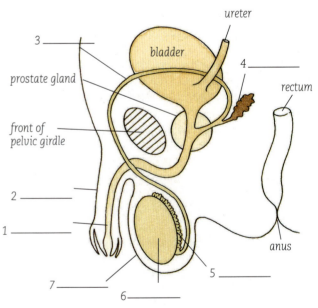

Male reproductive system – side view

living things | reproduction and growth | how do we reproduce?

How are gametes formed?

Sexual reproduction (page 38) involves the production of male and female gametes.

Male gametes The **sperm** are produced inside the testes which hang outside the body in the **scrotal sacs**. Sperm are first produced at puberty (pages 48–9). The release of fluid (**semen**), containing the sperm, may occur at night and take the form of wet dreams during adolescence.

The sperm have a head and a tail (see below). The head is 40 times smaller than the female egg. Sperm pass from the testes into a coiled tube, the **epididymis**, and then along the **vas deferens**. They are stored in the **seminal vesicles** until they pass out through the **penis**.

Female gametes The eggs are produced inside the **ovaries** which are found on either side of the **uterus**. After puberty (pages 48–9), one egg ripens each month and is shed from the ovary. This happens at about day 14 of the menstrual cycle, halfway between two periods.

The egg is about the size of a full stop. It passes along the **fallopian tube** (oviduct) towards the **uterus**.

How are the gametes brought together?

Stimulation of the penis causes it to become filled with blood and erect. During sexual intercourse the penis is placed inside the **vagina** of the female. Sexual stimulation eventually leads to **ejaculation** and the release of semen into the female. This may be accompanied by pleasurable feelings of orgasm.

If intercourse takes place within a few days before or a few days after the female has shed an egg, then **fertilisation** is possible. Millions of sperm are ejaculated during intercourse and they swim through the uterus into the fallopian tubes. The head of one of these sperm enters the egg and a **fertilised egg** is formed which can grow into a new individual (pages 62–3).

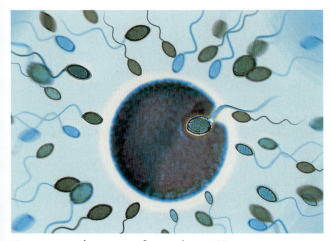

Human egg at the moment of conception × 400

Associated health aspects

D and C (Dilatation and curettage) This is an operation on the uterus. If a woman has heavy periods or is having an abortion in hospital, a D and C may be done. The neck of the uterus is widened (dilatation) and the inside is scraped (curettage). This removes the lining of the uterus wall. The bleeding stops after a few days and the tissues heal.

Prostate gland In a man the prostate gland surrounds the urethra where it leaves the bladder. In older men the prostate often enlarges, but it only causes serious problems in some cases. If the prostate is hard and inflexible it can interfere with the flow of urine from the bladder. A part of the gland may have to be cut away in an operation.

Cancer Cancer is when cells multiply out of control (page 47). Early detection and treatment can save lives.

Cancer of the cervix There is a bad-smelling discharge, and blood may be passed out between the periods or after intercourse. If left untreated, the cancer can spread to nearby organs.

It can be detected early by the use of a cervical **smear** or **Pap test**. A smear (a few cells) is taken from the cervix. From these the technician or doctor can say whether the person is likely to develop cervical cancer. She should then be treated and not run any risks. Adult women should have a Pap test at least every five years.

Cancer of the breasts This is felt first as a hard lump in the breast. If the lump is **malignant** it will be removed in an operation before it can spread to other parts of the body. Only special tests can distinguish it from a harmless **benign** tumour. Women should do a regular monthly examination of their breasts to feel for any unusual lumps. They should also have a breast screen every five years. This should be done more often if other female members of the family have suffered from breast cancer.

Cancer of the testes The testes are within the sac or scrotum. There is one lump, the epididymis, at the top and back of each testes. If there are other lumps, heaviness in the scrotum, or pain in the testes, the person should go for a check-up. Men should do a regular monthly examination of their testes to feel for any unusual lumps. Early detection and treatment can save lives.

Questions

1. List, in order, the structures a sperm passes through on its way from the testis, to fertilise an egg in the fallopian tube of a female.
2. How can cancer of the cervix, breast or testes be detected early? Why is this important?

What are sexually transmitted diseases?

Sexually transmitted diseases (STD) are infections which can be passed on during any form of sexual contact or intercourse. Some STDs, such as AIDS, can also be passed from one person to another by infected needles.

Symptoms of various STDs are described here. It is important you understand how they are passed on, and that you visit a clinic at once if you think you may have caught one of them. This is important because:
1. If treatment is begun at once, it is much easier to cure the infection.
2. If the person continues to have sex while he or she is infected, the disease will be passed on to other people.
3. If certain STDs are not treated then the person may become sterile, or a mother may pass the disease on to her babies.
4. The person may not have an infection, and can stop worrying about it.

The treatment given is usually antibiotics or other drugs. These should be taken exactly as prescribed. If the complete course is not taken then the infection may not be cured properly. In addition the infection may have a chance to become resistant to these drugs and so it will be very much more difficult to cure at another time.

If someone is suffering from a STD he or she must tell anyone with whom they are in sexual contact, so that the other person can be treated as well. If the sexual partner is not treated and cured, then there can be re-infection the next time they have sex. **Using a condom (rubber) can help a lot to reduce the risk of transferring STDs.**

A major problem with STDs is that symptoms may not be easy to see. For example, gonorrhoea and syphilis are easy to identify in a man. But in a woman the symptoms may be inside the reproductive system and not easily seen. The blisters of herpes are easily identified in both men and women, but the disease of AIDS may show no symptoms for many years.

Vaginal discharges

A woman may have an unpleasant vaginal discharge and irritation around the vulva. It is not dangerous to health, but is unpleasant.

Thrush The discharge is thick and white, and there may be pain during intercourse. It is caused by a fungus *Candida albicans*. Precautions are to wear cotton underwear and not to use vaginal deodorants or douches. Treatment is with anti-fungal suppositories placed in the vagina, and with anti-fungal cream.

Trich The discharge is greenish-yellow. It is caused by a protist, *Trichomonas*. Treatment is by a short course of tablets for the woman and her sexual partner.

Gonorrhoea

This is caused by a bacterium *Neisseria gonorrhoeae*.
Incubation period One to two weeks from infection.
Symptoms Man Some discomfort on passing urine. Slight discharge of pus from the tip of the penis.
   Woman May cause no visible symptoms. Sometimes there is an increase in vaginal discharge.
Diagnosis Samples are taken of the secretions. These are examined for the presence of gonorrhoea bacteria.
Treatment Infected people are given a course of antibiotics such as penicillin. They must complete the whole course.

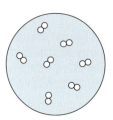
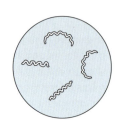

Bacteria that cause gonorrhoea – they are spherical and occur in pairs

Bacteria that cause syphilis – they are spiral-shaped

Syphilis

This is caused by a bacterium *Treponema pallidum*.
Incubation period Nine to ninety days from infection.
Symptoms A hard sore or bump develops on the penis in the man (or the anus in homosexuals) and on the area around the vulva in women. The disease has three stages:
   Primary stage The hard sore develops. It is painless but very infectious. There may also be enlargement of nearby lymph glands (page 128). The sore may heal in a few weeks – but the disease is not cured unless the correct course of antibiotics is taken.
   Secondary stage Occurs several weeks later. There may be a rash which covers the body, swelling of the lymph glands, and infectious sores around the anus. Antibiotics can still be given to cure the disease.
   Tertiary stage If untreated the disease goes to the third stage. The person is no longer infectious, but may suffer deterioration of the circulatory, skeletal and nervous systems. At this stage the illness is very difficult to treat.
Diagnosis Blood samples and samples from the sores are examined for syphilis bacteria.
Treatment Early treatment is essential – this involves a full course of antibiotic injections.

AIDS

AIDS is Acquired Immune Deficiency Syndrome which is caused by a virus, the Human Immunodeficiency Virus (HIV). In people with AIDS the production of white blood cells (lymphocytes) is upset.

The AIDS virus infecting a lymphocyte (white cell in the blood)

The immune system is damaged and people cannot protect themselves against infections such as tuberculosis (TB). TB can cause death – about a third of the total AIDS deaths world-wide are due to TB. For the first few years of the AIDS epidemic most suffering were homosexual and bisexual men, people who used intravenous drug injections and people who had been given blood transfusions from infected donors. Now AIDS is also common in the heterosexual community.

Incubation period Several months to many years. This is very dangerous because a person can infect others before they know they have the disease.

Symptoms Swelling of lymph glands all over the body, a rare type of skin cancer – Kaposi's cancer, fever, excessive tiredness, loss of mass, and respiratory disease.

AIDS can be passed on when infected semen enters the bloodstream through breaks in the anus, vagina or mouth of the sexual partner. It can be passed between drug users by infected needles, and by blood transfusions from infected persons. AIDS is not just a sexually transmitted disease. Parents can spread AIDS to an unborn baby. A mother can also spread AIDS to her child during birth. If a mother is infected, there is a course of drugs she can take before the birth which reduces the likelihood of this happening. There is no evidence that AIDS can be passed on by casual contact with an infected person.

Treatment There are several combinations of drugs that can slow the development of the disease. But the virus actually becomes part of the blood cells of the infected person, so it is extremely difficult to cure. New treatments might use genetically engineered HIV and a special drug which would kill cells if they became infected by introduced HIV. Scientists are in the process of developing vaccines.

AIDS in the Caribbean Figures from the PAHO in Barbados give estimates for the percentage of adults and children living with HIV/AIDS in 1999: Bahamas > 4%, Barbados > 1%, Belize 2%, Guyana 3%, Haiti >5%, Jamaica and Trinidad about 1%. In 1999 in the Caribbean there were 360 000 people reported infected. Causes are likely to be early sexual activity, having several sexual partners, and not using condoms.

What can be done? The best advice to anyone is to only have sex with one partner whom you believe is free of the disease, and **always to use a condom when having intercourse**. In addition, stay away from drugs and **never** use a needle which has been used for injecting drugs. If you believe you may have contracted the disease, there are tests you can have done. But, because the disease takes a long time to develop you may be told you are clear of the disease, but might still develop it later. The best protection is to make every effort not to become infected.

Herpes

This is caused by a virus *Herpes genitalis*.

Incubation period Four to seven days.

Symptoms There is an itchy feeling on the penis or around the vulva. This is followed by small blisters which break open to leave red, moist, painful ulcers which slowly crust over. The person may have a high temperature and feel unwell.

The sores heal in about two weeks. Half of the people who are infected, develop resistance to the disease and have no further attacks. The other half may have return attacks when they are under stress or feeling run-down.

Diagnosis There is a test for herpes which anyone can take, and especially for mothers before labour.

Treatment There are antiviral ointments that can be put onto the sores to help them heal. There is no simple treatment.

Danger If a mother has genital herpes, her baby should be delivered by Caesarean section. Forty per cent of babies who are infected during birth develop serious brain damage or blood poisoning, or they may even die.

Questions

1. There has been a gradual increase in the number of cases of STDs especially since the use of the contraceptive pill (page 57). Suggest reasons why.
2. In your country try to find out:
   (a) which is the most common STD
   (b) which age group is most affected.
   Suggest reasons for your findings.
3. Some people try to get drugs to treat STDs by seeing their 'friends' at the pharmacy. But they may get the wrong drugs, or take the drug incorrectly. What are the dangers of doing this?
4. The likelihood of infection with STDs is increased by having more than one sexual partner, having sex with casual acquaintances and not using condoms. What are the reasons for this?
5. What advice would you expect a doctor to give about the dangers and treatment of STDs?
6. What are the special problems associated with the (a) spread (b) diagnosis and (c) treatment of AIDS?

Is there a population problem?

Before the end of 1999 there were 6 billion (6×10^{12}) people on Earth. Look at the graph below. It is estimated that in the year 2020 there will be 8.5 billion people, and by 2100, 10 billion people on Earth. Every year the world population increases by 75 to 80 million people, of which 65 million are in developing countries. The world population is presently doubling every 50 years.

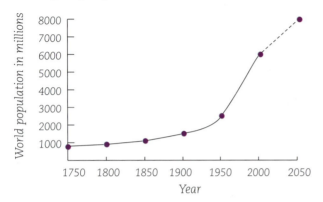

Projected increase in world population

But, are more people a problem, or an asset? Let us look at this in more detail.

Why do populations increase?
The population increases if there are more people born (**birth rate**) than die (**death rate**).
Birth rate Birth rate is the number of people born for every 100 people in the population. An annual birth rate of 3% means that for 100 people at the beginning of the year, there would be 103 at the end of the year.
Death rate In a similar way, an annual death rate of 1% means that for every 100 people at the beginning of the year there would be 99 alive at the end of the year.
Population growth Population growth is the difference between these two rates. If the annual birth rate is 3% and the annual death rate is 1%, then the annual population growth will be 2%. This means for every 100 people at the beginning of the year there would be 102 alive at the end of the year. But how many of these people are in a particular country depends upon immigration and emigration.
Immigration and emigration If more people leave the country (**emigrate**) than come into the country (**immigrate**) then the population size will decrease.

What is the effect of medical progress?
Until the beginning of the 1900s high birth rates tended to be offset by high death rates. Families had many children because of a lack of contraceptives (pages 56–9) and because they expected to lose several of their children at birth or when they were very young. Death rates were also very high, because of famine, malnutrition, plague and war.

But death rates are being reduced. There have been campaigns against malaria, smallpox, yellow fever and cholera so fewer children die (lower **infant mortality**). There have also been improvements in clinics and hospitals, clean water supplies, nutrition and education – though more needs to be done, especially concerning malaria.

In **developed** or **highly industrialised** countries these improvements in medical care, and so on, have taken place gradually, and have occurred along with a decrease in the birth rate due to effective contraception. In some European countries the population is now actually decreasing.

In **developing** countries the decrease in death rate has in general not been linked to a corresponding decrease in birth rate. The average rate of population growth is 2.1% per year. Parents in developing countries may want to have a large number of children as a symbol of their fertility, to help them in their work and to provide for them in their old age.

Recently, infection with HIV and death caused by AIDS and other diseases such as TB has affected population growth. Over 95% of AIDS-infected people live in the developing world. In nine countries in Africa it is projected that life expectancy in 2010 will be only 47 years.

Developed and developing countries
The developed countries, with a third of the world's population, consume more than three-quarters of the world's resources of food, energy and natural resources, such as minerals.

Many developing countries grow crops or export natural resources in the form of raw materials such as bauxite or petroleum, which account for 60 to 70% of their income. The prices are determined by the developed countries. If the crop fails or the price drops then the developing country may suffer great economic problems and lack of foreign exchange. This in turn may force the developing country to seek loans from developed countries or international bodies, such as the World Bank. But these have to be repaid, with interest.

Question

1 Do you think there is or will be a population problem in the Caribbean, and in particular in your country? Look through the arguments put forward on the next page to support two different points of view. Discuss these ideas with a few other students.

Population growth should be decreased

Population growth is seen as a problem. Some arguments which may support this are:

1. *More people means more mouths to feed* The greater the population the more food is needed. The soil may be poor and not able to produce. The cost of fertilisers and pesticides is high. Overpopulation leads to famine. More people also means there are extra demands for water and a greater likelihood of pollution.

2. *High birth rates mean lots of young children* In developing countries nearly half the population consists of children under 15. These children depend on their parents for their everyday needs and for education. It would be better to have fewer children and educate and care for them better, especially if the mother has the major responsibility. For example, Guyana has many young children and a 3% reported rate for AIDS. So there may be a future problem of children without parents. In many African countries there is already a 'missing generation', with grandparents bringing up their grandchildren, and no parents available to work.

3. *We will run out of natural resources* The more people there are the quicker we will use up our petroleum and mineral resources.

4. *More people means more pollution* This is more pollution from household wastes and from industries producing materials for the population.

5. *Cities become overcrowded* People tend to leave the land and move to the cities to look for work. The people often cannot afford to live in proper housing or to eat properly.

6. *A higher population means more crime and unemployment* The more people there are, the more likely there will be ones who cannot find jobs. If people are unemployed then they may turn to crime in order to get money. Drug problems can also make this worse.

Population growth should be increased

Population growth is seen as an asset. Some arguments which may support this are:

1. *More people means more hands to work* Work on the land or in factories needs people. For example, Africa's ten richest countries have virtually the same population growth rates as the ten poorest.

Population growth of itself does not necessarily cause problems. What is also important are the natural resources of the country and how able the country is to supply its own food and materials.

2. *More children means a young population* If we cut down population growth, then the present young people will grow old but not be replaced by more young people to carry on the work of the country. It is estimated that in the year 2025 about a quarter of the people in developed countries will be over 60 years old. This will cause heavy costs for pensions and health care, and may mean that the retirement age is raised.

3. *We will find more natural resources* As we notice that a natural resource looks as though it may run out, we try other methods to extract it, or we find substitutes to use. By the time oil runs out, we will have developed alternative energy sources. But water reserves are more of a problem, as these need careful management.

4. *In many countries there are under-populated places* People tend to go to the cities and it is these which are overcrowded, but in the rural areas there may be an urgent need for more people. If people could be spread out over the country it would be a good thing.

5. *More people can create more jobs* Many of the present developed countries achieved their economic growth during periods when they had high population growth. Many of these countries which now have a very low or zero population growth, such as West Germany and Japan, are now giving incentives to families to have more children.

> ### Question
> 2 Some Caribbean countries, such as Guyana, have quite low populations – especially in the countryside. Others, such as Barbados, are overcrowded – especially in the cities. What is your country like? What might be done about any problems?

A crowded scene in the Caribbean during a festival. Population growth may cause problems, or it may be an asset. There are two points of view

How can we control population numbers?

In many Caribbean countries there may be a need to limit population numbers (pages 54–5). Each couple will need to consider how many children they want to have or already have. The size of their families can be controlled by using methods which prevent fertilisation from occurring. Which method a couple chooses will depend upon their own personal beliefs.

The rhythm method

This is the only method allowed by the Catholic Church. It does not involve the use of chemicals nor mechanical barriers of any kind, and it does not upset the natural menstrual cycle (page 48). It is, however, not a very effective method of birth control. It needs careful attention to the particular days in the woman's menstrual cycle and can only work if the couple does not have sex on the 'unsafe' days each side of **ovulation** (egg release).

If the woman has a regular cycle of 28 days, she can find her 'safe' and 'unsafe' days as shown in the table below.

Rhythm method for a 28 day cycle		
Days 1–9	No egg present	} safe days
10–11	Live sperm may fertilise egg	
12–16	Ripe egg may be released	} unsafe days
17	Egg may still be present	
18–28	No egg present	} safe days

However, if the woman has irregular cycles, of other than 28 days, she should record their lengths for a year, and take away 18 and 11 like this:

Shortest cycle, for example, 25 days − 18 = 7

Longest cycle, for example, 30 days − 11 = 19

The unsafe days are then from day 7 to day 19 counting day 1 as the first day of her period.

The effectiveness of the rhythm method can be improved by the woman taking her temperature each day for three months and recording it. There is a drop and then a rise in temperature at ovulation so she can identify when, in her menstrual cycle, ovulation occurs. The four days *before* and three days *after* this date would be her 'unsafe' days.

The **Billings method** is a natural birth control method. Expert advice is needed to help the woman to record changes in the nature of mucus secretions from her vagina. A woman could get advice from a birth-control clinic. This helps to predict when ovulation is expected and intercourse can be avoided at these times of the month.

Spermicides

Spermicides are chemicals that kill sperm. They are available in the form of creams, jellies, foams or tablets (pessaries) which are inserted into the vagina shortly before intercourse. The time of insertion will be marked on the packet. A new amount of spermicide must be inserted for each occasion of intercourse.

Spermicide in the form of a jelly. The jelly can be inserted into the vagina using an applicator or some can be put onto a diaphragm before use. A condom is most effective when used together with a spermicidal jelly or cream. It gives some protection against STDs

Spermicides are not very reliable when used on their own. This is because a single ejaculation may contain millions of sperm, and it only needs one of these to escape to fertilise the egg. Spermicide should always be used together with a 'barrier method' such as a condom or diaphragm. These give physical protection to keep sperm away from the egg.

Condom

The condom or 'rubber' is a sheath made of strong but thin rubber. It is put onto the erect penis before intercourse. When sperm are produced during ejaculation they are caught in the closed end of the condom. So the sperm are prevented from getting into the vagina to cause fertilisation. When removing the condom, the open end should be held carefully so that sperm do not spill into the vagina.

The condom is fairly reliable when used on its own. But it is much more reliable if used with a spermicidal cream. It has the disadvantage that putting it on may interrupt the process of intercourse, and there may be an attitude, especially amongst men, against its use. But the condom has the great advantage of being easily available and giving some protection against STDs (pages 52–3).

There is also a female condom which is a sheath which fits inside the vagina and stops sperm getting into the vagina.

Diaphragm or cap

The diaphragm is a rounded rubber cap with a stiff outer edge. It is squashed and inserted through the vagina and placed over the **cervix** or neck of the uterus. The outside edge springs into place and so keeps the diaphragm securely over the cervix. This forms a barrier to stop sperm from entering.

This is how a diaphragm is inserted

A doctor or nurse will first identify the correct size for the woman and show her how to insert it. Before use it should be covered with spermicidal cream. It can be inserted some time before intercourse occurs. It must be left in place for at least six hours after intercourse and should then be cleaned carefully. It should be reinserted with more spermicide on the next occasion.

There is also a smaller cap which fits over the cervix and is used together with spermicide. The diaphragm and cap may protect against some STDs.

The contraceptive pill

Contraceptive pills contain hormones similar to those produced by a woman during her menstrual cycle. The most common and effective ones are combined pills containing oestrogen and progesterone. These pills prevent the monthly release of eggs but they do not prevent the woman from having periods, although there will be less blood loss than before. These pills are not suitable for smokers, or for women who are breast-feeding. Contraceptive pills may lose their effectiveness if the woman is vomiting, has severe diarrhoea or is taking certain medication such as antibiotics. At these times the couple should use an additional contraceptive method.

The contraceptive pill must be taken regularly for 21 days, preferably at the same time of day. During the first two weeks when a woman starts her first packet of pills she is not protected and should use an additional contraceptive method. If she forgets a pill she must take it as soon as she remembers and also the pill that should be taken on that day. If a pill is forgotten for 12 hours then the woman is no longer protected and should use an additional contraceptive method.

After the 21 days there are seven 'pill-free days' when the woman will have her period. She should begin the next packet of contraceptive pills on the fifth day after her period started. (See below.)

The yellow pills are the contraceptive pills, which contain hormones that prevent the monthly release of eggs. The first pill (at the top left) should be taken on the fifth day after the period starts. These pills are taken for 21 days. The red pills do not contain hormones, but provide iron to replace that which is lost during menstrual flow

The pill may cause side-effects such as nausea (feeling sick), breast discomfort, spotting of blood between periods and high blood pressure in some women. A woman may be advised by her doctor **not to use the pill** if there are circulatory diseases (pages 96–7) in the family, or if she has sickle-cell anaemia (page 61), jaundice, or diabetes (pages 127 and 129). There is also a correlation between use of the pill and cervical cancer (page 51), especially if the woman also smokes, but the combined pill does give some protection against cancer of the ovary and uterus.

Pills are of different kinds. A woman should choose a pill (with the help of her doctor) which does not cause her side-effects and which is not too strong (that is, contains more hormones than she really needs). A progesterone-only contraceptive pill can be used by women who smoke or are breast-feeding. The pill is a very reliable contraceptive.

There is also a contraceptive ring which can be put into the vagina by the woman and which fits in place around the cervix. The ring releases contraceptive hormones for three weeks until the period is due. The ring is then removed, and a new one put in for the next month. The hormone used is oestrogen, but the amount needed is only half of that used when taken as pills.

How can we control population numbers?

Intra-uterine devices
The commonest intra-uterine device (IUD) is a coil or T-shaped device made of plastic and copper. It is inserted by a doctor or nurse into the uterus where it is held securely. It has a thread attached to it which hangs out of the cervix.

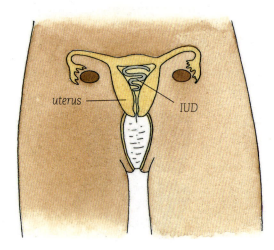

Because the IUD can be kept in place there is no preparation needed before intercourse. It is a reliable contraceptive method. It is believed that it irritates the lining of the uterus and so prevents implantation of the fertilised egg. The woman continues to ovulate and have her periods. If she wishes to become pregnant a doctor or nurse can remove the coil.

There are disadvantages with the coil: there may be heavier periods, and possible pelvic infections which could cause infertility.

A similar device, but which contains progesterone hormone, can be used. This is called an IUS (intra-uterine system). It has the advantage that periods are not so heavy. Both IUDs and IUSs can be left in place for five years and are 99% effective as contraceptives.

Sterilisation
If a couple is sure, after careful discussion, that they do not want any more children then one of them can have a sterilisation operation. This is usually irreversible, so it is a permanent method of contraception.

In the man, this is called a **vasectomy**. The doctor makes a small cut in the scrotal sac so that the vas deferens can be cut and tied up (see next column). This will stop sperm getting into the semen. The male hormones are not affected and so there is no change in masculinity, orgasms, etc. A vasectomy is usually done under a local anaesthetic and takes only about 15 minutes. After the operation the man may feel discomfort for a few days. He will still have sperm in

his semen for a few months and so should use another method of contraception during this time.

In the woman, sterilisation is called a **tubal ligation**. The operation is done under a general anaesthetic and takes about 20 minutes. A small cut is made in the abdominal wall through which the surgeon seals off the fallopian tubes. This is done with heat treatment or with small clips, or the tubes are cut and tied (see below).

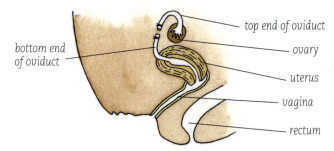

Under certain circumstances, for example, if a woman develops cancer of the uterus, the uterus, fallopian tubes and/or ovaries will be removed (**hysterectomy**). This will cause the periods to stop and the woman will become sterile.

Counselling is recommended before people undertake being sterilised. They have to consider that they may think differently later on. It is a permanent method, though sometimes it can be reversed. The couple must be very sure that they never want to have any, or more children.

This table shows the effectiveness of contraceptive methods when instructions are followed completely.

Number of pregnancies, on average, for 100 couples using the method for one year	
No contraceptive method	90
Spermicide alone	25
Rhythm method	20
Condom alone	15
Diaphragm alone	15
Condom with spermicide	2
Diaphragm with spermicide	4
IUD	1–2
IUS	1
Contraceptive pill	less than 1

Part 2

New methods of contraception

Injectable hormone contraceptives An injection of progesterone can be given which gives contraceptive protection for eight weeks (Noristerat) or 12 weeks (Depo-Provera). The effect is not reversible during these times, and regular periods may take a year or more to return after stopping the injections. They may give some protection against cancer of the uterus, but their long-term effects are not known.

A male contraceptive A male contraceptive is being developed in trials. Injections of chemicals that stop the production of sperm have been tested, but they had unpleasant side-effects. It might be that the men would not want to take the pills. Such a pill might also reduce the use of condoms, which in turn might lead to a greater spread of STDs.

Contraceptive under the skin Under local anaesthetic, small plastic tubes containing progesterone are placed under the skin in the upper arm.

The plastic allows the contraceptive to escape slowly into the body. They can prevent pregnancy for three to six years. The implant can be felt but not seen. When removed the woman's fertility is immediately restored.

'Morning-after pill' This is a pill, prescribed by a doctor, to be used by a woman within the first two days after she thinks she may have become pregnant. If fertilisation has in fact occurred, this kind of pill prevents the fertilised egg from becoming implanted in the uterus wall. An IUD fitted within five days of intercourse has a similar effect.

Some important issues

There are some issues related to the beginning and ending of life on which people hold different opinions. You can discuss these matters in small groups.

Abortion Abortion is the artificial ending of a pregnancy. (A miscarriage is a natural ending of a pregnancy.) An abortion is safer and easier and less of a health risk to the mother if it is done in the first three months of pregnancy. An abortion should only be done by qualified personnel under safe conditions such as in a hospital where adequate treatment is available should there be any complications. In some countries abortion is illegal, while in others it is allowed.

People have different ideas about abortion. Some think it should not be done because they feel the developing foetus is already a person. Other people claim the foetus only becomes a person at birth.

Some people say abortion should never be allowed, while others say the woman should have the choice to ask for an abortion. What do you think?

'Test-tube babies' Doctors can take an egg from a woman's ovary and mix it with sperm in a test-tube to achieve fertilisation. The fertilised egg can then be put back into the woman's uterus where it will develop normally.

But how far could this idea be taken? Could a foetus be grown outside the uterus? Could scientists change the genes in an egg and so change its characteristics? What about surrogate mothers having a baby for someone else? What about using the sperm, not of your partner, but of a close relative or of a stranger? What about using genetic material from three parents to make a child? And should you tell the child how he or she was conceived?

How much control should we have over genetic characteristics, fertilisation and development? With new developments in cloning it is possible to take a woman's egg, remove its nucleus and put in a nucleus from a cell of her partner. The baby who would be born would have *identical* genes to those of the father. What do you think?

When does a person die? It used to be easy to say when a person died – they just got old and their various organs stopped working. But now it is becoming much more difficult to decide.

It is now possible for a person to have a transplant of various organs such as the kidneys, liver, heart and lungs to replace damaged parts of his or her body. Or the person may be connected to a machine which can take over the functions of the heart, lungs or kidneys.

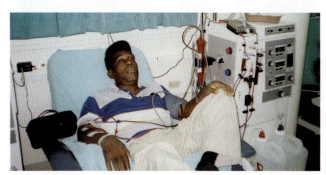

Kidney disease can cause death. This man is connected to a dialysis machine which cleans and filters the blood (see page 113)

Some people who are unconscious (in a coma) are being kept alive for years just by machines. Should the machines be switched off, and who should make the decision?

Euthanasia is the voluntary ending of life – it is legal in some countries. Do you think it should be?

Questions

1. For each contraceptive method described on pages 56–9, say how fertilisation is prevented.
2. What factors would a person consider when deciding which method of birth control to use?

Why are you just like you are?

If you look at yourself and at your parents you will see several similarities. Your nose may look like that of your mother and your ears may look like those of your father. But you will also look different from your parents, and from your brothers and sisters; you will have some different **characteristics**.

Experiments were carried out on the inheritance of characteristics in plants and animals. Gregor Mendel (1822–1884) did important work in finding out how characteristics were inherited in sweet peas. Other experimenters have found that the chromosomes (page 10) inside the nuclei of our cells contain **genes** which determine our characteristics and how we pass them on to our children.

The photograph below shows the variety of colours, shapes and sizes of pigeons. In every kind of plant or animal there is **variation** like this. A great deal of the variation is due to the particular genes (page 10) which are carried on the chromosomes.

Activity | Looking at characteristics
1. Observe and measure characteristics such as height and mass in members of your class. How do these characteristics vary?
2. Observe and record the colour of the eyes in each class member. How do these characteristics vary?

How do cells divide?
When organisms grow, more and more cells are made. One cell divides into two, and each of these later divides into two more and so on. When a cell is about to divide, each chromosome in the cell splits in half and each half becomes a whole new chromosome. The original cell has 23 pairs of chromosomes and as the cell divides in ordinary cell division (**mitosis**) the two new cells will also have 23 pairs of chromosomes. Each cell is identical to the parent cell. This is the basis on which asexual reproduction (page 39) produces offspring like the parent.

Mitosis occurs in all organs of the body except the reproductive organs where gametes are formed (page 51). Why is this? Think for a moment. If the egg your mother produced had 23 **pairs** of chromosomes and the sperm your father produced had 23 **pairs** of chromosomes, then when you were conceived (by fertilisation) you would have had 46 **pairs** of chromosomes. That is 92 instead of only 46 chromosomes! And in the next generation there would be 92 **pairs** and so on.

Instead of this happening, a special kind of reduction division called **meiosis** occurs in the formation of gametes. The result is that the number of chromosomes is halved. The parents each have 23 **pairs** of chromosomes (i.e. 46 chromosomes) but the egg and the sperm each have only 23 chromosomes (22 ordinary chromosomes and a 23rd, which is either an X or a Y chromosome). Remember from page 10, females have 22 pairs of chromosomes and two long sex chromosomes: XX. Males have 22 pairs of chromosomes and a long and a short sex chromosome: XY. The mother can only make eggs containing an X chromosome, whereas in the father half of the sperm have an X chromosome and half have a Y chromosome.

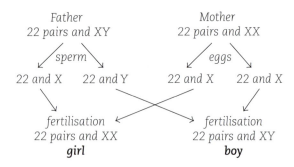

Meiosis and the formation of a girl or a boy

When an X-carrying sperm fertilises the egg the offspring has 22 **pairs** of chromosomes and XX – this grows into a girl. When a Y-carrying sperm fertilises the egg the offspring has 22 **pairs** of chromosomes and XY – this grows into a boy.

During meiosis one of each pair of chromosomes goes into a particular gamete. There is therefore a mixing-up of chromosomes. Also, the parts of the chromosomes in a pair can wind around each other (**cross-over**) and so parts of one chromosome may end up on the other chromosome of that pair.

This means that the gametes contain slightly different chromosomes and genes from each other and from the parent. This makes a variety of gametes. When these gametes fuse with each other in sexual reproduction (page 38) they give rise to offspring that, although having family characteristics, are in many ways unlike the parents. This is why *you* are just like you are – you have a unique mix of genes on the chromosomes and this gives you your unique characteristics.

Can you roll your tongue?

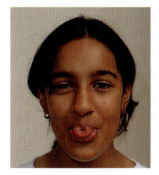

This girl can roll her tongue. Can you?

Tongue rolling is determined by just one pair of genes. Each gene can either be T (able to tongue roll) or t (not able to tongue roll). If there is at least one T gene in the pair then the person will be a tongue roller.

On the pair of chromosomes the person may have either T and T (this person will be a tongue roller), or T and t (will also be a tongue roller), or t and t (not a tongue roller).

When the T gene is paired with a t gene, the T will show its effect; we call the T a **dominant** gene. The t gene cannot show its effect when it is with a T gene and so we call the t a **recessive** gene.

When gametes are formed and fertilisation occurs it may be that recessive genes that were hidden in the parents will both come together in one of their children. For example, two tongue rollers (both Tt) could have a non-tongue rolling child (tt). The child shows a 'new' characteristic and in this respect will be different from his or her parents.

What is sickle cell disease?

This disease is caused by a gene which leads to damaged haemoglobin in the red blood cells. The cells become twisted when the amount of oxygen in the blood is reduced and this causes a **crisis** resulting in joint pains and fever.

A person with sickle cell disease has two recessive genes SS. A person with one A (normal adult haemoglobin) gene and one S (sickle haemoglobin) gene (AS) is said to have the sickle cell trait and may sometimes have minor problems. This shows what happens when the mother and father are both AS.

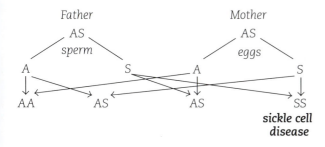

Roughly one in ten people in the Caribbean has the sickle cell trait (AS) – especially those of African and Indian descent. In the past, when drugs against malaria were not available, people with AS had an advantage against malaria and so the sickle cell gene is still in these populations. As you have seen, if two people with the sickle cell trait have children, one in four of them are likely to have sickle cell disease (SS). Because of this risk some couples with sickle cell trait go to a counsellor to talk over the situation before having children, though there are now medical advances for its treatment.

What is the Rhesus factor?

The Rhesus factor is a chemical (an **antigen**, page 161) which most people have on their red blood cells. These people are said to have Rhesus positive (Rh+) blood. People who do not have this antigen have Rhesus negative (Rh−) blood.

A problem can arise with a Rh− mother and a Rh+ father. The baby that is being carried may be Rh+ (having inherited the gene from its father). If any blood from the Rh+ baby gets into the mother there will be problems. The mother is Rh− and is not used to Rh+ blood and so she makes **antibodies** (page 161) to try and kill off this strange substance. These antibodies stay in her bloodstream. If she becomes pregnant again with a Rh+ baby then these antibodies may attack the baby's blood and destroy it. With each pregnancy there will be more and more antibodies and these may cause a baby's death. If a baby survives, it has to be given a big blood transfusion at birth to completely change its blood.

These problems **can be avoided**. At the beginning of pregnancy a woman should have a blood test. If she is Rh− and her partner is Rh+ she should have additional blood tests to be sure she has not made antibodies. When the baby has been born the mother will be given a **serum** called Anti-D. This serum destroys any Rh+ blood cells in the mother so she will not make any antibodies during future pregnancies.

What of the future?

The full set of genes in humans will be known through the Human Genome Project. Increasingly, it will be possible to transfer healthy genes into people to replace or supplement damaged ones and help overcome diseases such as cystic fibrosis and diabetes.

'Dolly', the sheep produced from some of her adult cells making an identical 'clone', opens up other possibilities. Cells from humans can now be removed, and cloned to make embryos with the same genes as the adult. The embryos are then grown for a short time in the laboratory. Cells from the embryo can be grown to make new tissues and organs. Some of these could then be transplanted back into the adult to cure such diseases as Parkinson's and Alzheimer's. How far should these developments be taken? What do you think?

What were you like before you were born?

A baby is **conceived** when a sperm fertilises an egg. The baby grows for about 38 weeks from this moment of conception until it is fully formed and ready to be born.

In order to calculate the probable birth date we count 40 weeks from the date of the start of the last period of the mother. This is because her last period would have been about two weeks before she conceived or became pregnant. Until about the twelfth week of pregnancy the developing baby is called an **embryo**. From 12 weeks until birth it is called a **foetus**.

Look at the pictures of the developing embryo. In the first few weeks it looks more like a fish than a human being, pictures (a) and (b). Only bumps show where the legs and arms will be. But in a very short time, the general shape of the limbs can be seen (c), and then the fingers and toes develop (d). A special machine using high frequency sound waves (ultrasound) can be used to show an image of the developing foetus.

The signs of pregnancy

1. The mother-to-be will miss her periods. This is because as soon as she becomes pregnant her periods stop until some time after her baby has been born. If she misses two periods she is almost certainly pregnant. She can find out with a pregnancy test, or a visit to the clinic (page 64).
2. Her breasts may feel heavy and tender. This is because of the hormones that are being produced by her body.
3. She may need to urinate more often. This is because the uterus is close to the bladder where the urine is stored. As the uterus begins to grow it presses on the bladder.
4. The mother may have 'morning sickness' in the first few months of pregnancy. This is most common early in the morning and it is reduced by eating something before getting out of bed. It is caused by changes in the body hormones.
5. She will notice a thickening of the waist before her abdomen starts to increase in size.

The fertilised egg divides into a ball of cells which becomes implanted in the uterus wall. This becomes the embryo which is attached to the placenta by the umbilical cord. The embryo develops extremely rapidly. At five weeks it is about the size of a grain of rice, but by twelve weeks it is about 60 mm long.

At 28 days the largest, most developed organ is the heart. The limbs first develop as buds; the nervous system, eyes and ears are present by six weeks. The proportions of a developing embryo are very different from those of an adult human being.

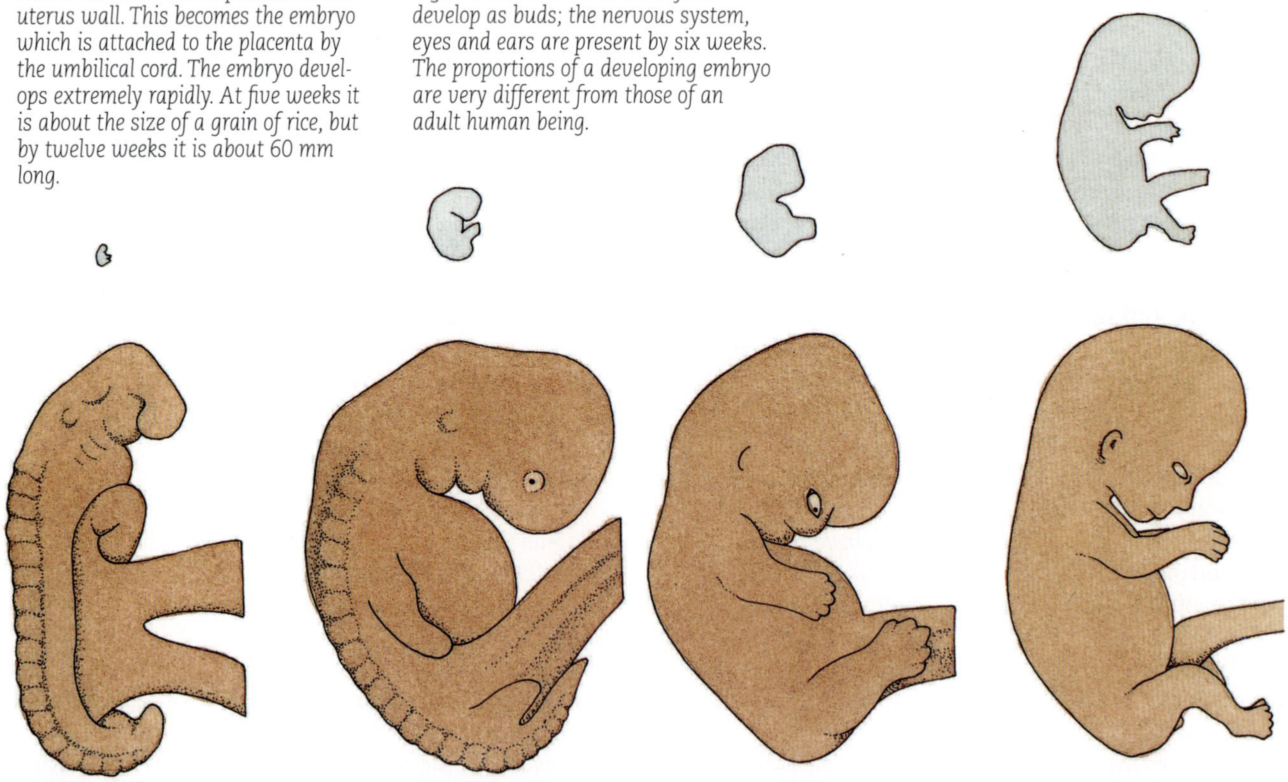

Time since last period — (a) 6 weeks — (b) 7 weeks — (c) 9 weeks — (d) 10 weeks

The developing embryo and foetus – (a) to (d) show different stages. The outline above each illustration shows the actual size

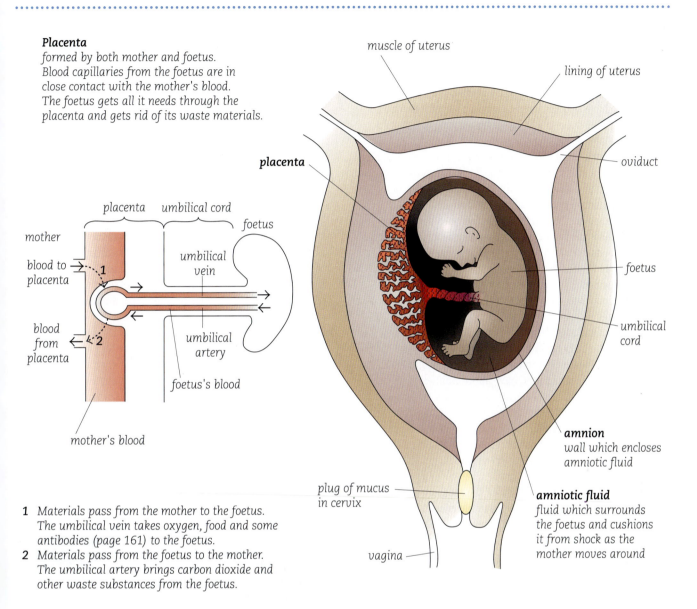

Placenta
formed by both mother and foetus. Blood capillaries from the foetus are in close contact with the mother's blood. The foetus gets all it needs through the placenta and gets rid of its waste materials.

1. Materials pass from the mother to the foetus. The umbilical vein takes oxygen, food and some antibodies (page 161) to the foetus.
2. Materials pass from the foetus to the mother. The umbilical artery brings carbon dioxide and other waste substances from the foetus.

The developing foetus inside the uterus and the functioning of the placenta

Development of the embryo and foetus

1. Conception and implantation in the uterus.
2. First three months: embryo develops basic structures and is very sensitive to drugs, germs, etc.
3. About four months: the foetus is moving actively in the uterus.
4. Fourth to eighth months: foetus continues to grow.
5. In the eighth month: the uterus is quite high up near the diaphragm.
6. In the ninth month: the baby's head usually drops down ready for birth.

Effect of development on mother-to-be

1. Mother misses her period.
2. Mother has increased frequency of urination, tender breasts, morning sickness and increase in size of waist.
3. Mother feels the movements of the foetus in the uterus ('quickening').
4. Mother's abdomen gets larger and larger.
5. Mother may suffer from indigestion ('heartburn') and feel uncomfortable.
6. Mother experiences dropping sensation ('lightening') and also increased frequency of urination.

living things | reproduction and growth | what were you like before you were born?

Why are pre- and post-natal care important?

Pre-natal care

Before a woman even knows she is pregnant, her baby's brain and heart have started to form. If you and your partner are trying to have a baby, it is important that both of you cut down on alcohol and smoking. The woman is also recommended to have supplements of folic acid for three months before and after conception. The pregnant woman should visit a pre-natal clinic to get advice.

Nutrition Nutrition of the mother is very important. She has to eat the right kinds of food (pages 74–9). The baby relies upon her for the protein, vitamins and minerals that are needed to build up its developing skeleton, brain, blood, muscles and so on. The mother will also need some additional carbohydrates and fats to supply her with the energy that she needs to carry around the increasing mass of the baby and the tissues inside the uterus.

Drugs Some drugs that the mother takes can easily travel from her blood into the foetus. This is especially dangerous in the first three months of pregnancy.

The drug thalidomide which was used in the 1960s caused some babies to be born without proper arms and legs. Mothers who smoke are more likely than non-smokers to lose their babies in a miscarriage, have abnormalities of the placenta, and produce smaller babies. Alcohol may also damage the developing baby if the mother drinks during her pregnancy. Hard drugs such as cocaine are very dangerous. Even the use of aspirins has been connected with certain birth defects. The best advice is do not drink, smoke or use drugs (see also pages 80–1).

Diseases Some diseases travel from the mother to the foetus. For example, if a mother has German measles in the first month of pregnancy there is a 50% chance the baby will be born deaf or with a heart disease. If the mother has had the disease already or has been vaccinated (page 161) there is unlikely to be a problem.

If the mother is suffering from syphilis, herpes or AIDS (pages 52–3) the baby may catch it and become seriously ill. These diseases must be treated.

X-rays X-rays can cause damage to the developing tissues of the foetus. They have been largely replaced by the use of ultrasound which works like an echo and from which the doctor can see the developing baby.

Labour and birth

Labour In pre-natal classes the mother-to-be will receive advice and be given exercises to do to help with labour. Many women continue to work throughout their pregnancy. Towards the end of the pregnancy the woman may become tired easily and may have backache.

Labour begins as the muscles in the uterus wall (page 63) begin to contract. The contractions are felt as irregular pains in the abdomen, which become more regular and closer together. The muscles open up the **cervix**: the exit from the uterus. The amnion surrounding the baby will burst to release the amniotic fluid (page 63). This is called the 'breaking of the waters'. At least by this stage the mother-to-be should be under the care of a midwife. Labour is often longer with first babies than with subsequent babies.

When the cervix is fully open (dilated), the contractions become more regular and powerful as the muscles of the uterus push down on the baby. The baby's head appears at the exit of the vagina, followed by the rest of its body.

After the baby has been born the **placenta** is also expelled. The placenta is called the 'afterbirth'.

The new baby As soon as the baby has been born it has to begin breathing for itself. A newborn baby often cries as it takes its first breath. This expands its lungs. If there is any problem the doctor or midwife will clear out the baby's breathing passages, hold it upside down, and may give it a slap on its behind.

Tying the cord When the baby is born it still has the umbilical cord attached to its abdomen by which it was attached to the placenta inside the uterus (page 63). The doctor or midwife ties the cord 10 cm and 15 cm from the baby's abdomen and makes a cut in between (see below). The stub of the cord which remains on the baby is kept clean and dry until it shrinks and drops off and the wound heals over as the navel.

Immediate care As soon as the cord has been cut, the baby is wrapped up and given to the mother to feed it. This first feed is good for the baby and also helps the mother's uterus to contract.

A clamp has been applied to the cut end of the umbilical cord of this newborn baby. Notice also that the baby's head has been moulded during its passage through the vagina. All babies are pale when they are first born. Their skin colour develops as they are exposed to light

Twins The mother may carry two babies at once (twins). They may have developed from one fertilised egg, and so be **identical twins** with similar chromosomes and of the same sex. If instead they have developed from two separate fertilised eggs they will be **non-identical twins** who may be of different sexes, and will not look any more alike than ordinary brothers and sisters.

One baby is born and then the other one, usually after only a few minutes. Then the afterbirth or placenta is expelled. When there are more than two babies, each one will tend to be smaller than if only one baby had been developing inside the uterus. With multiple births, because the uterus will be very large, there is a risk of the mother going into labour before the babies are fully developed (**premature labour**).

Whenever a baby is born early it needs to be kept warm in an incubator and may need a special oxygen supply.

Post-natal care

The baby After the baby is born it has to live on its own instead of being supplied with all its needs inside the uterus. This table compares a baby's life before and after birth.

	Before birth	After birth
Keeping warm	Kept at constant temperature by amniotic fluid	Must keep its own temperature constant
Getting oxygen and food	Brought from the placenta	Has to breathe and feed for itself
Getting rid of carbon dioxide and wastes	Taken away by the placenta	Has to breathe, defecate and urinate for itself

Keeping warm is a problem because a newborn baby has little fat under its skin. It is hard for very young babies, even in the Caribbean, to keep their body temperatures constant at 37 °C and so they should be kept warm.

Breast milk is the best food for a young baby. It is the right composition and is delivered at the right temperature, is free from germs and contains antibodies (pages 160–1) which help to protect the young baby from some diseases. This table compares human breast milk with cow's milk.

	Constituent (g/100 cm^3)			
	Sugar	Fat	Protein	Minerals
Human milk	7	4	1.2	0.4
Cow's milk	4.7	3.8	3.3	0.75

Gastro-enteritis (diarrhoea) occurs in many small babies, especially those who are bottle-fed. This is due to infection because the water for the feed has not been boiled or because the bottle has not been cleaned properly. It can be a cause of death. The baby has watery stools and may become dangerously dehydrated. The baby should be taken to the clinic.

The mother's milk becomes more concentrated as the baby grows older. But after a few weeks it can be supplemented with solid food such as baby cereal as well as orange juice and vitamins.

The mother Breast-feeding can begin soon after the birth of the baby. During the first few days a special liquid, **colostrum**, is produced, and then the regular supply of milk is made. The mother must make sure she has enough water and milk to drink and follows a nutritious diet. The baby's sucking action encourages more milk to be made, and breast-feeding is an enjoyable experience for both the mother and the baby.

Emotional reactions After the birth the new mother may have a fit of depression and feel like crying. She may feel unable to take proper care of her baby. These feelings are normal and are related to changes in her hormones and should soon disappear.

A **post-natal check-up** should take place six weeks after the birth. At this time both the mother and the baby will be checked to see if they are in good health. The mother should continue visiting the clinic to get advice with feeding and caring for her baby.

The role of the father

The father can help a lot before the birth by caring for the mother and helping her to have a good diet, and to visit the clinic regularly. As she becomes larger, she will need more help around the house. Fathers may also be present at the birth of their babies.

After the birth the mother will need a lot of love as she settles into her new role. The father can help to take care of the baby. The mother must be careful to still show love and understanding to the father who may feel jealous of the attention given to the baby.

Questions

1. Describe how each piece of advice on pre- and post-natal care of the mother and baby will help to develop a healthy child.
2. What problems might a newborn baby have, and how can they be dealt with?
3. 'Two people make a baby, and two people are needed to look after it!' What do you think?
4. Make a list of the costs of having a baby – both before and after it is born, until the teenage years.

How are plants able to make food?

Activity | Testing a leaf for starch

Starch is food. Let us see if it is present in leaves.

Starch turns iodine solution from yellowish-brown to blue-black. But leaves are green, and this green colour would get in the way of the test. So we first have to remove the chlorophyll (i.e. decolourise the leaf) by boiling it in alcohol, and then we can test for starch.

1 *Dip leaf in boiling water for 20 seconds to soften it*

2 *Turn off the Bunsen burner. Put the leaf into a test-tube of alcohol and stand it in the beaker of boiling water. Leave for 10 minutes*

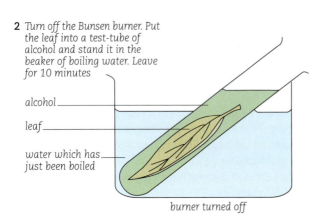

3 *Wash decolourised leaf in water*
4 *Test decolourised leaf with iodine solution*

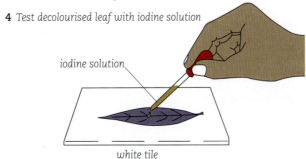

How to test for starch in leaves

You can test the leaves of several plants. Choose fairly soft leaves such as *Hibiscus* or balsam, rather than tough leaves. The starch test will show that leaves contain starch. But do the plants make it and under what conditions?

To find out if plants make starch, we first have to remove the starch which is there, and then see if the leaves can make some more.

Activity | Removing starch from a leaf

Use a plant in a pot, for example balsam or coleus.
1 Leave the plant in the dark for 24 hours. During this time the starch in the leaves will change to sugar and will be distributed around the plant.
2 How do you know if all the starch has gone? You break a leaf off the plant, decolourise it and test for starch. When there is no starch left we call the plant a **de-starched** plant.
3 Now, can the plant make some more starch? Put the potted plant in the light and leave it for 2 or 3 hours. Break off a leaf, decolourise it and test for starch.

Plants can make their own food

You will find that the plant has made some more starch. How useful it would be if animals could do this as well! But they cannot.

In what ways are plants different from animals? An important difference is that plants contain chlorophyll. This is important in making food, a process which we call **photosynthesis**.

Plants can carry out photosynthesis and do not need to eat energy-containing foods. They are self-sufficient (**autotrophic** or 'self-feeding'). They still need minerals from the soil, and certain raw materials for photosynthesis to occur, but these do not have to contain energy. Plants can use simple inorganic compounds for this process.

Animals, on the other hand, rely on plants or animals which have eaten plants, for food. They are **heterotrophic** or 'other-feeding'. They need complex energy-containing organic compounds.

What do plants need to carry out photosynthesis? You will do some experiments to find out. In each case you will have to use a **control** (page 33).

You may think that **light** is important. In this case you will need to put part of a leaf in the light, and another part in the dark and see what happens. In a similar way we can test if **chlorophyll** and if **carbon dioxide** are important. The plants you use in the tests must otherwise be identical to each other – the same kind and size. Only the one variable you are interested in should be different.

In each of the Activities you must start with a de-starched plant. Why? Because otherwise you will not be able to tell if the variable you are testing for affects the production of starch. The starch you might find could have been there from before.

Activity | Is light necessary for photosynthesis?
Use a de-starched plant. The plan is to leave part of a leaf in the light and part in the dark. Then compare the places where starch has been made.
1. Make an aluminium foil cover and cut out a shape or some letters from one side. Carefully attach the foil over a leaf growing on a plant, and hold it on with paper clips like this.

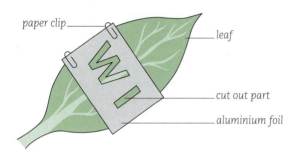

2. Leave the plant in the light for 2 or 3 hours. Then remove the leaf, decolourise it, and test it for starch.
3. Compare the parts which (a) received light and (b) did not receive light. What do you conclude?

Activity | Is chlorophyll necessary for photosynthesis?
Use a de-starched plant. Choose one which is **variegated**: partly green (with chlorophyll) and partly white (no chlorophyll). *Tradescantia* (Wandering Jew) or variegated *Hibiscus* would be suitable.
1. Draw the pattern of the green and white parts of your leaf like this.

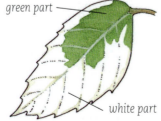

A variegated *Hibiscus* leaf

2. Leave the plant in the light for 2 or 3 hours. Remove the leaf that you drew. Decolourise it and test for starch.
3. How does the pattern of starch production compare to the distribution of chlorophyll? What do you conclude?

Activity | Is carbon dioxide necessary for photosynthesis?
Use a de-starched plant. Some leaves should receive carbon dioxide and others should receive none. Sodium hydroxide pellets can be used to absorb carbon dioxide.
1. Put some sodium hydroxide pellets into a dish. Put the dish inside a plastic bag and fix the bag around a leaf. Use another plastic bag without the pellets to enclose another leaf like this.

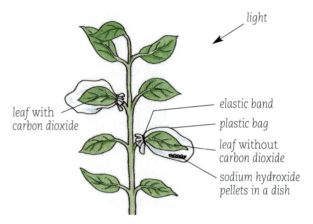

2. Leave the plant in the light for 2 or 3 hours. Then remove a leaf that was (a) with, and (b) without carbon dioxide. Decolourise and test both leaves for starch. What do you find? What do you conclude?

What happens in photosynthesis?
Scientists have shown that water is also needed for photosynthesis. Water and carbon dioxide are the two compounds which react together in photosynthesis. The chlorophyll is needed to help speed up the reaction. We call it a **catalyst**. Light provides the energy which is built up into the food. The products of photosynthesis are oxygen, and glucose which is quickly changed to starch and other compounds (page 72).

Photosynthesis can occur in any green part of plants, but most of it occurs in the leaves which are especially well-adapted for this job (pages 70–1).

Questions
1. Why do we need to remove chlorophyll from a leaf before we test for starch?
2. What is a 'control'? What is the control in each of the Activities on light, chlorophyll and carbon dioxide?
3. What is a 'de-starched' plant? Why do we need to use one when finding out what is necessary for photosynthesis to occur?
4. Why are animals unable to make their own food?

How do substances move?

On page 67 we saw that during photosynthesis water and carbon dioxide are used up, and food and oxygen are produced.

$$\text{Water + Carbon dioxide} \xrightarrow[\text{chlorophyll}]{\text{light}} \text{Food + Oxygen}$$

How do water and carbon dioxide enter the plant and how does the plant get rid of oxygen?

Water is taken in by the roots and transported up the plant in long tubes called **xylem vessels**. Some of it is used in photosynthesis and some is lost from the leaves as water vapour, in the process of **transpiration** (page 91).

In photosynthesis, *carbon dioxide* has to enter through the leaves, and *oxygen* has to pass out. The leaf will also be carrying out **respiration** in which oxygen is used and carbon dioxide produced, so both these gases will have to travel in and out of the leaf.

The structure of the leaf is adapted to help the movement of water vapour, oxygen and carbon dioxide (pages 70–1).

What principles underlie the movement of substances?

1 Substances consist of tiny particles such as atoms and molecules and it is these particles which move (pages 16–18).
2 Particles in gases and liquids are constantly moving at random.
3 As a result of their movement the particles tend to spread themselves out evenly in any space which is available to them. This is called **diffusion**.
4 Particles move, **diffuse**, from a place where they are numerous (in *high* concentration) to a place where they are less numerous (in *lower* concentration). We say they are moving along their **diffusion gradients**.

We can illustrate this as follows:

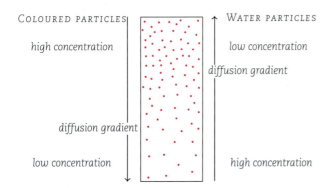

Coloured particles and water particles move along their diffusion gradients

Activity | Demonstrating diffusion

1 Your teacher will use a bag made of special material. Inside the bag is a mixture of starch particles and glucose solution.

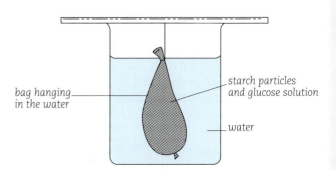

2 Before hanging the bag in the water your teacher will test the water for starch and glucose (page 75).
3 After one hour, your teacher will test the water again for starch and glucose. You will see that there is now glucose in the water, but no starch.

Explanation We can imagine that the wall of the bag (**membrane**) has small holes in it. The particles of starch are too large to go through the membrane. But the particles of glucose can pass through, along their diffusion gradient, from the place of high concentration (in the bag), to the place of lower concentration (in the water). The membrane is said to be **partially permeable**, because it allows glucose to go through but not starch. We can show this on a diagram:

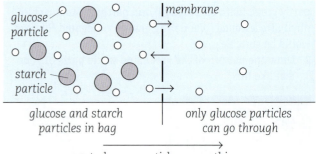

Diagram illustrating diffusion

 FACT: Diffusion is the movement of particles along their diffusion gradients. Particles tend to diffuse from a place where they are in high concentration to a place where they are in lower concentration, unless they are stopped from doing so by a membrane through which they cannot pass.

Activity | Investigating osmosis

1. Make two 'potato cups' from two halves of a potato by removing the skin and scooping out the centre parts. Make the bottoms flat.

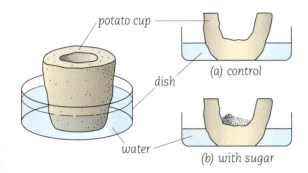

2. Place the cups in water in separate dishes. Put some cane sugar (sucrose) into one of the cups (b).
3. After 1 hour look inside the cups. You will find that in (a) (the control), there is no change. But in (b) there is a lot of liquid in the cup.

Explanation We can imagine the potato has membranes with holes that are too small for the sugar particles to go through. But the water particles can go through. The liquid in the dishes is entirely water, which is a higher concentration of water than in the sugar inside the potato cup. So water particles move along their diffusion gradient into the sugar solution. This movement of water (the solvent) is called **osmosis** and the membrane is said to be **semi-permeable**. This is illustrated below.

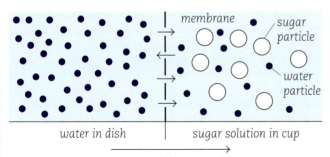

Diagram illustrating osmosis

 FACT: Osmosis is a special case of diffusion. Osmosis is the movement of solvent particles (usually water) from a place where they are in high concentration to a place where they are in lower concentration, through a partially or semi-permeable membrane which stops the passage of the large solute particles.

Where do diffusion and osmosis occur in living things?

Diffusion and osmosis are occurring all the time. The substances that enter and leave living cells do so through the cell membrane (page 10). This cell membrane is partially permeable and so allows only some substances to pass through – it controls the movement of substances.

The cell membrane will only function properly when the organism is alive. The surfaces through which movement occurs are also large and this helps diffusion and osmosis (see pages 286–7).

Plants Water enters plant roots (page 90) by osmosis. It also travels from cell to cell by osmosis. In the leaf, excess water diffuses from the cells into the air spaces, and then diffuses (still along its diffusion gradient) out into the air (pages 70–1).

Mineral nutrients in the soil diffuse into the plant roots when there is a diffusion gradient. The plant may also be able to take up important minerals by 'active uptake' (page 90).

When the plant carries out photosynthesis, carbon dioxide diffuses into the leaf along its diffusion gradient, and oxygen diffuses out. When the plant respires, oxygen diffuses into the leaf and carbon dioxide diffuses out (pages 70 and 108).

Animals When food has been broken down in the gut (page 84) it is changed from large particles (such as starch), into small particles (such as glucose) that can diffuse through the partially permeable membranes found in the walls of the gut.

In the body, oxygen diffuses along its diffusion gradients: it passes from the blood where it is in high concentration, into the cells where it is in lower concentration (page 128). In the lungs oxygen passes from the air spaces of the lungs into the blood (page 100).

In a similar way, carbon dioxide diffuses along its diffusion gradients: out of the body cells and into the blood, and out of the blood and into the air spaces of the lungs.

Questions

1. What do the following terms mean?
   (a) High concentration (b) Low concentration (c) Diffusion gradient (d) Diffusion (e) Partially permeable membrane (f) Osmosis.
2. Give two reasons each why diffusion is important to (a) plants and (b) animals.
3. In an Activity about osmosis, some students first boiled one of the potato cups for 10 minutes. They then put sugar into it and placed it in water. They found that no water entered the cup. Suggest a reason for this.
4. You have some Koolaid crystals and water. You want to make a drink and sweeten it with sugar. How will diffusion be involved?

What is a leaf like inside?

Activity | What are the parts of a leaf?
Look at a leaf such as *Hibiscus*.
1. Look at how the leaf is connected to the stem.
2. Find the parts that are labelled in this photograph.

Questions
1. What colour is the leaf? Why?
2. Is the blade thin or fat? What do you think is the reason for this?
3. Look at the veins, especially on the lower surface. What do you think are the functions of the veins?
4. How is the structure of the leaf suited to its functions?
5. Collect a wide range of leaves. How are they (a) similar and (b) different?

Activity | What is a leaf like inside?
1. Collect the following things: a container (such as a beaker) with some hot but not boiling water in it, a piece of sponge, a leaf, and a pencil or paper clip.
2. Put each object into the water, one at a time.
   (a) What do you see in each case?
   (b) How would you explain your observations?
   (c) Do you think the leaf is more like the pencil, or more like the sponge?
3. You will have seen air bubbles coming out of the sponge and out of the leaf. Look at the picture on the opposite page which shows the inside of a leaf. Which part of the leaf do you think contains the air spaces?
4. A leaf is green because it has chloroplasts which contain chlorophyll. Look at the picture again and make a list of the cells which contain chloroplasts.
5. You have seen that the veins support the leaf blade. Look at the picture and list the tissues that are found in the vein. What are the functions of these tissues?

Activity | How do gases enter and leave the leaf?
You will look at the cells on the under surface of a leaf.
1. Choose a *Hibiscus* leaf. Fold it in half with the pale under surfaces of the leaf together.
2. Slide the two halves past each other, as in this picture, so as to strip off a piece of the transparent under surface (called the **epidermis**).

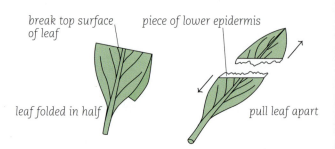

3. Put a small piece of the epidermis on a slide in a drop of water and cover it with a cover-slip (page 34). Observe it under the low power of the microscope. The epidermis consists of ordinary epidermal cells, together with pairs of cells, **guard cells**, on either side of small holes called **stomata**. Find the parts labelled in this diagram.

Stomata

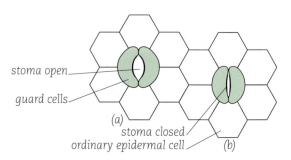

The stomata are open (a) when water enters the guard cells by **osmosis** (page 69) and makes them swollen or **turgid**. The stomata are closed (b) when water is drawn out of the guard cells so they are limp or **flaccid**.

Day and night, water is evaporating into the air spaces in the leaf. There is more water vapour inside the leaf than outside and so there is a **diffusion gradient** (page 68) and water diffuses out. Photosynthesis which occurs in the daytime, uses up carbon dioxide and produces oxygen. During both the day and night respiration uses up oxygen and produces carbon dioxide. Depending upon the balance between these two processes (page 101) there are diffusion gradients set up, and oxygen and carbon dioxide diffuse into or out of the leaf.

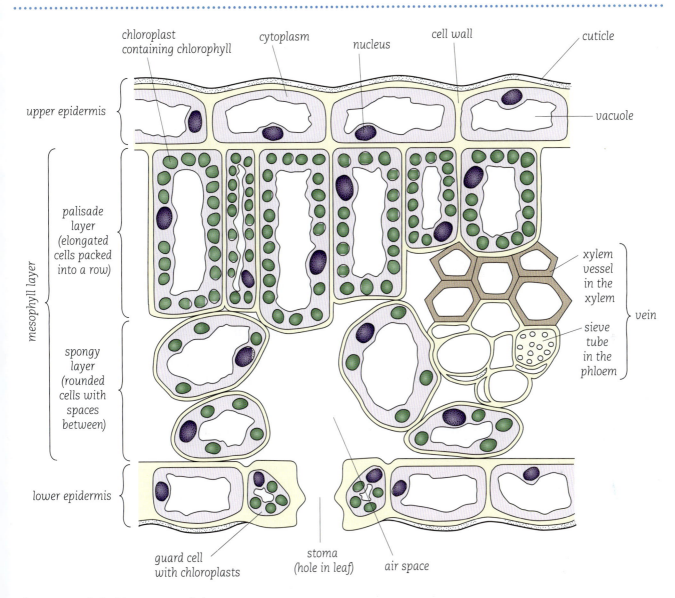

The structure of a leaf (transverse section)

How leaf structure is important for photosynthesis

1. On a plant there are usually many leaves, arranged so that they do not overlap each other. They are held at an angle to the sun so that they receive the maximum light for photosynthesis.
2. There is a main vein and a branching network of veins which give support to the softer tissues of the leaf blade.
3. The leaf blade is thin and flat which provides a large surface area to volume ratio (page 286) for the diffusion of gases.
4. The cells of the leaf contain chlorophyll which is needed for photosynthesis to occur.
5. The upper epidermis is only one cell thick and the palisade mesophyll cells are elongated, so that light can enter the leaf easily without having to pass through too many cell walls.
6. The spongy mesophyll cells have large air spaces between them which allow water vapour to evaporate, and oxygen and carbon dioxide to move freely into and out of the leaf.
7. The lower epidermis has many stomata enclosed by guard cells which control the movements of water vapour, oxygen and carbon dioxide.
8. The veins contain xylem vessels which bring water to the leaf, and phloem sieve tubes which take food away from the leaf (page 90).

How is food used by the plant?

We have seen that carbon dioxide, water, chlorophyll and light are necessary for photosynthesis to occur, and that oxygen and food are produced (page 67).

A chemical equation for photosynthesis

We can describe what happens in photosynthesis by using symbols to represent the compounds involved. For more on chemical equations see pages 22–3.

The starting substances (reactants) are water (H_2O) and carbon dioxide (CO_2). The products are oxygen (O_2) and glucose ($C_6H_{12}O_6$). The catalyst which speeds up the reaction is chlorophyll, and light provides the energy to make the reaction possible.

We can summarise the reaction as follows:

$$\text{Water} + \text{Carbon dioxide} \xrightarrow[\text{chlorophyll}]{\text{light}} \text{Glucose} + \text{Oxygen}$$

$$6H_2O + 6CO_2 \longrightarrow C_6H_{12}O_6 + 6O_2$$

Activity | The production of oxygen

You will need a water plant such as Canadian pondweed (*Elodea*).
1. Put some of the pondweed under an upside-down glass funnel in a beaker of water.
2. Fill a test-tube with water, and put your finger over the open end. Carefully upturn the test-tube over the stem of the funnel like this.

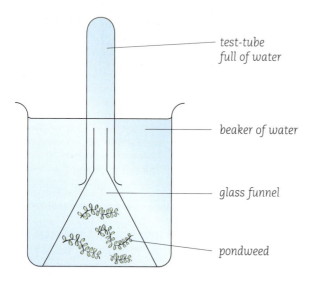

3. Observe what happens when the pondweed is left in the light.
4. Test the gas which collects in the test-tube with a glowing splint. It should relight. This shows the presence of oxygen.

The production of food

The glucose produced in photosynthesis is built up into starch which is stored in the plant. It is this starch which we test for to show that photosynthesis has occurred (page 66). Glucose can also be built up into cellulose which forms a major part of cell walls.

The mineral salts, such as nitrates and sulphates, which are taken in from the soil are combined with substances produced from glucose to make amino acids. These are built up into proteins (page 74) which are needed for the growth of the plant, and can be stored, especially in seeds and fruits.

Fatty acids and glycerol are also formed and are the building blocks of fats and oils (page 74). They are mainly used for food storage in seeds and fruits.

Every day the plant stores food. When needed some of this food is broken down and respired to release energy. Animals eat the plant food, digest it and also carry out respiration to release the trapped energy.

Building up molecules

Small molecules such as glucose are joined together to make complex molecules such as starch. At each link position a molecule of water is removed. This kind of reaction occurs wherever food is built up for storage.

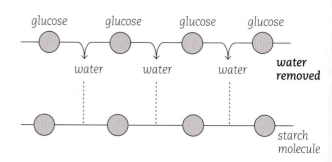

Breaking down molecules

Complex molecules such as starch are broken down into simple molecules such as glucose. At each link position a molecule of water is added. This kind of reaction occurs wherever food stores are broken down.

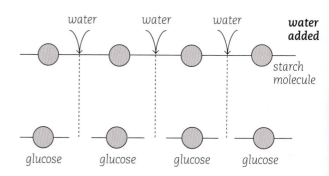

Activity | Plant storage organs

Excess food which the plant makes is stored for future needs and accumulates in particular parts such as roots or stems. These parts become fat and swollen and are called **storage organs**.

Select a variety of plant storage organs such as those shown below. In each case identify the particular part which is storing food.

Questions

1. Write a word equation and a chemical equation to describe photosynthesis.
2. (a) How are starch, cellulose, proteins and fats made by the plant? (b) Of what use are these substances to plants? (c) Of what use are these substances to animals?
3. How are foods (a) built up and (b) broken down?

A variety of storage organs. Name each one

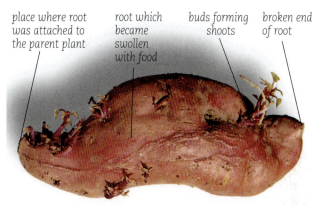

Sweet potato – a root tuber

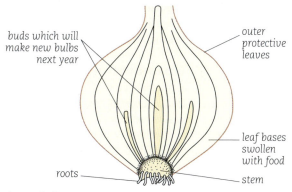

Onion – a bulb

Peas – seeds inside the fruit (pod)

How do plants use the stored food?

When stored food is needed it is broken down to the simple building blocks: glucose, amino acids, fatty acids and glycerol. These substances then travel around the plant. They are respired (combined with oxygen) to release their energy for everyday needs. The stores are also useful for energy release and growth:

1. during bad conditions such as drought when plants may not be able to make enough food
2. for the growth and development of the plant
3. for vegetative reproduction as buds grow out to make new plants (see also page 39)
4. for the developing embryo, from food stores in the seeds and fruit.

How do animals use plant food?

Animals, including humans, eat plants (especially plant storage organs) to obtain nutrients and roughage (page 74). The foods are broken down during digestion, and are later combined with oxygen to release energy.

1. Storage crops such as cassava, yam, eddo, dasheen and sweet potato are a major source of starch.
2. Grains (cereals) such as rice, maize, oats and wheat (which is used to make flour) also provide animals with starch and some vitamins.
3. Seeds such as beans and peas provide animals with starch and protein.
4. Fruits such as mango, tomato, pawpaw and citrus provide us with sugars, roughage and vitamins.

What are you eating?

Your food consists of the storage organs and other parts of plants, and of parts of animals or products made from them. Plants and animals supply your basic nutrients.

Proteins

The individual building blocks of proteins are **amino acids**. Two amino acids joined together are called a **dipeptide**. More than two amino acids joined together are called a **polypeptide**. A large polypeptide is called a **protein**.

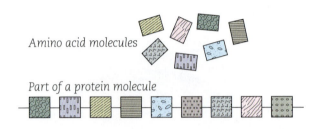

Proteins contain carbon, hydrogen, oxygen and nitrogen, and often sulphur and phosphorus as well. We use proteins for growth and repair and to make substances such as enzymes, antibodies, nails, and haemoglobin in the blood.

Foods rich in protein are meat, fish, eggs, milk, peas and beans.

Fats

Fats contain two parts: glycerol and fatty acids.

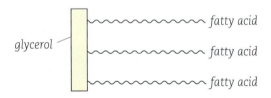

Different kinds of fatty acids make different kinds of fats. Fats from animals are mainly **saturated** (the fatty acid parts have as many hydrogen atoms as possible) – these are found in red meat, butter and full-fat milk and cheese. Excess intake of these can raise the level of **cholesterol** (page 97) in the blood which can cause atherosclerosis. Fats from plants (for example, nuts), and most of those in fish and chicken are mainly **unsaturated** (the fatty acid parts could contain more hydrogen atoms). The best fats for health are **polyunsaturated**, for example sunflower, corn, soya-bean and olive oil.

Animal fats are usually solid and plant fats (oils) are usually liquid at room temperature. Fats contain carbon, hydrogen and oxygen only. Fats are used for insulation and energy. Removing visible fat from meat and skin from chicken, using cooking methods other than frying, and drinking semi-skimmed milk can help reduce our fat intake.

Carbohydrates

Carbohydrates contain carbon, hydrogen and oxygen. They are built from small molecules such as glucose.

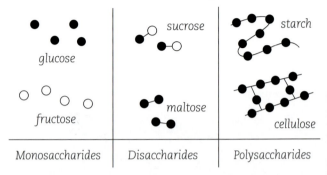

Sugars Sugars are either **monosaccharides** or **disaccharides**. They usually taste sweet. They are used in our bodies as a source of energy.

Some sugars give a red solid (or precipitate) when heated with solutions called Benedict's and Fehling's solutions. These sugars are called **reducing sugars**. All monosaccharides are reducing sugars. Some disaccharides, such as maltose, are also reducing sugars.

Sucrose is a disaccharide which is a **non-reducing sugar**. It does not give a red precipitate when heated with Benedict's or Fehling's solution. Sucrose first has to be broken down (by boiling it with acid) into its simpler parts which will then give a positive test.

Sucrose is the sugar extracted from sugar cane which is used in sweets, cakes, biscuits and so on. Glucose and fructose are found in fresh fruit. Glucose is also the end product of the digestion of starch. It is better for health to take our sugars in fresh fruit than as refined sugar.

Starch Starch is a **polysaccharide** made of many glucose building blocks. It is the storage substance used in our bodies to provide glucose as a source of energy.

Foods rich in starch are plant storage organs such as yams, cassava, dasheen, Irish and sweet potatoes, some grains such as rice and maize and also peas, beans and foods made from flour, such as bread and cakes. These foods also provide us with fibre (see below).

Cellulose (dietary fibre) Cellulose is made of glucose building blocks. It makes up the cell walls of plants. It is found in fruits, vegetables, storage organs and wholemeal bread.

We cannot digest cellulose. Its function in our body is as roughage to give bulk to the material in the alimentary canal and help prevent constipation.

Water (page 138), minerals and vitamins (pages 78–9) are also an important part of our diet.

The food tests for proteins, fats, starch, reducing and non-reducing sugars are described opposite.

Activity | Testing for protein (Biüret test)

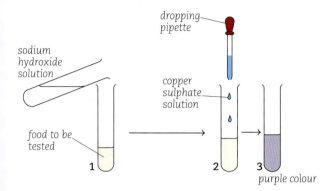

First test milk, then try other food items.
1. Put 2 cm³ of the food in a test-tube and add the same amount of dilute sodium hydroxide solution.
2. Add 1% copper sulphate solution from a pipette, drop by drop. Shake the test-tube.
3. A purple colour shows the presence of protein.

Activity | Testing for fats (emulsion test)

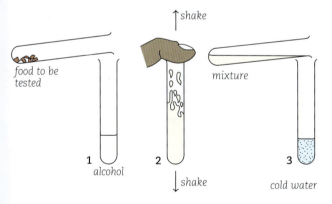

First test margarine, then try other food items.
1. Add 1 cm³ of the food to 2 cm³ of alcohol (for example, ethanol) in a test-tube.
2. Put your thumb or a cork over the top of the test-tube and shake the tube thoroughly.
3. Add the mixture to a test-tube with 2 cm³ of water. A milky appearance (drops of fat) indicates fat is present.

Activity | Testing for starch (iodine test)

First test cut-up rice grains, then try other food items.
1. The food can be either solid or liquid. Add a few drops of iodine solution.
2. The straw-coloured iodine solution will go blue-black to show the presence of starch.

Activity | Testing for reducing sugars (Benedict's test)

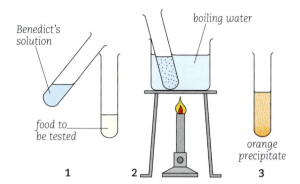

First test glucose, then try other food items.
1. Add 2 cm³ of Benedict's solution to 1 cm³ of the food in a test-tube.
2. Put the test-tube into a beaker of boiling water for about 5 minutes.
3. An orange-red precipitate shows the presence of a reducing sugar.

Activity | Testing for non-reducing sugars

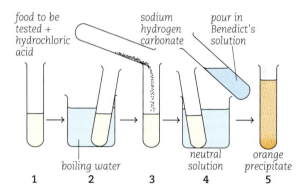

First test sucrose, then try other food items. The substance to be tested must have given a negative result in the test for reducing sugar.
1. Add a few drops of dilute hydrochloric acid to 1 cm³ of the food in a test-tube.
2. Put it into a water bath of boiling water for about 3 minutes to break down any sugar into reducing sugars.
3. Cool. Then add a little sodium hydrogen carbonate until the fizzing stops. (This is to neutralise any left-over acid.)
4. Add 2 cm³ of Benedict's solution and put it back into the water bath for about 5 minutes.
5. An orange-red precipitate shows there is now a reducing sugar. This means the food originally contained a non-reducing sugar.

living things | food and nutrition | what are you eating?

What are you really eating?

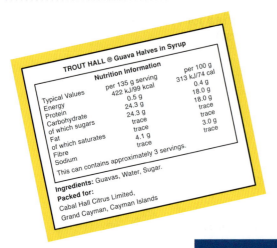

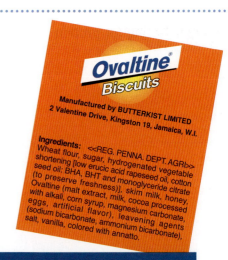

Labels from some foods on sale in the Caribbean

Activity | What is in our food?

1. The picture above shows labels taken from different foods. Examine them carefully. The ingredients can be divided into two groups, the **natural products** (as described on page 74, such as energy-providing foods, minerals and vitamins) and the **food additives** (such as food colourings and preservatives).
2. Make a full list of the natural products: the substances which occur naturally in these foods.
   (a) Energy-providing foods such as green peas, guavas, oats, brown sugar, fat.
   (b) Minerals such as salt, iron.
   (c) Vitamins such as vitamins A and C.
   (d) Natural flavourings such as salt, pepper and spices.
   (e) Indicate whether there is any water present. If no water is present we say that the food has been dried or **dehydrated**.
3. Make a full list of the food additives: the synthetic compounds added to these foods.
   (a) Food colouring such as yellow colouring.
   (b) Preservatives such as sodium nitrite.
   (c) Flavour enhancers such as monosodium glutamate.
   (d) Gelling agents such as gelatine and agar which help food to 'set'.
   (e) Emulsifiers such as lecithin which keep fat particles evenly mixed in the food.
   - Repeat steps 2 and 3 with other foodstuffs.

Taking care what we eat: natural products

These are the carbohydrates, fats, proteins, vitamins, minerals, water, natural flavours and spices which make up the major part of the food, and are listed in the ingredients.

Carbohydrates Too much starch and sugar can lead to obesity (page 47). Starches and sugary foods which stick to our teeth also contribute towards tooth decay (page 83). Cellulose (dietary fibre) can help to prevent constipation (page 85).

Fats Too much fat can lead to obesity. Animal fats (saturated fats) have been linked to the build-up of material (atheroma) in the arteries which may predispose certain people to heart attacks or strokes (pages 96–7). To help avoid these problems we should eat less fat and try to replace animal fats with unsaturated plant fats.

Protein There are nine essential amino acids which we need in our diet and cannot make for ourselves. Soya bean and most animal proteins contain all these essential amino acids, whereas different plant proteins lack one or more of them. A vegetarian must eat a wide variety of plant proteins to get these essential amino acids.

Taking care what we eat

Food additives These are mostly synthetic compounds added, in small amounts, for different reasons. Food additives are listed in the ingredients and many of them have E numbers.

Colours are mainly synthetic dyes which are added to most processed foods to restore or improve their appearance and so encourage people to buy them. However, tartrazine (a yellow food colouring used in products such as margarine and orange juice) has been withdrawn from use because it is carcinogenic (page 47).

Preservatives are added to most foods which are sold moist, because in the presence of water and oxygen bacteria can grow on the food (page 158) and cause decay. For example, processed meats such as bacon and corned beef may contain a chemical called sodium nitrite. This makes the food safe from botulism bacteria which could kill us. But, the nitrite produces chemicals in the body which can be carcinogenic. For this reason processed meats should never be given to babies and infants. To reduce our intake of preservatives, we should eat more fresh foods, and foods which have been preserved by drying rather than by the addition of chemicals.

Flavourings and flavour enhancers These are synthetic compounds which are added to improve flavour and make food more attractive. An example of a flavouring is saccharin which was added to increase the sweetness of certain foods and bottled drinks. In large quantities it was found to be carcinogenic in mice, and is no longer used on a large scale. People can, however, still use it in safe quantities to sweeten tea and coffee.

An example of a flavour enhancer is monosodium glutamate which is used in many Chinese dishes. This causes a reaction in some people who are sensitive to it.

Other concerns There has been recent discussion on two other issues.

Organic food All food is organic in that it consists of energy-containing carbon compounds (pages 24–5). But when people talk about *organic food* they usually mean 'food produced without the use of synthetic fertilisers or pesticides, and without the widespread use of drugs and antibiotics for rearing animals'. However, the cost of producing these foods is higher. Practical suggestions are to wash food well before eating, and to grow as much of your own food as possible.

Genetically modified (GM) foods These are foods where genes have been modified or new genes added to food plants, animals or micro-organisms. Genetic modification, especially of crops, allows disease resistance, and an increase in size and rate of growth. It is very unlikely that the foods would be harmful to humans, but it might be possible that some of the changes made to domestic plants could escape into wild plants. Careful monitoring of tests is necessary to avoid possible problems.

Food contaminants

These are organisms or chemicals which have got onto or into our food by accident. We do not usually know that they are there, unless they make us ill!

Micro-organisms Micro-organisms may have got onto the food from flies or when the food was handled (pages 154 and 158). Cleanliness, care in handling, use of refrigeration, eating fresh food and careful cooking can all help to avoid problems (page 153).

Pesticides Pesticides may be applied to crops at too high a dose, or too close to harvest time. Pesticide residues may then be found on or inside our food. Some insecticides are very resistant to being broken down and they accumulate, especially in the fatty part of meat, fish, poultry and dairy products.

EDB (ethane dibromide) is a pesticide, used in the growing of grain. EDB has been found to cause cancer. EDB residues were found in flour, peanuts and cornbread mix in the USA and 77 products were withdrawn from the stores because of this.

Animal hormones Animal hormones, for example DES (diethylstilbestrol), have been fed to beef cattle and chickens to improve growth. DES is related to the female hormone oestrogen and when it is found in the food we eat, it can cause the development of breasts in young children of both sexes. It may also cause cancer.

Pollutants Pollutants such as lead (from gasoline), sulphur dioxide (from burning coal) and caustic fumes (from processing bauxite) can end up on our food. This is one reason why it is important to use lead-free gasoline.

Fertilisers Fertilisers added to the land may be washed into rivers which are used for irrigation and for drinking purposes (page 147). Fertilisers and pesticides are also sprayed onto fruit and vegetables. Unfortunately we cannot see these when we buy the food.

What can we do about food contaminants?

There are advantages in using pesticides, fertilisers and hormones for the growing and rearing of food, but do the dangers exceed the benefits?

We do not know if a food contains or has a food contaminant on its surface. But we can wash all fresh fruit and vegetables before eating or cooking them. This removes any germs, fertilisers or pesticides. We could also remove the outer skin, but sometimes, for example with Irish potato, we may be removing valuable nutrients. We can also trim fat from meat, and eat as much fresh food as possible. Growing our own food without using chemicals is also a good way to avoid problems. Dried food can be made up with water when we need it. This reduces the problem of the foods going bad. For more advice on food handling and cooking see page 153.

What is a balanced diet?

A balanced diet supplies all our needs.

An ideal balance of nutrients	
55% carbohydrates	Energy-giving
15% fats and oils	
20% protein	Body-building
10% water, minerals vitamins and fibre	To maintain health

People of different ages, sexes and occupations need different amounts of energy. This is measured in kJ/g (page 105).

Energy requirements in kJ	
Child 2–3 years	5 900
Child 5–7 years	7 500
Girl 12–15 years	9 600
Boy 12–15 years	12 000
Sedentary job: woman	9 000
man	10 000
Active job: woman	12 000
man	13 000
Pregnant or nursing woman	13 000
Very active job	16 000

An unbalanced diet

Malnutrition Malnutrition can mean eating too much fat and carbohydrate so we become obese (page 47). It is also a diet with insufficient proteins, vitamins and minerals especially in young children, pregnant and breast-feeding mothers. Before birth and in the first two years, protein is especially important for brain development.

A disease called **kwashiorkor** may develop in children without sufficient protein. Their bellies become swollen, their skin scaly and their hair may fall out.

Starvation Starvation occurs when, over time, the body needs more energy than is being supplied. At first the fat stores are used up, and the person becomes skinny. If the food is still inadequate, the proteins which make up the muscles of the body are digested to provide energy. This state of starvation (**marasmus**) is especially common in children. The child will have a small body in relation to the size of its head, with thin limbs and a swollen belly.

Slimming Slimming is like starvation, but the person intentionally eats less food for a short period of time, so the fat stores are used up and the person becomes slimmer. Sometimes this is taken to extremes and the person becomes ill with **anorexia**, which is the excessive loss of mass, and may have to be treated in hospital. The person may need help to re-establish eating habits and body mass.

The vitamins and minerals we need are shown in Tables 1 and 3. Table 2 gives nutritional information about Caribbean foods.

Table 1 The vitamins we need

Vitamins	Some sources	Why they are needed	Results of deficiency
A	Carrots, mangoes, spinach, red peppers, liver, margarine butter, milk	Aids growth and helps maintain healthy tissues in throat and lungs. Part of pigment in rod cells of retina	Infections of respiratory system. **Night blindness**, from lack of production of pigment in retina
B complex	Wholemeal, cereals (including brown rice), milk, liver, meat, green vegetables	Help in chemical reactions such as respiration. Also for healthy skin and muscles and prevention of anaemia	Tiredness, loss of energy and mouth sores. **Beri-beri** (muscular weakness and paralysis) where white rice is major part of diet
C	Fresh vegetables and fruit especially citrus fruits (oranges, limes, grapefruit)	For healthy skin and gums. Helps cuts and sores to heal properly. May help combat colds	Bleeding gums and slow-healing sores. **Scurvy** (pains in joints, bleeding gums) with lack of fresh fruits and vegetables
D	Fish oils, butter, eggs. (Made in the skin in sunny conditions like the Caribbean)	Helps body absorb calcium from our food for strong bones and teeth. Especially important during growth	Poor bones and teeth. **Rickets** (soft leg bones which curve) found in children with deficiency
E	Wheat germ, eggs, dark green vegetables	Probably helps protect cells from damage	Deficiency unlikely

Table 2 Nutritional value of common Caribbean food

Constituents of food/100 g edible portion

Name of food	Protein (g)	Fat (g)	Carbo-hydrate (g)	Calcium	Iron	Vitamin A	Vitamin B complex	Vitamin C	kJ
Banana, plantain	1	–	14	+	+	+	+	++	260
Bread (brown)	8	1.1	49	+	++	–	++	–	1 000
Cabbage	1.5	–	3	+	+	++	+	+++	84
Carrot	1	0.3	10	+	+	+++	+	++	202
Coconut flesh	4	38	12	+	++	–	+	–	1 697
Groundnut	26	46	10	++	++	–	++++	–	2 344
Irish potato	2	–	19	+	+	+	++	++	353
Kidney bean	24	2	48	+++	+++	–	++	–	1 285
Maize grains	10	4.5	70	+	++	+	++	–	1 461
Oranges	0.8	–	10	+	+	+	+	+++	185
Pawpaw	0.6	–	9	+	+	++	+	++++	160
Pumpkin	1.2	0.1	7	+	+	++	+	++	147
Rice (unpolished)	8	2.0	76	+	++	–	++	–	1 487
Soya bean	35	18	12	+++	+++	–	+++	–	1 470
Spinach	4	–	5	+++	++	+++	++	++++	168
Sugar (brown)	0.2	–	96	+	++	–	+	–	1 634
Sweet potato root	1.8	0.7	27	+	+	+++	++	++	508
Sweet potato leaves	2.3	–	4	+++	++	+++	+	++++	118
Tomatoes	1	–	4	+	+	++	+	+++	88
Yam	2	–	25	+	++	–	+	++	454
Beef (lean)	22	8	–	+	++	–	+++	–	672
Butter	0.6	82	0.4	+	–	++	+	–	3 108
Chicken	19	7	–	+	++	–	+++	–	584
Eggs	13	11	–	++	++	++	+	–	634
Fish, sea-water	19	1.0	–	+	++	+	+++	–	340
Fish, dried (salt)	42	4.0	–	+++	++	+	+++	–	857
Liver	18	5	–	+	++++	++++	++++	–	491
Milk (cow's)	4	4	4.6	+++	–	+	+	+	294
Pork (lean)	14	35	–	+	++	–	++	–	1 558

Key – = none present or not known   +, ++, +++ and ++++ = increasing amounts of particular constituent

Table 3 The minerals we need

Minerals	Some sources	Function in the body
Sodium chloride (salt)	Seafood, processed food, especially meats	Part of blood plasma. Needed for digestion and passing of nervous impulses
Potassium	Banana, avocado, citrus fruits, milk, green vegetables	Necessary for healthy skin, normal growth and passing of nervous impulses
Phosphorus	Milk	Part of bones and teeth. Needed in chemical reactions of respiration
Calcium	Beans and peas, spinach (calaloo), milk	Part of bones and teeth. Deficiency causes soft bones and tooth decay
Iron	Liver, spinach, beans and peas	Part of haemoglobin. Deficiency causes **anaemia** (weakness and tiredness)
Iodine	Seafood, table salt	Part of thyroxine. Deficiency causes goitre (swelling in the neck)

Questions

(Also refer to pages 74–5.)

1. What are the components of a balanced diet? Give an example of a food which contains each component.
2. How would you test for the presence of each of the energy-containing foods?
3. How do energy requirements differ with sex, age and occupation?
4. Why are vitamins important in the body? Choose three vitamins and for each one name some foods in which it is contained, why it is important, and what problems a lack or deficiency might cause.
5. Choose two minerals and for each one name some foods in which it is contained, why it is important, and what problems a deficiency might cause.
6. Which groups of people might need to supplement their diet with vitamin pills and minerals, and why?
7. What are the causes of (a) kwashiorkor (b) marasmus and (c) anorexia?
8. How are starvation and slimming (a) similar and (b) different?
9. What are the (a) advantages and (b) disadvantages of a vegetarian diet?

Use and abuse of drugs

How can drugs be useful?

Drugs are chemical substances used as medicines. Some drugs can be useful, such as antibiotics, insulin and painkillers. All drugs should be taken exactly as prescribed.

Antibiotics can be used to treat infections caused by bacteria, such as gonorrhoea and syphilis. Antibiotics are **not** effective against viral diseases such as AIDS. However, they can be used against some of the infections, such as TB, that can infect people with reduced immune system protection. What is important is that the full course of antibiotics be taken, otherwise resistant strains of TB can develop.

Insulin, for the treatment of diabetes (pages 113 and 127), is a life-saving drug which allows diabetics to lead a normal life. Measured amounts of insulin have to be given with care and in response to the testing of the patient's blood.

Painkillers also have to be treated with care. Aspirin and paracetamol tablets should not be given to children under 12 years of age. Instead, they can be given small amounts of paracetamol elixir. Everyone should avoid having too much paracetamol as it can damage the liver, and large doses of aspirin can cause death. Special 'enteric-coated' aspirin are prescribed long-term for people who have had a heart attack or stroke, or as a preventative measure. These pills help reduce the risk of heart attacks, clots and strokes.

How are drugs harmful?

Other substances that are not prescribed medicines are also called drugs. These have an effect on the body – usually a short-term pleasurable effect – which is often followed by a bad reaction. These drugs can have a harmful effect on your health if they are used in excess. For example:

- **Nicotine** in cigarettes. This is an addictive drug which raises the heart beat and blood pressure. In pregnancy it can cause a miscarriage, or an under-weight baby. The greater the number of cigarettes smoked, the greater the danger of contracting bronchitis, lung cancer and having a heart attack (pages 102–3).
- **Caffeine** in coffee, tea, cocoa, and cola drinks. Caffeine stimulates the nervous system, makes a person more energetic and causes more excretion of water (it is a diuretic). More than five cups a day can cause restlessness and trembling. Caffeine should be avoided in the first three months of pregnancy.
- **Alcohol**. Some of the short-term effects are described on page 253. Alcohol is a relaxant, and as with caffeine, it is a diuretic. Drinking large amounts of alcohol can cause loss of concentration and coordination, slurred speech, blurred vision and vomiting. Heavy long-term drinking can damage the liver (cirrhosis), and the heart and lead to general ill health. Other problems include disturbance of the general diet and lifestyle, and possible violence.

Dangers of drug abuse

- There is *no* safe dose nor safe number of times you can use a drug. Many people, often teenagers, have died after their first try. You *cannot* experiment and be safe.
- The 'down' that comes after a hit or a trip is much worse than you felt before using the drug.
- The pleasurable effects of the drugs are short-lived, and as the body's senses have been affected you are at greater risk from accidents.
- Many drugs 'hook' you so you need to keep taking the drug to get the effect, and usually you need more of the drug to get the same effect.
- Drugs are expensive, so a habit can lead to stealing in order to get money for the habit.
- It is very dangerous to mix drugs, or to take them with alcohol – the combined risks are worse.
- Injected drugs, such as heroin, carry the additional health risk of hepatitis or HIV/AIDS through the use of infected needles.
- Drugs purchased may be mixed with other substances to increase the dealer's profit, but these can cause even worse health risks for the user.
- Using or supplying illegal drugs can lead to fines or imprisonment, and may bring other problems.

Avoiding drugs

Coming off drugs causes unpleasant 'withdrawal' symptoms. Some drugs are *very* difficult to give up. It is better *never* to take drugs so that there is no risk of becoming addicted.

- It is *not* necessary to do as your friends do. Find someone else like yourself who wants to keep control of their lives and start a **S**ay **N**o to **D**rugs club (SND). Make slogans and posters and get support from teachers and parents to set up club activities and encourage other students to join.
- Tell everyone you meet that drugs *don't* solve problems. Most countries have groups which supply information to help prevent drug abuse. Collect this information and ask if people could come and speak at your school.
- Explore other ways to spend your time such as: start a new hobby; join a science, church or adventure club; take part in games or learn a new language or musical instrument. If these facilities aren't available, again talk to teachers and parents and see what can be set up. It is so much more worthwhile to do useful things with your life.

FACT: *If you start using drugs, it means you are giving up the control of your life. You risk damaging your health and your family, and you could be on the path to ending your life. Take good care of your life, say NO to drugs – they just aren't worth it.*

A few of the drugs shown in this table can be prescribed medicinally, but most have a bad effect on health.

Types of drugs	Use and effects	Long-term problems
Stimulants ('pep' pills)		
▪ Amphetamine (speed, uppers)	Powder sniffed or injected. Increases physical and mental processes. Weight loss. Dilated (enlarged) pupils, sleeplessness	To maintain the effect users take increasing amounts. High doses can give panic attacks and the body sometimes needs two days to recover. Feel depressed and very hungry. Resistance to disease is lowered, with serious effects on health
▪ Cocaine (coke, snow)	Powder sniffed or injected. Stimulates and heightens sensations. Dilated pupils. Short 'high' leads to addiction	Cocaine, including crack (more dangerous), gives an exhilarating effect which comes quickly and goes quickly. The drug then has to be taken more often to maintain the high and this can lead to dependence, or going on to other drugs, such as heroin. Over time, exhilaration is replaced by sickness, sleeplessness and weight loss. Sniffing cocaine can damage the nose membranes which is very painful
▪ Crack cocaine (rock)	As with cocaine. Made into crystals. Smoked, or heated and vapours inhaled. Hard come-down	
▪ Ecstasy (E, MDMA)	Tablets, liquid or powder. Gives feeling of energy and friendliness, followed by a feeling of misery	Large amounts cause anxiety, confusion and lack of coordination, making driving dangerous. May damage certain brain cells and the liver. Can cause death
Depressants		
▪ Barbiturates, for example Phenobarbitone (page 253)	Sedative. Extreme tiredness, lack of coordination	Sleeplessness, double vision, possible death from overdose especially if used with alcohol. Ulcers at injection site
▪ Tranquillisers, for example Benzodiazapines, such as Valium, Librium, Mogadon	Sedatives. Relieve anxiety. Can be prescribed as medicines in small amounts. Very dangerous with alcohol or if injected. Users drowsy and forgetful	Some tranquillisers that are intended for short-term support, become addictive and cause more problems. Can cause temporary memory loss and users trying to quit may suffer panic attacks. Withdrawal has to be very slow over a long period of time
Narcotics		
▪ Opiates, for example opium, morphine, Methadone, Pethidine	Prescribed as short-term medicines for relief of pain. Can become addictive	Small amounts can be prescribed: morphine to relieve pain, for example in cancer patients; Pethidine during child birth; Methadone helps addicts kick heroin addiction. Opiates can become addictive drugs
▪ Heroin (smack)	Injected, sniffed or smoked. Gives 'high' but bad come-down	Some addictive users may need the drug just to feel normal. Overdose can cause coma and death. Injecting can damage veins and users risk infections of hepatitis or HIV/AIDS
Hallucinogens		
▪ Cannabis (ganja, dope, grass, weed, marijuana)	Smoked as joints. Relaxant, mood swings, dilated pupils, red watery eyes, lethargy Hash is concentrated cannabis usually smoked in a pipe	Takes effect quickly. Impairs learning and concentration – therefore dangerous to drive. Long-term smoking of cannabis is more dangerous than tobacco (page 102). Users are tired, lacking energy, anxious. Not considered addictive, but users come to rely on it to feel more relaxed socially
▪ Hashish (hash)		
▪ LSD ('acid')	On squares of paper dissolved on the tongue. After about an hour have a long 'trip' with distorted vision and depression	'Trip' involves distorted vision and hearing or feeling of being outside the body. Bad trip can be terrifying but can't be stopped. Later you may relive parts of trips as flashbacks. Use complicates mental health problems
▪ Magic mushrooms	From special wild mushrooms. Raw, dried, cooked. Quicker effect and shorter trip than LSD	Problem to distinguish edible mushroom from poisonous ones can cause death. Cause vomiting, diarrhoea, stomach pains. Use complicates mental health problems
Solvents		
▪ Solvents sniffed as vapour, for example glue, paint thinner, aerosols, dry cleaning solvents, lighter fuel, gasoline	Vapours sniffed. Causes vomiting, nose bleeds, fatigue, lack of appetite and coordination. Effect similar to alcohol. Can become unconscious and choke on vomit	Effects short but users repeat the dose. When affected, accidents can occur. Vomiting, black-outs and heart problems can be fatal. Squirting gas down the throat can also cause a reaction which floods the lungs and causes death. Long-term abuse damages brain, liver and kidneys

living things | *food and nutrition* | *use and abuse of drugs*

How do we use our teeth?

Our teeth are important in eating our food

Activity | Eating a cracker
Each person will need two small crackers.
1. For the first cracker pretend that you do not have any teeth.
   (a) Put the cracker into your mouth.
   (b) Now try to eat it using only your tongue and the saliva in your mouth.
   - Can you break up the cracker?
   - Would you be able to swallow it easily?
2. Now take the second cracker.
   (a) Bite off a piece.
   - Which teeth did you use?
   (b) Continue to eat the piece of cracker.
   - Which teeth do you use now?
   - What do they do?
3. In what ways do our teeth help us to eat our food?

Activity | Different kinds of teeth
1. Open your mouth and use a mirror to look at the teeth in your bottom jaw.
   - Do you have any **fillings** in your teeth?
   - Do you have any decayed or broken teeth?
2. Look at the picture on the right which shows the teeth an adult should have on each side of the mouth. Identify your **incisors, canines, premolars** and **molars**.
   - Do you have the same numbers of each kind of tooth as are in the picture?
3. If you do not have the full number of adult teeth, there are two possible reasons.
   (a) Look and see if there are spaces between your teeth. This may be where you have had a tooth removed (**extracted**).
   (b) You may not yet have got all your wisdom teeth (the molar teeth at the back of the mouth). These often come out after you are 20 years old.

What do our teeth do?
Our teeth are partly buried in our gums. The part of the tooth we cannot usually see is called the **root**. The root anchors the tooth firmly in the gums. It is coloured dark in the picture below.

The top part of the tooth is above the gums. This part is called the **crown**. The crowns of the different types of teeth have different shapes because they do different jobs. Look at the different shapes shown in the drawing below.

Incisors These are biting teeth. They are used for cutting our food into small pieces. These teeth have a narrow ridge along the top of the tooth. The incisors in the top and bottom jaws work against each other rather like the blades of a pair of scissors. We use our incisors mainly when we eat plant food, like cabbage, lettuce or spinach, where we cut off pieces bit by bit.

You can see very well-developed incisors in a rabbit (**herbivore**) which lives on plant food.

Canines These are tearing teeth. They are used for tearing pieces off our food. These teeth have a pointed top to the crown and are longer than the other teeth.

We use our canines mainly when we eat meat. The sharp point is used to tear off a piece of the meat.

You can see very well-developed canine teeth in a dog (**carnivore**) which tears meat.

Cheek teeth These are the molars and premolars. They are chewing teeth. They have a wide flat top which has small bumps and hollows on it.

When the food has been cut or torn off it is pushed to the back of the mouth. The cheek teeth in the top and bottom jaws then work against each other to grind the food into smaller pieces.

Cheek teeth are especially well developed in herbivores such as cows, sheep and goats which have a lot of plant food to grind.

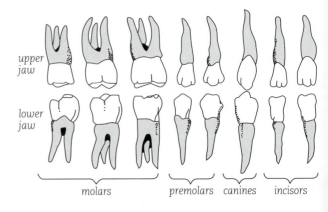

Human teeth on one side of the mouth

Mechanical digestion Our teeth break down the food we eat into small pieces. We call this **mechanical digestion**. This allows us to swallow food easily. It is important because it increases the amount of surface area which can be acted upon by the enzymes in our alimentary canal (pages 84–7).

Taking good care of our teeth

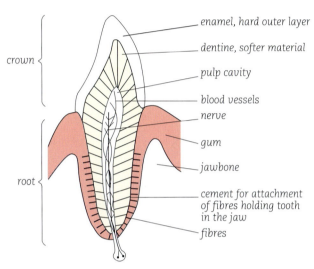

Inside a tooth

Cleaning When we eat, some of the food gets between our teeth and forms a coating called **plaque** (page 209). This plaque is acted on by bacteria which produce acids that can attack the enamel. We need to clean this away. Cleaning materials we can use are described on page 209.

What we eat Some foods are good for our teeth. For example, milk and cheese contain calcium and phosphorus which are needed to build strong teeth. We also need vitamin D which helps us to use the calcium in our food, and vitamin C for healthy gums. Crisp foods such as carrots, apple and otaheite apple help to dislodge food particles from between our teeth.

But starchy and sugary foods are bad for teeth. They can get stuck between the teeth and the bacteria will act on these food particles to release acids that attack the enamel of the teeth.

Looking after our teeth When you were a baby your first teeth grew gradually through your gums. These were your **milk teeth**. It was important to clean these milk teeth carefully because they were keeping a space for your adult teeth. From about six years onwards you began to lose these milk teeth. As they fell out, more teeth grew out from the gums. These are your **adult teeth**.

After the adult teeth have grown, there are no more teeth waiting in the gums, so we must care for our adult teeth so that they do not decay.

Tooth decay

Acids from food decay can wear away the **enamel** and then the **dentine**. This exposes the nerve which is inside the **pulp cavity**. When we bite on this tooth, or if hot or cold food or drink gets near to the tooth, we feel a **toothache**. This is because the nerve has been stimulated and we feel pain.

We need to visit a dentist, who will drill away the decayed part of the tooth and fill up the hole with a **filling**. If the decay has gone too far the dentist will have to remove or **extract** the tooth.

Fluoridation

Fluoridation is the addition of fluorides to toothpaste or drinking water. Fluorides make teeth harder so the acid cannot decay them so easily.

Fluoride toothpaste contains, for example, sodium mono-fluorophosphate. It may also contain calcium compounds. Fluoride toothpastes have a limited effect and harden only the very outer surface of the teeth. The teeth become more resistant to decay by 15 to 30%.

Fluoride can also be added to drinking water. This is more effective than toothpaste. It builds up in the teeth which become more resistant to decay by 54 to 84%. This graph shows how different concentrations of fluoride in the water reduce the number of decayed teeth. But there are also dangers and side-effects from having too much fluoride in the water.

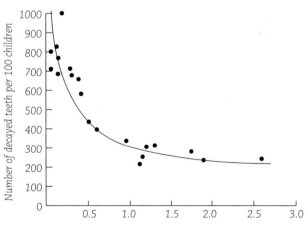

Questions

1. Describe how the different types of teeth are adapted for their job.
2. What causes tooth decay and what can we do to prevent it?

living things | *food and nutrition* | *how do we use our teeth?*

How do we digest our food?

The lining of our gut (**alimentary canal**) is like a sieve with very small holes. The food we eat is made up of large pieces. **Digestion** is the process by which large particles of food are broken down into very small particles which can pass through the gut lining into the bloodstream.

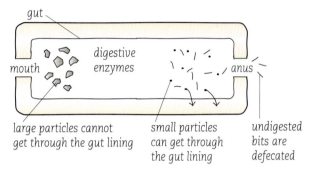

Digestion and defecation

Look back to page 68 where you did an Activity on diffusion using starch particles and glucose solution. The bag which you used is like the gut, with very small holes. The starch particles were too big to go through the holes, but the glucose particles could pass through. The bag and the gut lining are partially permeable membranes. If we are to make use of the starch we eat, we first have to digest it to make glucose.

You can also refer to the Activity on page 82 with the cracker. What changes did your teeth make to the cracker, and what changes did the saliva make? You are now going to find out.

Digestion occurs in two stages

1 Mechanical digestion This occurs mainly in the mouth (page 82). It is the cutting up and chewing of large pieces of food into small pieces – as with chewing a cracker. The food itself is not changed.

2 Chemical digestion This starts in the mouth when food is mixed with saliva and continues in the different parts of the alimentary canal. The food is changed from insoluble molecules such as starch into soluble ones such as glucose. Enzymes (pages 86–7) bring about these changes by breaking down the food into simpler substances.

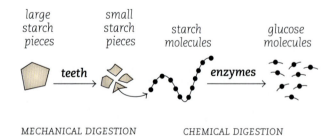

Why is the structure of the small intestine important?

The small intestine is very long, about 5 m in an adult, and this gives plenty of opportunity for digestion of the food to be completed. The food completes its final stages of digestion into small soluble molecules such as glucose (from carbohydrates), amino acids (from proteins) and glycerol and fatty acids (from fats).

The inner wall of the small intestine has many finger-like projections called villi which absorb the digested food, as you can see in the picture below. The villi greatly increase the surface area through which **absorption** can occur. The wall of the villus is only one cell thick so that digested food can easily diffuse across it.

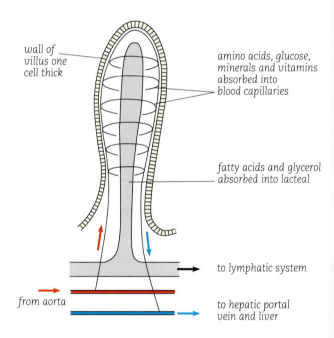

How a villus works

Glucose and amino acids enter the blood system and travel first to the liver where their concentration in the blood is adjusted (page 129). The glycerol and fatty acids enter the lacteals which are part of the lymphatic system (page 128).

What happens to undigested food?

Not all of the food that we eat can be broken down into soluble molecules and absorbed. Some of it, such as **roughage** (dietary fibre, cellulose), is largely unaffected by the digestive process and passes on down to the large intestine.

In the large intestine water is removed from the waste, and the waste material then accumulates in the **rectum**. About once a day it is passed out of the **anus** as **faeces**.

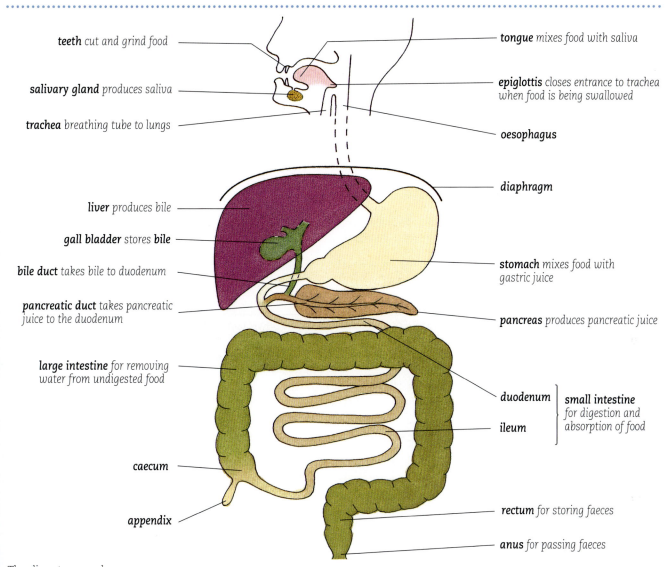

The alimentary canal

Upsets in the alimentary canal

Vomiting Vomiting (being sick) occurs when some of the contents of the stomach are sent back up, and out of the mouth. You vomit because your body has been upset in some way. This upset might be because of an infection, drinking too much alcohol, stress, poisoning, or hormone changes due to pregnancy (page 62). Vomiting is preceded by **nausea** (feeling sick).

Diarrhoea This is the passing of a large amount of very runny faeces. Some upset in the large intestine means that water has not been removed from the faeces. Diarrhoea can be brought on by an infection, stress, or by certain spicy foods. If it does not stop within a few days it is best to visit a doctor or else the diarrhoea could cause you to become dangerously dehydrated.

Constipation Constipation (hard stools) is the passing of hard faeces, which may cause pain. If it goes on for a long time it can cause stretching of the large intestine, intestinal diseases and possibly cancer. It is prevented by a diet high in dietary fibre (roughage, cellulose) such as cereals, wholewheat bread, bran, fruit and vegetables.

Questions

1. What are (a) mechanical and (b) chemical digestion? Why is each one important?
2. What are the main parts of the alimentary canal? What happens in each part?
3. Describe what is meant by the following: (a) absorption (b) defecation (c) diarrhoea (d) constipation.

How do enzymes help in digestion?

Enzymes are chemical substances which speed up the breakdown of food in digestion.

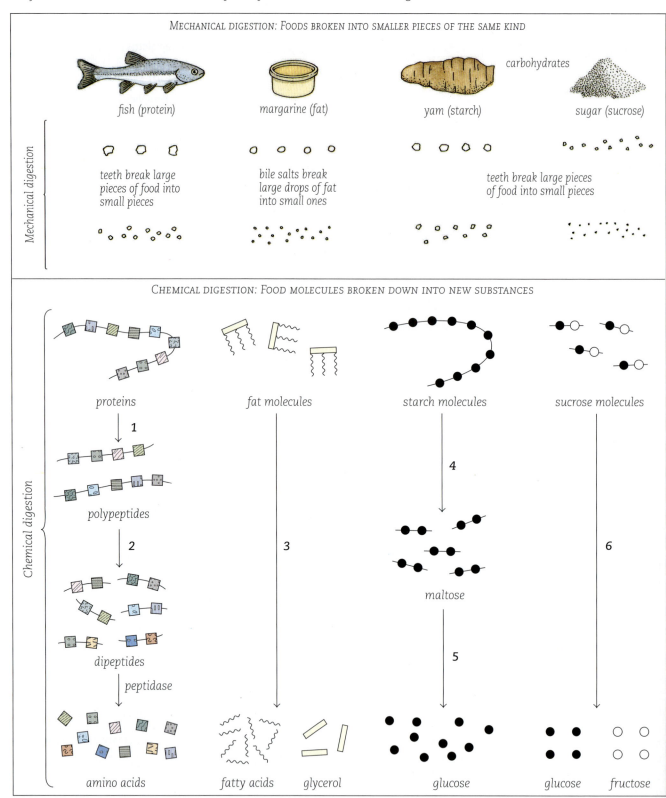

What do the enzymes do in the alimentary canal?

Enzyme activity in different parts of the alimentary canal

Part of canal	pH	Digestive juice	Enzymes present	Substrate	End products
Mouth	Slightly alkaline	Saliva (from salivary glands)	Salivary amylase	Cooked starch	Maltose
Stomach	Acid	Gastric juice (from stomach wall)	Pepsin	Proteins	Polypeptides*
Duodenum	Alkaline	Bile (from liver) (not an enzyme)	Bile salts (not an enzyme)	Fats	Emulsified fats (small fat droplets)
	Alkaline	Pancreatic juice (from pancreas)	Amylase	All starch	Maltose
			Lipase	Emulsified fats	Glycerol and fatty acids
			Trypsin	Polypeptides	Dipeptides*
Ileum	Alkaline	Intestinal juice (from intestine wall)	Maltase	Maltose	Glucose
			Sucrase	Sucrose	Glucose and fructose
			Lipase	Emulsified fats	Glycerol and fatty acids
			Peptidase	Dipeptides	Amino acids

*See page 74.

Digestive juices

Digestive juices are produced in the mouth, stomach and pancreas, and also in the ileum, which is part of the small intestine.

The juices contain the enzymes which help break down the food into simpler substances. Each enzyme works best at a particular pH, and the different digestive juices have different pHs which are most suited to the enzymes at that particular part of the gut (page 88).

Bile is produced in the liver, and stored in the **gall bladder** until it passes along to the duodenum.

Bile does not contain any enzymes. It contains sodium hydrogen carbonate which neutralises the acidity from the stomach, and helps to make the conditions in the duodenum become alkaline. It also contains bile salts which break up (**emulsify**) the fat into small droplets so that they are more easily digested by the lipase in the duodenum and ileum. The bile is therefore important in mechanical digestion.

As the food passes down the alimentary canal it is very thoroughly mixed with the digestive juices.

In the mouth, the teeth and tongue help to mix the food with the saliva. In the stomach, the muscular walls squash and squeeze the food so that it is mixed up with the gastric juices. All along the alimentary canal the walls of the canal alternately squeeze and relax on the food (**peristalsis**). In this way the food is mixed with the juices and pushed along to the next part.

The intestines are very long and have folded walls. This assists in the mixing of the food with the digestive juices.

As a result, the enzymes in the digestive juices are able to break down the complex food molecules into simple soluble molecules. These are then absorbed into the bloodstream or lacteals (page 84) for use by the body.

Questions

1. In the chart opposite are six places marked 1 to 6. These show the positions of six different enzymes.
   (a) Look at the table above and see if you can find the names of each of the enzymes 1 to 6.
   (b) Also write down the place in the alimentary canal where each of the enzymes is found and what it does.
2. Imagine you have just eaten a meat patty or beef pie.
   (a) What groups of food were present in your patty or pie?
   (b) Describe the steps by which each type of food was digested.
3. Imagine you have just eaten a calaloo or potato patty. Repeat steps (a) and (b) from Question 2.
   (c) What else was present in this patty and what will happen to it?
4. Which of the following actions are mechanical digestion and which are chemical digestion?
   (a) The biting and chewing of our food.
   (b) The action of pepsin on protein.
   (c) The action of bile salts on fats.
   (d) The churning action of the stomach.
   (e) What happens to protein in the mouth.
   (f) The action of lipase on emulsified fats.
5. Describe what is meant by: (a) peristalsis (b) emulsify (c) substrate (d) end product.

How do enzymes work?

Enzymes in everyday life

We use enzymes in many ways, for example:

- Papain is an enzyme in pawpaw leaves, which can be used to tenderise meat. The process of making the meat tender is like digestion.
- Biological washing powders contain enzymes which can remove stains made from foodstuffs. The enzymes break down the food stains so that they can be washed away. The temperature of the washing cycle should be kept low (you will find out the reason for this shortly).
- Bioyoghurt contains living bacteria whose enzymes may help improve the digestion of our food.
- Fruit drinks, such as those from banana, contain enzymes which can aid in digestion.

Activity | Breaking down starch

1. Use a piece of cracker or bread of about 1 cm². Break it up with a knife or pestle and mortar, and mix it with a little water. This is your sample of cooked starch.
2. Divide your sample equally between three test-tubes.
   (a) Add a third of a test-tube of dilute hydrochloric acid to the first test-tube and boil it for 2 minutes. Allow it to cool. Add solid sodium hydrogen carbonate until the fizzing stops (to neutralise the acid).
   (b) Add saliva from your mouth to the second test-tube and leave it for 5 minutes.
   (c) Leave the third test-tube as a control.
3. Use a dropping pipette to take out a little of the contents from each test-tube. Test them with a few drops of iodine solution to see if they still contain starch.
4. Record your results.

Digestive enzymes

You will find that the starch is broken down by boiling it with hydrochloric acid. The enzyme amylase present in saliva also breaks down starch. Enzymes are useful because they speed up reactions in the body without the need for high temperatures or boiling with acid.

During digestion (page 87) large molecules of proteins and starch are broken into their smaller units: amino acids and glucose. Fats are broken down into glycerol and fatty acids.

During the breakdown, a molecule of water is added between each pair of units, so that the units break apart (page 72). A large molecule is broken down to make many small molecules of a new substance. These small molecules are soluble and are absorbed into the body.

Activity | How does pH affect salivary amylase?

1. Prepare a sample of cooked starch as in the previous Activity.
2. Divide your sample equally between three test-tubes.

Add saliva from your mouth to each test-tube.
(a) Add dilute hydrochloric acid to the first test-tube.
(b) Add dilute sodium hydroxide solution to the second test-tube.
(c) Leave the third test-tube at neutral pH.

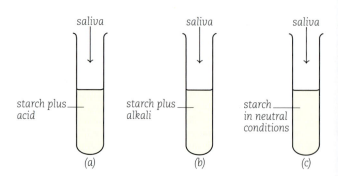

3. Leave the test-tubes for 5 minutes. Use a dropping pipette to take out a little of the solution from each test-tube. Test the solution with iodine solution to see if it still contains starch.

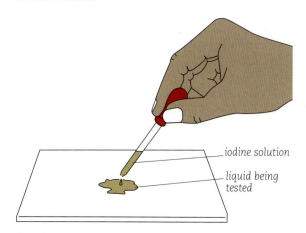

Testing whether starch has been broken down

Questions

1. What happened in very acid conditions (a)?
2. What happened in very alkaline conditions (b)?
3. What happened in neutral conditions (c)?
4. Of the three conditions provided, in which one does salivary amylase work the best?

In the body salivary amylase works best in the slightly alkaline conditions which are found in the mouth. In very acidic or very alkaline conditions it does not work well.

Activity | How does temperature affect salivary amylase?

1. Prepare a sample of cooked starch as in the first Activity.
2. Divide your sample equally between four test-tubes.
   (a) Add two drops of saliva to the first test-tube (a) and boil the contents for 3 minutes.
   (b) Put the second test-tube into a beaker containing ice, or put it in the refrigerator.
   (c) Put the third test-tube into a beaker containing water at room temperature.
   (d) Put the fourth test-tube into a beaker containing water at about 37 °C (body temperature).
3. Add two drops of saliva to tubes (b) to (d). Note the time.

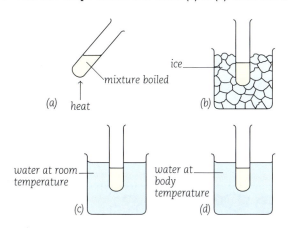

4. At 1 minute intervals, use a dropping pipette to take out a little solution from each test-tube. Test it with iodine solution on a spotting tile.

spotting tile	(a)	(b)	(c)	(d)
1 minute	O	O	O	O
2 minutes	O	O	O	O
3 minutes	O	O	O	O

Questions

5. What happened when the enzyme was boiled (a)?
6. What happened in very cold conditions (b)?
7. What happened at room temperature (c)?
8. What happened at body temperature (d)?
9. At which temperature did the salivary amylase work most quickly to break down the starch? What would be the advantage of this?
10. Describe in your own words the way in which temperature affects the action of the enzyme.

Enzymes in living things

Enzymes are important in living things because they **catalyse** certain reactions. This means that they make the reactions occur more quickly, but they are not themselves used up in the reaction. The enzymes act as a 'go-between' to bring together particular molecules which are going to react. After the reaction, the enzyme is ready to 'bring together' another set of molecules of the same kind. Enzymes cannot make reactions occur that would not usually happen.

If we burn sugars such as glucose with a flame the sugars combine with oxygen and release energy. In the body, glucose combines with oxygen in reactions controlled by respiratory enzymes. The enzymes allow the reactions to occur step by step and at a lower temperature.

All of the reactions in living things, such as photosynthesis, respiration, digestion and making proteins for growth, are controlled by enzymes. Enzymes help to build up, or break down molecules in a step by step way.

The characteristics of enzymes

Characteristics	Notes
1 They are proteins	Amino acids from our food are used to make enzymes
2 They are destroyed by heat	This is because they are proteins which cannot work when heated above 40 °C
3 They do a particular job	Each enzyme is important for one special reaction
4 They can be re-used	They catalyse the reaction but are not changed by it
5 They are sensitive to pH	Different enzymes work best at different pH values
6 They are sensitive to poisons	For example, arsenic and cyanide destroy respiratory enzymes and so kill the person
7 They are helped by vitamins and minerals	For example, the B vitamins help the respiratory enzymes

Questions

11. Which of the characteristics of enzymes would explain each of the following observations?
    (a) Boiling enzymes will stop their activity.
    (b) Pepsin breaks down protein, but not starch.
    (c) Enzymes are only needed in small amounts.
12. Look at the table on page 87. If the enzyme pepsin had been used with the protein in the Activity to find the effect of pH, what result would you have expected?

How are substances moved around in plants?

Why do substances have to move?

Many plants are quite large, and some of them, such as trees, are very big indeed. They therefore need some kind of **transport system**. (See also page 286.)

The roots take in water and mineral salts from the soil and these substances must be carried to other parts of the plant. Carbohydrates and other foods are made in the leaves and have to be carried to the rest of the plant.

How do substances enter a plant?

On page 69 you did an experiment with a scooped out potato cup. You put sugar inside the cup and placed it in some water in a beaker.

After a while you found that water had moved from the beaker into the cavity of the potato cup. The water had passed through the potato cells. Water was at a high concentration in the beaker and a very low concentration in the potato cup. Water had moved in along its **diffusion gradient**. This process is called **osmosis**.

This is also what happens when water enters a living plant through its roots.

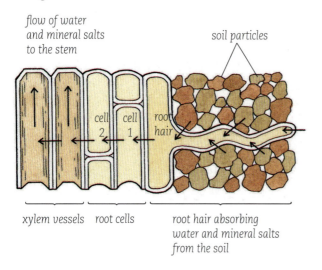

Absorption of water and mineral salts

Root hairs have a semi-permeable cell membrane. There is a higher concentration of water in the soil than in the root hairs. Water can therefore enter into the root hair along its diffusion gradient. In the same way water can pass, by osmosis, from the root hair into the next cell, and so on.

Mineral salts may enter the plant in a similar way along their diffusion gradients. But the concentration of a particular mineral salt is often higher in the root cells than in the soil. So, to absorb these, the plant uses energy for **active uptake** to absorb salts *against* their diffusion gradients.

What are the substances transported in?

Inside a plant is a system of tubes, rather like the blood vessels inside an animal. The tubes are arranged in **vascular bundles** or veins.

Xylem lies towards the inside of the vascular bundle. It consists mainly of **xylem vessels**. The xylem vessels are empty pipes which have walls thickened with woody tissue which may simply have small holes in it, or may be arranged as rings. Look at the picture below.

When water and dissolved salts enter the root hairs they pass into the xylem vessels, and are transported up the plant and into the veins in the leaves. Some of the water is used in photosynthesis, and the rest is lost by transpiration. Mineral salts are used in making proteins (page 72).

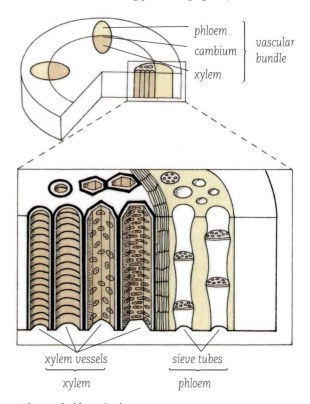

Xylem and phloem in the stem

Phloem lies towards the outside of the vascular bundle. It consists mainly of **sieve tubes**. The sieve tubes are made of cells which are separated from each other by walls with holes in them – sieves – hence their name.

When food has been made in the leaves it passes into the sieve tubes in which it is transported around the plant. The food is combined with oxygen during respiration to provide energy (page 106) or it is stored in fruits, seeds or storage organs (page 73).

Activity | How do substances rise in plants?

1. Your teacher will set up a cutting of a small plant such as balsam in some water which has been coloured with a red dye, such as eosin.

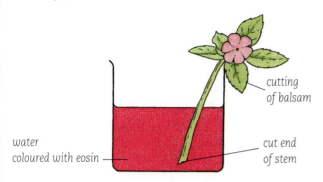

Balsam in coloured water

2. Observe the plant carefully to see what happens to the coloured water.
3. Use a razor blade to cut thin sections across the stem at three different places.
4. Look at the cut edge of your sections with a hand lens. What do you notice?
5. The irregular holes in your section are cuts through the xylem vessels which run up the plant. From your observations, does it look as though the coloured water has travelled up the xylem vessels?

Water and its dissolved salts travel up the xylem vessels in the same way as the coloured water has done.

How does transport occur?

Diffusion The movement of mineral salts into the plant, and the movement of oxygen, carbon dioxide and water vapour (pages 68 and 70) are examples of diffusion.

Osmosis The movement of water into the root hairs, and to the centre of the plant are examples of osmosis.

Capillarity This is the rising of liquid in very narrow tubes. If we observe glass tubes of different internal diameters we find that liquids rise up highest in the narrowest tube. (See also page 130.)

In plants, the xylem vessels are extremely narrow and so the water (and dissolved mineral salts) rise up to a certain height by capillarity.

Transpiration pull In the xylem vessels there are long tubes of water which are in contact with the water in the leaf. As water is lost by evaporation from the leaves (**transpiration**) more water is drawn up from the xylem vessels to take its place. So water is pulled up through the plant.

Root push/pressure As water enters through the roots, the water which is already there is pushed up. This is called root push, or root pressure. Water is pushed up through the plant.

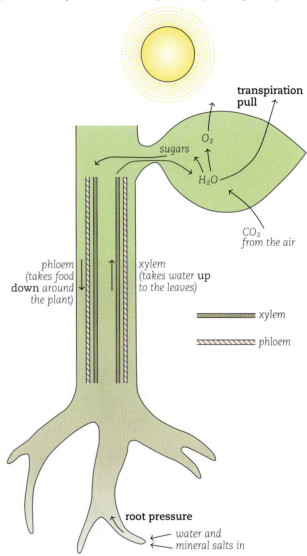

Summary of how substances move around in plants

Questions

1. List the various substances which enter and move around in the plant. Briefly describe how each one moves.
2. What do you understand by the following? (a) Diffusion (b) Osmosis (c) Capillarity (d) Transpiration pull (e) Root push or pressure
3. What are (a) xylem and (b) phloem? Why is each of these important to the plant?

What is our blood made of?

Part of the blood	What it is	What it does
Plasma	The liquid part of the blood. It is water which contains salts, proteins and antibodies.	It transports carbon dioxide, food substances and wastes such as urea. It also distributes heat, hormones and antibodies. It contains substances used in blood clotting.
Red blood cells	Red discs which have a hollow on each side. They do not have a nucleus. They contain haemoglobin. They are made in the bone marrow.	They carry oxygen from the lungs to all parts of the body. They also carry some carbon dioxide from the body cells to the lungs.
White blood cells	Two kinds, phagocytes and lymphocytes. They are colourless and each has a nucleus.	They protect the body from infection by germs (bacteria and viruses).
	Phagocytes The nucleus is made of several lobes joined together. The cytoplasm has granules in it. They are made in the bone marrow.	Phagocytes (cell eaters) attack and engulf any germs that enter the body. They are the chief defence against disease-causing micro-organisms.
	Lymphocytes The nucleus is roughly circular. The cytoplasm is clear. They are made in the lymph glands.	Lymphocytes protect the body by producing antibodies which either kill the microbes, or make them clump together so that they can be removed in the lymph glands.
Platelets	Very small pieces of cells from the linings of the blood vessels.	They gather where a blood vessel has been cut, and they plug up the hole as the first part of the blood-clotting process.

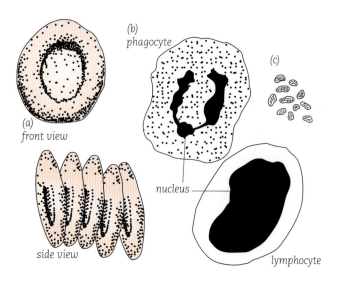

Blood cells – (a) red blood cells (b) white blood cells (c) platelets

How the red cells carry oxygen

The red cells contain the red pigment, **haemoglobin**. When blood passes through the lungs the haemoglobin picks up oxygen, to form **oxyhaemoglobin**. When this blood reaches the cells of the body, oxygen is released:

Lungs	→	Blood	→	Cells
Oxygen picked up		Oxyhaemoglobin in red cells		Oxygen released

The red cells then circulate in the blood back to the lungs, and the process is repeated.

How does blood clot?

If you cut yourself a blood vessel may be damaged and blood will come out. The body has a way to deal with this problem. First the platelets help to plug up the hole. Then the damage to the blood vessel causes the production of a substance called **thrombin** which interacts with **fibrinogen** in the blood to form a mesh of fibres (**fibrin**) over the cut.

Red blood cells and platelets are caught up in these fibres to form a **clot** which seals the blood vessel and stops further bleeding. This clot should be left undisturbed. (If it is picked off, then germs could enter the wound.) The tissues under the clot then begin to heal.

Questions

1. List the various things which are transported by our blood. In each case say which part of the blood is responsible.
2. List three ways in which red and white blood cells are different.
3. (a) What is oxyhaemoglobin? Why is it important?
   (b) What problems would a person have if their haemoglobin did not work properly? (See page 61.)
4. Why is it important that blood should be able to form a clot?

Activity | Looking at blood

Your teacher will give you a slide which has a blood smear on it. This is better than making your own slide which can be dangerous because of AIDS. Also, a prepared slide will be stained so as to show the cells more clearly.
1. Look at the slide under the low power of a microscope, then under high power.
2. Try to identify the cells shown below.

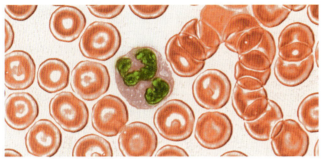

Human blood cells × 500. The nucleus of the white cell has been stained green

Blood donors

A person who is in an accident may lose a lot of blood and may need to be given some extra blood. We call this a **blood transfusion**.

Blood is stored in a **blood bank** in a hospital, in case of emergency. People are needed to give blood to the blood bank. We call these people **blood donors**.

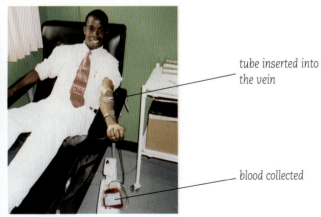

A blood donor

Donating blood is not painful. Blood is collected from the person through a tube, as in the photograph. Over the next few weeks the person makes more blood. Donating blood can help to save someone's life.

In most countries, donated blood is treated so that any viruses which could cause hepatitis or AIDS are killed.

What are blood groups?

Red blood cells contain substances called **antigen**s. Blood plasma contains substances called **antibodie**s. People have different antigens and antibodies so humans are divided into four different blood groups: A, B, AB and O, like this:

Blood groups	A	B	AB	O
Antigens on red cells	A	B	A and B	none
Antibodies in plasma	anti-B	anti-A	none	anti-A and anti-B

Notice, for example, that a group A person has A antigens and anti-B antibodies. If, instead, a person had A antigens and anti-A antibodies these could react together and the blood cells of the person might clump together or **agglutinate**.

This gives us a way of finding out the blood group of a person. Serum, which is plasma without the clotting agents, is used. The person's blood is added to anti-A and anti-B serum.

 Fact: *If the blood clots with anti-A serum only, the person is group A.*
If the blood clots with anti-B serum only, the person is group B.
If the blood clots with both anti-A and anti-B, the person is group AB.
If the blood clots with neither anti-A nor anti-B, the person is group O.

Visit your local hospital or blood bank and try and find out your blood group. Try and find out the percentage of people in each blood group in your country by asking at the hospital or blood bank.

Why are blood groups important?

A person who needs a blood transfusion must be given blood of the right group; it must be properly **matched**.

If the blood that is given contains antigens which interact with antibodies in the person receiving blood, then clumping together or agglutination of the blood will occur.

If the blood cells clump together this will clog the blood vessels, and stop the blood from flowing properly. If blood does not flow properly it could cause serious problems. For example, the brain may suffer from a lack of oxygen and this in turn can lead to shock and unconsciousness. Or the kidney vessels may become blocked. Either event could cause death. This is why it is so important to match blood groups.

How does our blood circulate?

On page 68 you saw that particles can move from one place to another by diffusion. This is important for the movement of substances over short distances (for example gaseous exchange, page 100) but a large organism such as an animal cannot rely only on diffusion (pages 286–7).

To move substances around inside our body we need a system of vessels (the blood vessels) and a pump (the heart) to push the blood around. In this way food and oxygen can get to our cells, and carbon dioxide and other wastes can be removed.

What are our blood vessels like?

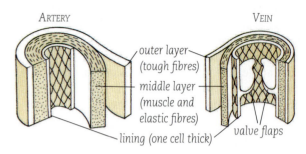

- Arteries have a thick layer of muscle and elastic fibres because blood is forced along them at a high pressure.
- Arteries have no valves as blood is pumped along them in only one direction.

- Veins have a thinner layer of muscle and elastic fibres because the blood returning to the heart is at a lower pressure.
- Veins have valves to stop blood from flowing backwards.

Arteries divide up into **capillaries** to supply the tissues. Capillary walls are only one cell thick, so food, waste and gases pass through easily. Capillaries join to make veins.

Activity | Measuring your pulse
As the heart beats it pumps blood into the arteries. This makes a **pulse** which you can feel.
1. Feel for your pulse in your wrist (as in the picture).
2. Count the number of pulse beats in 1 minute.

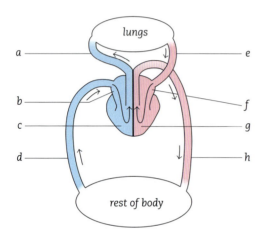

Simple circulation

Look at this very simple diagram showing our **circulatory system**.
- **Arteries** are blood vessels which take blood *away* from the heart, such as **(a)** in the diagram.
- **Veins** are blood vessels which take blood *towards* the heart.

Blood *enters* the heart at the **auricles** or **atria** (singular: atrium). Blood from the body enters the right atrium. Blood from the lungs enters the left atrium.

Blood *leaves* the heart from the **ventricles**. Blood to the body leaves the left ventricle. Blood to the lungs leaves the right ventricle.

Questions
Look at the diagram above.
1. Which are the arteries?
2. Which are the veins?
3. Which is the right atrium?
4. Which is the left atrium?
5. Which is the left ventricle?
6. Which is the right ventricle?

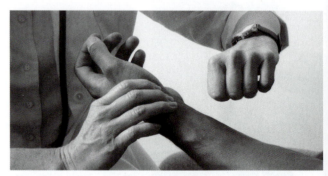

The pulse is taken while the person is relaxed

Questions
In the diagram opposite, which vessel takes blood:
7. from the right ventricle to the lungs?
8. from the left ventricle to the body?
9. to the liver?
10. from the alimentary canal?
11. from the lungs to the left atrium?

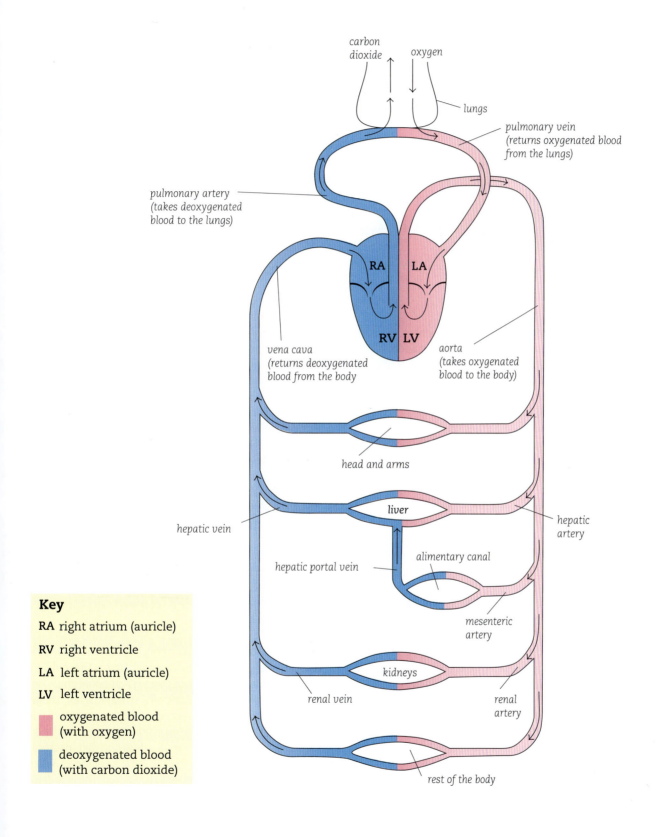

Problems with our circulatory system

What is the heart like inside?

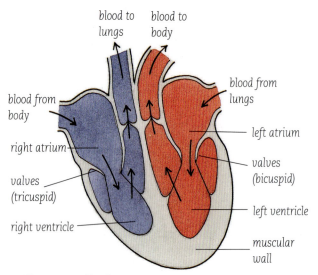

Inside a mammalian heart

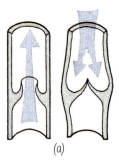
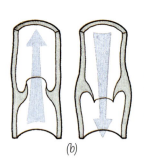

Varicose veins

In the picture you can see that the walls of the atria are fairly thin. This is because the atria only receive blood and pass it on to the ventricles. The walls of the ventricles are much more muscular. The left ventricle is the most muscular, as it pushes blood around most of the body, but the right ventricle only pushes blood to the lungs (page 95).

Between the atria and the ventricles there are valves which only allow blood to flow from the atria into the ventricles and not the other way around. They work like trap doors to stop blood from flowing back.

High blood pressure

Blood pressure is a measure of how hard the heart has to work to pump blood. A reading of 120/80 is considered 'normal' for a healthy adult. The '120' refers to the pressure when the ventricles contract, and the '80' refers to the pressure when the heart muscles relax.

A person with high blood pressure or **hypertension** is pumping blood very strongly. This may be because the arteries have become partly blocked up, or lost some of their elasticity. So the heart has to pump more strongly to force the blood through the arteries. Hypertension puts a strain on the heart and could cause delicate blood vessels to break.

Problems with circulation

Varicose veins Veins carry blood from the body back to the heart. They contain valves so that blood travels in one direction only: towards the heart (next column (a)). However, in some people who stand a lot, the veins become swollen and the valves do not close up properly, which means the blood can flow both ways (next column (b)).

Varicose veins become swollen and twisted and legs ache and may swell. The problem may be helped by exercise and the person should avoid standing still. When sitting the person should put his or her feet up, so blood circulates back to the heart more easily.

Problems with the arteries Arteries carry blood containing oxygen and food from the heart to all parts of the body. The arteries become narrowed when fatty deposits called **atheroma** are laid down on the inner walls. As more and more deposits are laid down, a large mass called a **plaque** is formed. This not only narrows the arteries but also makes them less elastic. Other substances are also laid down and this causes 'hardening of the arteries'.

Arteries which have become narrowed and hardened do not let blood pass through them easily. So the organs which they supply with blood may not get sufficient oxygen and food for their needs. Because the blood vessels are narrower, it is more likely that blood clots will form in them, and also more likely that blood clots will block them up.

The major problems are with the blood supply to the brain and the heart. If there is a blood clot in the brain arteries, the person will have a **stroke**, followed by paralysis and perhaps death. The heart is supplied with blood through the coronary arteries. If this blood supply is reduced it can cause **angina** or a **heart attack**.

Heart problems

Angina This is a pain in the heart felt across the middle of the chest. It may occur when there is additional strain on the heart as with exercise or emotional stress. It is not usually very severe and goes away when the person rests. It can be an early warning of a possible heart attack. It is caused by a slight narrowing of the coronary arteries so that insufficient blood gets to the heart during the exercise or stress. When angina is felt, special tablets can be sucked. These widen the blood vessels to the heart so that the heart receives more oxygen.

A heart attack This is a pain like a heavy weight across the middle of the chest. It can come on at any time. It is usually very severe and does not stop when the person rests. It can cause death. It occurs when the coronary arteries become so narrowed that blood cannot reach parts of the heart muscle. This is often because there is a blood clot (**thrombus**) in the coronary artery. A heart attack is therefore sometimes called a **coronary thrombosis** or just 'a coronary'.

It is usually treated by rest and drugs. Some drugs reduce the rate at which the heart beats, so that it does not need so much oxygen through the coronary arteries. Other drugs widen the coronary arteries so that more blood flows through them, while other drugs reduce the likelihood of blood clots forming.

Heart surgery After severe heart attacks the person may have to have a **coronary bypass** (see below). In this operation a new piece of blood vessel, taken from the person's leg, is used to renew the blood circulation to the heart muscles.

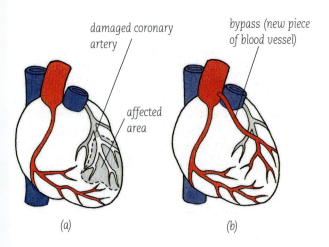

Coronary bypass operation – (a) before operation (b) after operation

Open heart surgery may be needed to correct the flow of blood through the heart. Most commonly, the valves are not working properly and new artificial ones have to be put in.

Factors linked with hypertension and heart attacks

Saturated fats Animals fed diets high in saturated fats develop hypertension. Countries where diets are high in saturated fats, such as the USA and UK, have a higher death rate from heart attacks than is found in the Caribbean. Hypertension in animals can be reduced by feeding them unsaturated fats to replace some of the saturated fats. Eating unsaturated fats has been found to reduce the likelihood of heart attacks in humans.

Cholesterol Cholesterol is a hormone found, for example, in eggs and fatty meat. Animals fed a high-cholesterol diet develop hypertension. People with a lot of cholesterol in their blood are more likely to develop hypertension and heart attacks. Cholesterol in the blood is reduced with a diet low in cholesterol and saturated fats.

Sodium Sodium is found in salt. Animals that are fed a high salt diet develop hypertension. These same animals when put on a low salt diet show a lowering of blood pressure. However, a similar lowering of blood pressure also results if the salt is kept high, but the animals are given more unsaturated fats.

People with a very high salt intake may reduce their hypertension by reducing their salt intake. But if their salt intake is average, then reducing it further will not reduce their blood pressure.

Stress Research shows that men who are aggressive, impatient, competitive and hostile (type 'A' behaviour) have twice the risk of heart attacks as men showing more relaxed, easy-going type 'B' behaviour. Aggressive behaviour causes increase in blood pressure and heart rate and secretion of the 'stress' hormone, adrenaline, which can lead to additional plaque formation in the arteries.

Inherited tendencies You may have a higher likelihood of heart disease if other members of your family have had it, or if you suffer from diabetes (page 113).

DIET
overeating
too many saturated animal fats
too much cholesterol
too much salt

CIGARETTE SMOKING
especially if also using the **contraceptive pill**

STRESS AND STRAIN

LACK OF EXERCISE

INHERITED TENDENCIES

Some factors which are linked with hypertension and heart attacks

How do we breathe?

Activity | What happens as we breathe?

Your teacher will set up this model to show what happens during breathing.

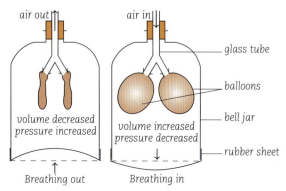

Breathing model

The balloons represent the lungs, the glass tubes represent the breathing tubes, the bell jar represents the chest cavity and the rubber sheet represents the **diaphragm** or sheet of muscles at the base of the chest cavity.

The left-hand diagram shows what happens during breathing out or **exhalation**. The diaphragm is pushed up and the volume inside the chest cavity is reduced. This increases the pressure in the chest cavity and so air is pushed out of the lungs.

The right-hand diagram shows what happens during breathing in or **inhalation**. The diaphragm is pulled down and the volume inside the chest cavity is increased. This allows the lungs to expand and air is pulled in along the breathing tubes.

This is a simplified model because it does not show the movements of the ribs, which form the framework of the chest cavity. Below is a side view of the ribs to show how they move during exhalation and inhalation.

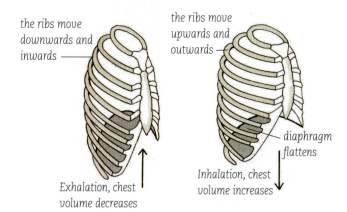

Chest movements

Activity | Carbon dioxide in inhaled and exhaled air

1. Set up two test-tubes as shown below.

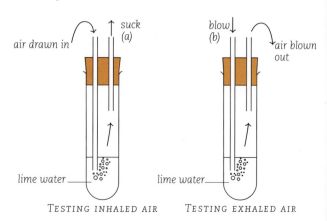

Testing for carbon dioxide

2. *Suck* air in through the *short* tube in (a). Notice that the tube does not reach into the lime water. Air from the atmosphere will be drawn in to bubble through the lime water, so you are testing inhaled air.
3. Now *blow* air out through the *long* tube in (b). This tube reaches down into the lime water. Your exhaled air will bubble through the lime water.
4. Describe what happens when inhaled air passed through the lime water. What does this tell you?
5. Describe what happens when exhaled air passes through the lime water. What does this tell you?

Activity | Temperature of inhaled and exhaled air

For this Activity you will need an ordinary thermometer and a clinical thermometer (page 120).

1. Use the ordinary thermometer to take the temperature of the air around you; i.e. the temperature of inhaled air.

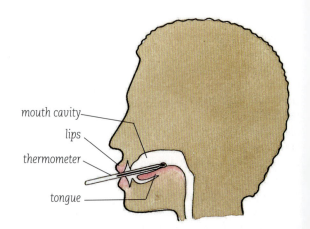

Taking the temperature of exhaled air

2. Now put the bulb of the clinical thermometer in the air space between your tongue and the roof of your mouth. Carefully close your lips. Wait for two minutes so as to take the temperature of the air you would be exhaling.
3. Record the temperatures of (a) inhaled and (b) exhaled air.

Activity | Water vapour in inhaled and exhaled air

For this Activity you will need a piece of glass, with *smooth edges*, or a mirror.
1. When the glass or mirror is lying on the bench it is in contact with the air we breathe in. Can you see any cloudiness which would mean there were water droplets in this inhaled air?
2. Now hold the glass or mirror *very close* to your mouth. Breathe some exhaled air onto it. Can you see some cloudiness or water droplets on the glass? You will find that exhaled air contains more water vapour than inhaled air.

Inhaled and exhaled air

Accurate measurements have been made to compare inhaled and exhaled air; the results are shown below.

Gas	Breathed in air (inhaled air)	Breathed out air (exhaled air)
Oxygen	20%	16%
Carbon dioxide	about 0.03%	4%
Nitrogen and other gases	about 80%	about 80%
Heat*	usually less	body temperature
Water vapour*	usually less	saturated air

* Note that actual figures in inhaled air will vary with climatic conditions.

Oxygen There is more oxygen in inhaled air than in exhaled air. Some of the oxygen from the inhaled air passes from the lungs into the blood and is taken to the heart and then to the rest of the body.

Carbon dioxide There is more carbon dioxide in exhaled air than in inhaled air. The blood which arrives in the lungs brings carbon dioxide, produced in respiration, which passes out in the exhaled air.

Nitrogen and other gases There is the same amount of these gases in inhaled and exhaled air. These gases just pass in and out of the lungs and are not used by the body.

Heat and water vapour Exhaled air is warmer and contains more water vapour than inhaled air. We shall see (page 104) that heat and water are produced in the cells as they respire.

Parts of our respiratory system

Air enters the respiratory system through the nose and mouth and goes to the back of the throat. It passes through the voice box into the windpipe (**trachea**) and the two **bronchi**. These all contain rings of cartilage which help to keep them open as air passes through them. The bronchi lead into the lungs where they break up into small **bronchioles**. The bronchioles break up into a mass of air sacs called **alveoli** where gas exchange takes place.

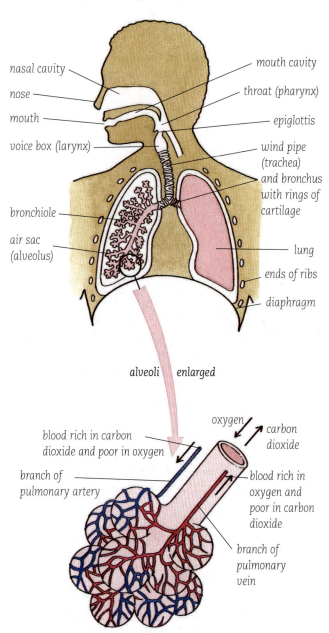

Respiratory system

living things | respiration | how do we breathe?

How are gases exchanged?

Animals and plants take in and give out gases: we call this **gaseous exchange**. It occurs by diffusion (page 68) through surfaces which we call **respiratory surfaces**.

Characteristics of respiratory surfaces
1. **Large and thin** Respiratory surfaces are large so that there is a large surface area (pages 286–7) through which gaseous exchange can occur. The surfaces are also very thin so that the gases only have a short distance to diffuse.
2. **Moist** Respiratory surfaces are moist because the gases have to dissolve in the moisture in order to diffuse across the surface.
3. **Supplied with gases** Gases are usually brought to and from the respiratory surface by some kind of breathing movement.
4. **Supplied with blood** Most respiratory surfaces in animals are supplied with blood vessels which bring some gases and take away others.

Gaseous exchange in mammals

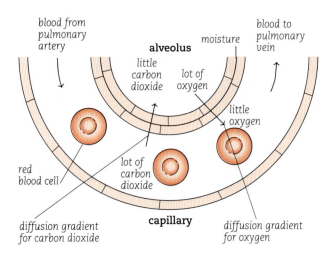

Respiratory surface of a mammal – the alveolus

There are very many alveoli in the lungs which give a large respiratory surface and their walls are only one cell thick. They have a layer of moisture in which the gases can dissolve. Blood containing a lot of carbon dioxide is brought close to the alveolus in a capillary from the pulmonary artery.

There is less carbon dioxide in the alveolus and so there is a **diffusion gradient** (page 68) and carbon dioxide diffuses out of the capillary into the alveolus. In the alveolus there is more oxygen than in the capillary so oxygen diffuses out of the alveolus into the capillary.

This leads into the pulmonary vein.

Gaseous exchange in fish
Water, containing dissolved oxygen, is drawn into the fish and passes out over its **gills**.

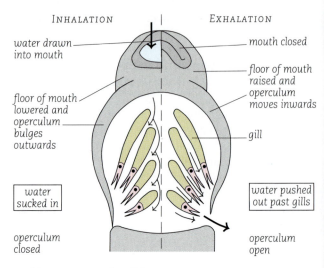

Breathing movements in a fish

The water comes into close contact with the gills which have a large surface area and are thin and well supplied with blood (page 143).

Blood with a large amount of carbon dioxide comes to the gills and is in contact with the water, which contains less carbon dioxide. There is a diffusion gradient and carbon dioxide passes out of the blood into the water. There is more oxygen in the water than in the blood so oxygen passes into the blood.

Gaseous exchange in amphibians
Amphibians, such as toads and frogs, have three different respiratory surfaces. The young tadpole has gills. The older tadpole and the adult amphibian can exchange gases through their skin and their lungs.

The *gills* are thin and feathery with a large surface area and are well supplied with blood. Oxygen and carbon dioxide can easily move between the water and the blood along their diffusion gradients.

The *skin* is kept moist by liquid secreted by glands and by the animal staying in moist places. The skin covers the whole body and it is well supplied with blood. Oxygen and carbon dioxide move between the air and the blood along their diffusion gradients.

The *lungs* are not used very much. They are simple bags which do not have the highly branched structure that is found, for example, in mammals. The lungs are thin-walled and supplied with blood vessels. Some gaseous exchange occurs in the lungs but the skin is more important.

Comparison of gaseous exchange in living organisms				
Organism	Large respiratory surface	Moist respiratory surface	Supplied with gases	Supplied with blood
Fish	**Gills:** divided into many small thin parts to give a large surface area	Gills surrounded by water	Breathing movements, so water comes in through the mouth and out through the operculum	Gases diffuse into and out of the blood supplied to the gills
Amphibian	**Gills:** feathery with large surface area **Skin:** large area all over the body **Lungs:** fairly small, not used much for gaseous exchange	Gills surrounded by water Animal stays in moist places to keep skin moist Lung cells secrete fluid	Water currents near gills Air movements near to the skin Breathing movements, so air enters and leaves lungs	Gases diffuse into and out of the blood vessels supplying the gills, skin and lungs
Mammal	**Alveoli** in lungs: very many make a large surface area	Cells in alveoli secrete fluid	Breathing movements, so air enters and leaves lungs (pages 98–9)	Gases diffuse into and out of the blood supplied to the lungs
Flowering plant	**Surface of cells:** inside leaf, large surface is in contact with the air	Cells with covering of water which is brought in xylem vessels	No breathing movements, gases just diffuse	No blood system Gases diffuse into and out of cells and stomata

Gaseous exchange in flowering plants

Animals take in oxygen and give out carbon dioxide all the time. But plants do not always do this, it depends on whether it is night-time or day-time. The gas exchanges which occur in the plant depend upon the balance between photosynthesis and respiration.

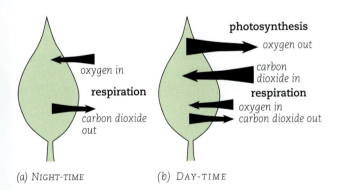

(a) NIGHT-TIME (b) DAY-TIME

Gaseous exchange in leaves

In the night-time when it is dark (a), the plant only carries out respiration. So it needs to take in oxygen and pass out carbon dioxide.

In the day-time (b), respiration still occurs, but so does photosynthesis. There is an *overall* intake of carbon dioxide and output of oxygen.

The respiratory surface in the plant is the large, moist surface of the cells in the leaf near to the air spaces connected to the stomata (page 71). Gases in high concentration in the air outside reach the leaf cells and diffuse into the air spaces along their diffusion gradients. Other gases diffuse from the air spaces into the leaf cells along their diffusion gradients. So, in the night-time oxygen diffuses in, and carbon dioxide diffuses out. In the day-time there is a net diffusion in of carbon dioxide and diffusion out of oxygen.

Unlike animals, there are no breathing movements and no blood to move the gases. The movement of the gases depends on diffusion and the movement in the xylem and phloem (page 90). This means that gaseous exchange is slower in plants than in animals. But plants are simpler and less active than animals, and have less need for rapid gaseous exchange.

Questions

1. What do the following words mean? (a) Diffusion (b) Diffusion gradient (c) Gaseous exchange
2. (a) What are the characteristics of respiratory surfaces? (b) How do the alveoli in the lungs of a mammal show these characteristics? (c) How are the respiratory surfaces in flowering plants simpler than those in animals? Why are they simpler?
3. Choose either a mammal or a fish and describe how breathing movements help in gaseous exchange.
4. Describe the diffusion gradients in a leaf (a) in the night-time and (b) in the day-time.

Problems with our respiratory system

We are at constant risk from dust particles, smoke, germs and other substances in the air which we breathe in.

If any part of the respiratory system is diseased, blocked or not working properly we cannot get sufficient fresh air. We may then cough, become wheezy or breathless, or develop more serious symptoms.

Smoking Tobacco smoke contains tar, nicotine and carbon monoxide. Tar condenses as a sticky syrup-like substance in the lungs. It can cause bronchitis, emphysema and lung cancer. Nicotine (page 253) is a drug which is absorbed from the lungs. It raises the heartbeat and increases the risk of blood clots. It can become addictive and is probably also responsible for some of the circulatory problems associated with smoking. The carbon monoxide can become combined with the haemoglobin in the red blood cells and reduce their ability to transport oxygen around the body. This can contribute to heart disease.

Statistics show that 40% of people who smoke more than 20 cigarettes a day die before retirement, compared to 15% of non-smokers.

Bronchitis Bronchitis is inflammation of the bronchi in the lungs. It is usually caused by cold and flu viruses but is made worse by cigarette smoke and pollution. If you have bronchitis, you will have a cough with a heavy fluid called **phlegm**, breathlessness and wheezing.

Emphysema This is a disorder of the alveoli of the lungs. They become stretched and lose their elasticity so gaseous exchange is less efficient. Emphysema often occurs after long-standing bronchitis in smokers, and in those living in polluted areas.

Lung cancer The cells that line the bronchi become damaged, and some of these may form tumours which lead to cancer in the lungs. Only about 3 in 1 000 cases occur in non-smokers.

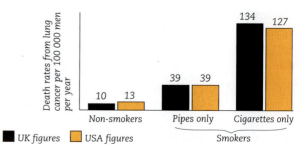

Lung cancer, pipes and cigarettes

Look at this bar chart. Describe how the annual death rate differs between non-smokers, pipe smokers and cigarette smokers. Also, smoking three to five marijuana joints a week is equivalent to smoking 16 cigarettes a day. Marijuana increases the rates of bronchitis, lung cancer and heart disease as well as deterioration in the nervous system.

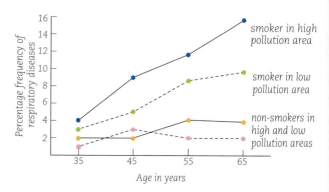

Respiratory diseases, smoking and pollution

Look at this graph. How does (a) smoking and (b) pollution affect the frequency of respiratory diseases?

In areas of high pollution, and in enclosed places where there is insufficient ventilation (page 135), 'passive smoking' occurs, so that non-smokers are exposed to the risks of developing respiratory disease.

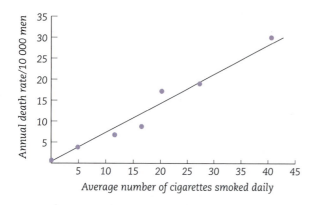

Lung cancer and cigarette smoking

Look at the graph. Describe how the annual death rate is related to the average number of cigarettes smoked.

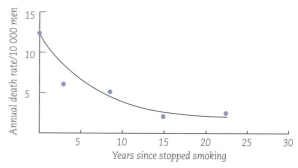

Giving up smoking

Look at this graph. What is the effect of giving up smoking?

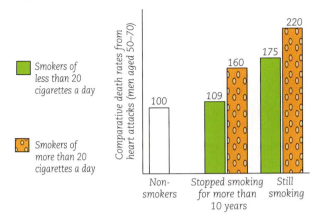

Smoking and heart attacks

Look at this bar chart. How do the numbers of people dying from heart attacks (page 97) compare for non-smokers and smokers of more than 20 cigarettes a day? Consider those who (a) have stopped smoking for at least 10 years and (b) are still smoking.

Pneumoconiosis Pneumoconiosis means 'dust in the lungs' and covers many diseases caused by inhaling dust particles, usually over a period of at least 10 years. It is associated with certain occupations such as quarrying, rock drilling, metal grinding and working with aluminium, asbestos, cement and certain synthetic fibres.

If these particles are inhaled, especially in enclosed areas with poor ventilation (page 135), they eventually cause little patches of irritation in the lungs. The resulting scar tissue makes the lungs less flexible and the person suffers from breathlessness.

Allergies

Allergies are reactions in the body caused by excess sensitivity (**hypersensitivity**) to substances inhaled, eaten or brought in contact with the skin. The substance causing the reaction is called an **allergen**. Some common allergens are pollen, dust and spores that are breathed in, drugs, and certain plants and foods.

People produce **antibodies** (page 161) to kill any invading germs and so protect the body from infection. This is a useful reaction. But a person who is allergic to a certain substance also produces antibodies against it even though it is not harmful.

In an allergic reaction irritating chemicals called **histamines** are released. It is these chemicals which cause the various symptoms of the attack.

Asthma Asthma often occurs as an allergic reaction in the lungs brought about by inhaling pollen, dust, fur from pets, or chemical substances. It can also be brought on by emotional stress.

In an attack of asthma the bronchi and bronchioles become partly blocked due to the contraction of their muscles, and the main symptom is breathlessness. In severe cases the face and lips turn grey because there is not enough oxygen in the blood. The attacks tend to become less serious as the person grows older and less sensitive to the allergen.

Hay fever This is an allergic reaction in the eyes, nose and throat. If the person sneezes and has a runny nose and red, watery eyes during certain months, he or she is allergic to certain kinds of pollen.

Other allergies People may be allergic to certain specific **foods**. The allergy may show up, for example, as a reaction in the intestine, such as diarrhoea after eating the food, or there may be a headache or a rash. A few people are allergic to nuts. If they eat nuts they have difficulty breathing.

People may also be allergic to certain **drugs**, for example the antibiotic **penicillin**. Such people develop a rash, an itchy skin and wheezing if they are given the drug.

The skin may become inflamed, itchy or develop small blisters when in **contact** with certain allergens. These are often plants, for example poison ivy.

Treatment of allergies

1 People having an asthma attack may use an **inhaler** to breathe in chemicals to help relax their breathing.
2 **Avoidance** of allergens is possible if the allergen is, for example, shellfish or penicillin, but impossible if it is pollen or dust.
3 **Antihistamine** tablets can be given to neutralise the effect of the histamines produced in the allergic reaction. These tablets help to prevent and stop attacks, but also make the person feel sleepy. It is therefore important not to drive or operate machinery while taking antihistamine tablets.
4 'Desensitising' can be done if the allergen has been identified. The person is given a very small dose of the allergen which does not produce a reaction, then slowly the quantity is increased up to normal levels.

Questions

1 (a) What reasons might a person give for *not* wanting to give up smoking?
   (b) What reasons would a doctor give to recommend that a person *should* stop smoking?
2 What are some of the contributory factors (possible causes) for the development of lung cancer?
3 Which of the respiratory diseases described on these pages would be more likely to occur in surroundings with poor ventilation?
4 If you had an allergy, but did not know what caused it, how would you go about trying to find the cause?

What is respiration?

Activity | Do respiring seeds release energy?

1. Your teacher will set up two thermos flasks with seeds like this.

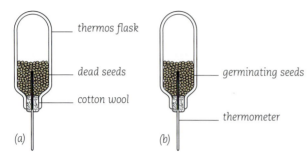

Seeds and energy release

2. Flask (a) contains seeds which have been boiled in water for 5 minutes to kill them. They were then washed in disinfectant to kill any bacteria on their surfaces.
3. Flask (b) contains seeds which were moistened and allowed to germinate. They were also washed in disinfectant.
4. Record the temperatures on the thermometers when the apparatus is first set up, and again after three days. What do you find?

Activity | Do respiring seeds release carbon dioxide?

You will need two test-tubes, two cloth bags, lime water, and dead and germinating seeds which have all been washed in disinfectant. Set up this apparatus.

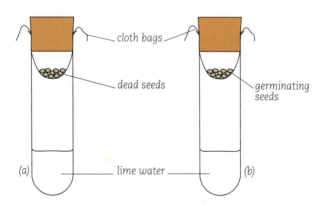

Seeds and release of carbon dioxide

Questions

1. Why were dead seeds as well as living germinating seeds used in the two Activities above?
2. Do respiring seeds release energy? Explain.
3. Do respiring seeds release carbon dioxide? Explain.

What happens when mammals respire?

This apparatus can be set up to find out if mammals release carbon dioxide when they respire.

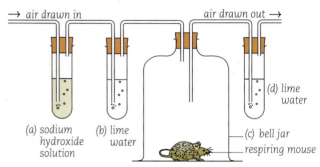

Mammal and release of carbon dioxide

Test-tube (a) contains sodium hydroxide solution. This is used to absorb carbon dioxide from the air which is coming into the apparatus.

Test-tube (b) contains lime water. If this goes milky it shows the presence of carbon dioxide. Test-tube (b) could instead have contained sodium hydrogen carbonate solution. This is also an indicator: in the presence of extra carbon dioxide it goes yellow (page 246).

The bell jar (c) contains a live, respiring mouse. Test-tube (d) contains lime water. Air, from the bell jar containing the mouse, is drawn through the lime water.

Questions

4. Why is test-tube (a) necessary?
5. Would you expect the lime water in test-tube (b) to go milky? If yes, why?
6. Would you expect the lime water in test-tube (d) to go milky? If yes, why?
7. If a plant was used in the bell jar, what precautions would have to be taken? What would you change?

What happens when plants respire?

Green plants carry out respiration all the time, to combine food with oxygen and release energy, carbon dioxide and water. If a plant were put in the bell jar above, the plant would also carry out photosynthesis which uses carbon dioxide and produces oxygen (page 101). To record respiration, the bell jar and plant should be covered with dark cloth.

Respiration is important because it is the process by which animals and plants release the energy they need for everyday activities. We can write a word equation for respiration:

Food + Oxygen → Energy + Carbon dioxide + Water

How much energy can be released?

We can find out how much energy there is in a food by burning it. The heat energy that is released can be used to heat a known volume of water. By the increase in temperature we can calculate the amount of energy released by the food. We use this information:

> **Fact:** The temperature of 1 gram (g) of water is raised by 1 °C by 4.2 joules (J) of energy.
> Note: In the past the calorie was used. We used to say 1 g of water was raised 1 °C by 1 calorie.
> 1 calorie (cal) = 4.2 joules (J)
> 1 kilocalorie (kcal) = 4.2 kilojoules (kJ)

Activity | Heat released from a peanut

1. Find the mass of a peanut and then stick it on a pin supported in plasticine.
2. Measure 20 cm³ of water into a test-tube.
3. Read and record the initial temperature of the water.
4. Light the peanut with a match so that it burns and hold the test-tube above the burning peanut to catch the heat.

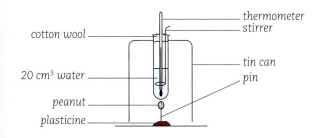

5. Stir the water.
6. Take the final temperature of the water.
7. Work out the energy released like this:

Mass of water (g) = 20 (mass of 20 cm³ water)
Temperature increase (°C) = Final − initial temperature
Energy released by nut in joules (J) = temperature increase (°C) × mass of water (g) × 4.2
To find the energy released per gram (J/g), divide the number of joules by the mass of the peanut.

Questions

The energy of the peanut is about 20 kJ/g.
8. How does your figure compare to this? Why?
9. How could you improve your accuracy?

The energy value of different kinds of food

We can use the apparatus in the last Activity to find the energy released from other kinds of food. Or we can use a calorimeter like this, which works in the same way but is much more accurate.

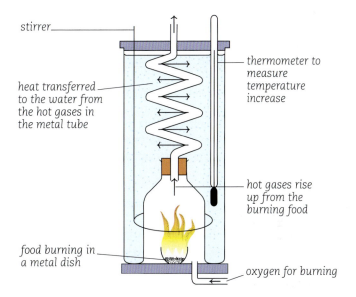

The food substance is placed in the metal dish and is set alight by a small electric current. It then burns and releases energy which heats up a known mass of water. The energy value in kJ/g is then worked out, as for the peanut.

The energy value of different food substances, found using a calorimeter, is given below:
- Carbohydrate: about 17 kJ/g
- Fat: about 39 kJ/g
- Protein: about 18 kJ/g

Respiratory substrates Any of these foods can be used as respiratory substrates to release energy for daily needs, and different groups of people will need different amounts of energy (page 78).

Carbohydrates such as starch and sugar are the best respiratory substrate. Their main use in the body is to release energy.

If we do not eat enough carbohydrates and *fat* in our diet, our body will begin to make use of the fat stored under the skin, which is supposed to be for insulation. This is the principle involved in slimming.

Proteins are usually used for growth and repair. They would be an expensive respiratory substrate. However, if insufficient carbohydrates and fats are available, the body will start to use up the protein in the muscles and other body organs. This is dangerous, and is what happens during starvation.

What are aerobic and anaerobic respiration?

Aerobic respiration

The normal process of respiration takes place in the presence of oxygen and produces a large amount of energy. It is called **aerobic respiration**.

In plants, substances such as glucose are transported to all the living cells. This food may have recently been made during photosynthesis, or it may be food brought out from storage organs.

In animals (for example in humans) the food which we eat is digested and then taken by the blood to all the living cells of our body.

In the cells of plants and animals, a series of chemical reactions takes place under the control of enzymes, in very small structures called **mitochondria** (page 10). Oxygen combines with the respiratory substrate such as glucose, releases a large amount of energy and produces carbon dioxide and water.

Glucose + Oxygen → Energy + Carbon dioxide + Water

$C_6H_{12}O_6 + 6O_2 \rightarrow$ Energy $+ 6CO_2 + 6H_2O$

Anaerobic respiration

Anaerobic respiration takes place when there is no oxygen. It also consists of a series of chemical reactions under the control of enzymes, but the respiratory substrate (glucose) is only partly broken down and releases only a small amount of energy.

Anaerobic respiration releases less energy from a given quantity of food than aerobic respiration.

In plants, such as germinating seeds, and in yeast, anaerobic respiration leads to the production of **ethanol**, carbon dioxide and a small amount of energy. Your teacher will show you that a little ethanol can be burned to release a lot of heat. This shows that the ethanol produced in anaerobic respiration contains trapped energy that has not been released.

In animals (for example in our muscles) when there is insufficient oxygen for aerobic respiration, anaerobic respiration produces **lactic acid**. This causes our muscles to feel fatigued. This is what happens when an athlete sprints for 100 m without taking a breath. Lactic acid builds up in the muscles and makes them ache.

In plants and yeast:

Glucose → Energy + Ethanol + Carbon dioxide

In animals:

Glucose → Energy + Lactic acid

Comparison of aerobic and anaerobic respiration

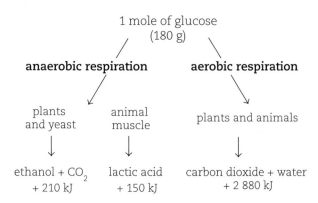

Aerobic respiration releases much more energy than anaerobic respiration.

Activity | Anaerobic respiration in yeast

Yeast is a fungus which does not carry out photosynthesis. It is therefore simple to set up some yeast with food (sugar) and find out how it respires in the absence of oxygen.

1. Boil half a test-tube of water to drive out the dissolved air (including oxygen).
2. Let the water cool to 35 °C. Mix into it a teaspoonful of dried yeast and a teaspoonful of sugar.
3. Add a layer of oil to stop air getting in, and set up this apparatus.

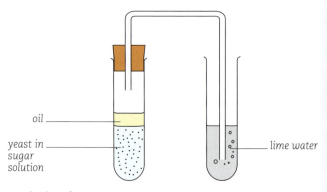

Respiration of yeast

Questions

1. What gas is produced as the yeast respires? How do you know?
2. What is the other product of the reaction? *Hint:* Smell the contents of the first tube.
3. How would you set up a control for this experiment?
4. What use is made of this reaction in everyday life?

Activity | Anaerobic respiration in germinating seeds

On page 104 we saw how germinating seeds released heat and produced carbon dioxide. The seeds were supplied with air, so they had enough oxygen to carry out aerobic respiration. But seeds can also carry out anaerobic respiration.
1. Fill a test-tube with vegetable oil and put it upside down in a dish also containing oil.
2. Push some germinating seeds up into the test-tube like this.

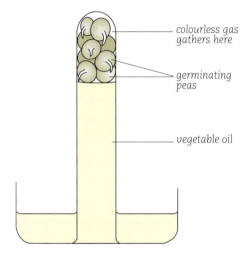

Respiration of germinating seeds

3. Set up another test-tube and dish but with seeds which you have killed by boiling them. At the top of the test-tubes, where the seeds are, there will be very little air.
4. After a few days examine your two test-tubes. What do you notice?
5. Put your finger over the mouth of each test-tube in turn and remove it carefully from the dish. Pour a little lime water into each test-tube. What do you notice?
6. How can you explain your results?

Uses of anaerobic respiration

Making bread In bread making we make use of the fact that in the anaerobic respiration of sugar, yeast produces carbon dioxide. It is the bubbles of carbon dioxide gas which makes the bread light and fluffy.

First, dried yeast is mixed with sugar and warm water. The yeast breaks down the sugar by anaerobic respiration or **fermentation** to release carbon dioxide, ethanol and energy. This mixture is then added to the flour to make a dough. The dough 'rises' as more gas is produced. The dough is then baked to make bread. The heating kills the yeast before very much ethanol has been made. The heating also makes the carbon dioxide gas expand so that the bread becomes fluffy.

Making alcoholic drinks When making alcoholic drinks we make use of the fact that in the anaerobic respiration of sugar, yeast produces alcohol (ethanol).

Depending upon the starting material and added flavourings, we can make alcoholic drinks with different flavours. For example, beer is made by the anaerobic respiration (fermentation) of hops, while red and white wines are made by the fermentation of different kinds of grapes.

The amount of alcohol in different drinks also varies. For example, the alcohol content of beers is usually 5 to 8% by volume, while that of wines is 10 to 14%.

The problem is that if the fermenting liquid contains more than about 14% of alcohol by volume, it kills the yeast. When rum is made, the sugar from the sugar cane is first fermented and then this liquid is distilled.

Distillation chamber. The water and alcohol mixture is heated and the alcohol vaporises at about 80 °C

The distilled liquid has a higher proportion of alcohol than the original liquid. Distilled spirits such as rum, whiskey and gin contain over 40% of alcohol by volume. Drinks which contain about 57% of alcohol by volume are called 'proof' and those with more than 57% are called 'overproof'.

These very strong alcoholic drinks have a bad effect on humans and can cause problems with our health as discussed on pages 80 and 253.

Questions

5. Give three similarities and three differences between aerobic and anaerobic respiration.
6. Why is anaerobic respiration important for germinating seeds?
7. When do you think anaerobic respiration might occur in our muscles?
8. How is anaerobic respiration (fermentation) used in (a) making bread and (b) making alcoholic drinks?

How do organisms get rid of waste?

Excretion and egestion

Getting rid of waste materials that have been produced by chemical reactions inside the body is called **excretion**. **Excretory products** include urine, sweat and carbon dioxide.

Getting rid of the remains of undigested food is called **egestion**. The material that is egested is called **faeces**. The diagram below illustrates the difference.

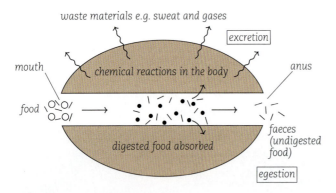

Simplified diagram of a longitudinal section of an animal to illustrate the difference between excretion and egestion

How do plants get rid of waste products?

A plant's main excretory products are *excess oxygen* from photosynthesis (in the day), and *carbon dioxide* from respiration (in the night-time). These gases pass out along diffusion gradients in the leaf (pages 70 and 101). Other wastes produced from the plant's activities collect in the leaves. These wastes are lost when the leaves fall off during **leaf fall**.

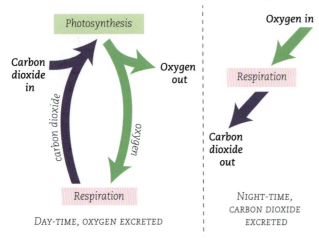

Plant excretion

What are the excretory products in a mammal?

Chemical reactions produce excretory products. One of the main chemical reactions which occurs in all living cells is respiration. Respiration produces excretory products of *carbon dioxide, water* and *heat*.

Another important chemical reaction is the breakdown of amino acids, if the body has more protein than it needs. (These excess amino acids are from proteins that were eaten by the animal but were not needed for growth and repair.) The breakdown occurs in the liver. The nitrogen-containing part of the amino acids is converted into a substance called **urea**. The urea circulates in the blood until it reaches the kidneys where it is excreted in the **urine** (pages 110–12). The remaining (carbohydrate) part of the amino acid molecule can be used to release energy.

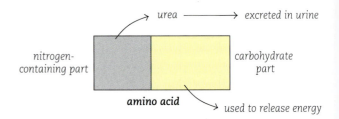

Breakdown of excess amino acids

The lungs, kidneys, and skin which get rid of excretory products, are called **excretory organs**. Of these, the kidneys (pages 110–13) are the most important.

Excretion from the lungs

In the lungs, carbon dioxide passes out from the blood into the alveolus along a diffusion gradient (page 100). This air is then breathed out and the mammal gets rid of carbon dioxide.

Most of the excess water in the body is lost through the kidneys, the skin and the faeces, and some water (as vapour) is lost from the lungs.

Average water balance over 24 hours			
Input/cm³		Output/cm³	
Food and drink	1 400	Lungs	400
		Skin (sweat)	500
From respiration	350	Urine	700
		In faeces	150

The energy released during respiration is partly used up in body activities. The remaining energy (as heat) is distributed by the blood to all parts of the body. When the blood reaches the lungs, the air in the alveoli is cooler than the blood and so some heat passes from the blood into the air and passes out with the breathed out air. Other heat is lost via the skin, as heat is needed for the evaporation of sweat and this is taken from the body.

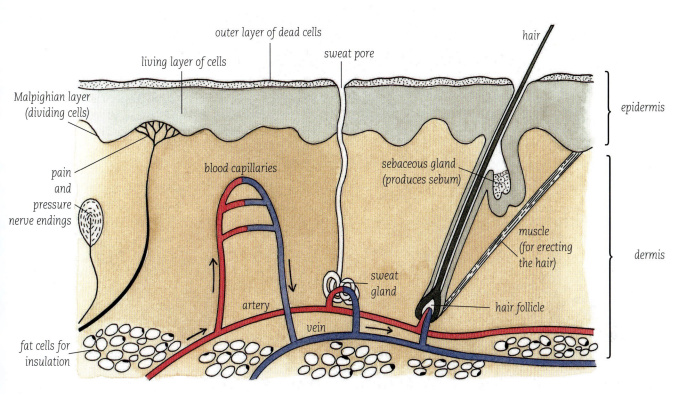

Vertical section through the human skin

The sweat glands

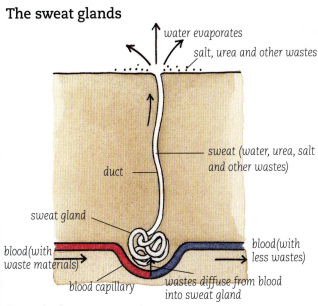

Sweat gland

Sweat is mostly water, with a small amount of urea (about 1.2%), and 1 to 2% of salt (sodium chloride). The main function of the sweat glands is to help keep the body temperature constant by the evaporation of water (page 117). But the loss of water also helps in the excretion of urea and salt.

Blood containing urea, salt and water passes close to the sweat gland. There is a higher concentration of these substances in the blood than in the sweat gland and so they diffuse along their diffusion gradients into the sweat gland. The sweat passes up the sweat gland duct to the surface of the skin. The water evaporates and the urea and salt are left on the skin.

The waste materials can accumulate on the skin and be broken down by bacteria to give rise to unpleasant odours. We all sweat more on very hot days as more water is lost to keep us cool, so more salt and wastes accumulate. This is one reason why we wash and keep clean (page 208).

Questions

1. Explain the difference between excretion and egestion, giving examples in each case.
2. Explain the following: (a) animals produce faeces but plants do not (b) our skin may taste salty.
3. What excretory products are released from: (a) the lungs (b) the kidneys (c) the skin (d) leaves in the day-time (e) leaves in the night-time?
4. Describe the excretion of one substance. Use these terms: concentration, diffusion, diffusion gradient.

What do the kidneys do?

The excretory system

The **kidneys** are the main excretory organs and are found inside the abdomen near to the back wall. They form part of the **excretory system**. Here is a diagram of the excretory system of a male.

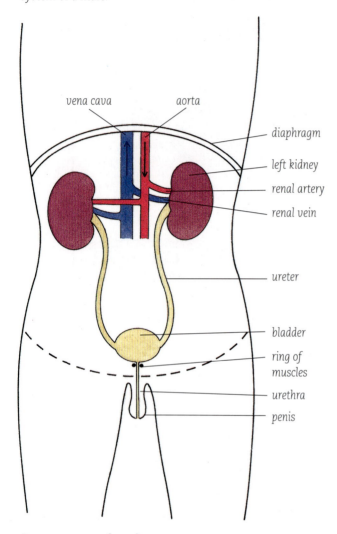

Excretory system of a male

A tube called the **ureter** runs from each of the kidneys. The ureters lead into a balloon-like structure called the **bladder**. The **urethra** is the passage that leads from the bladder to the outside of the body.

As urine is formed in the kidneys, it flows down the ureters and accumulates in the bladder, which increases in size. A ring of muscles keeps the opening of the bladder closed, until such time as the person **urinates** so that **urine** passes out of the body.

In the male the urethra runs down the middle of the penis, which is also part of the reproductive system (page 50).

How do the male and female systems differ?

In males the urethra is the passageway out of the body for both urine and sperm. However, both fluids cannot pass down the penis at the same time.

In a female (below) the urethra which carries urine is separate from the vagina, which is the female passageway from the reproductive system (page 50). Urine is produced in the same way and accumulates in the bladder before being passed out of the urethra.

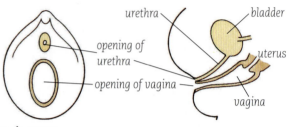

Female system

Having two exits is very important in the female because the vagina has other separate functions. During intercourse it has to receive sperm from the male and it also has to expand a great deal during the birth of a baby.

How is urination controlled?

Adults can keep the ring of muscles at the base of the bladder tight, so that they only urinate when it is convenient. But in a young baby, when its bladder contains about a cupful of urine, a reflex action (page 124) is triggered and the baby urinates. As the child grows older it gradually learns to overcome the reflex action and to control the muscles, although some bed-wetting may still occur. Older people, and women in late pregnancy, also may not be able to control the ring of muscles. They may therefore lose urine when they do not want to. This is called **incontinence**.

There is also frequency of urination which occurs in early pregnancy. As the uterus begins to grow it presses on the bladder so that the bladder has to be emptied more often. As the uterus grows larger it is higher up in the abdominal cavity and so does not press on the bladder. As a result of hormone changes the woman may also accumulate more water than usual in her body that she has to get rid of.

Increase in size of the prostate gland in older men may also cause more frequent urination.

Frequent urination is also a symptom of **diabetes** (page 113) where the body passes extra liquid to try and get rid of excess sugar from the blood.

The factors which affect *how much* urine we excrete will be dealt with on pages 112–13. We will now look at the simple structure and functioning of the kidneys.

Activity | What are the kidneys like inside?
1. Obtain a kidney of a mammal such as a pig or sheep.
2. With a sharp knife, cut the kidney along its length.
3. Look inside, identify the parts, and make a labelled diagram like this.

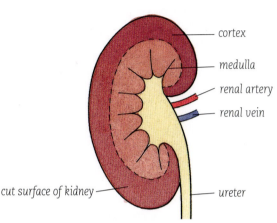

Cut open kidney

What do the kidneys do?
Urea, the waste product from broken down amino acids (page 108), is produced in the liver. It travels round the body in the blood. Some urea is lost in the sweat; most is lost in the urine.

Blood comes to the kidneys in the renal artery. This blood contains urea and salt not needed by the body. Small tubes, called **kidney tubules**, inside the kidney remove the urea and salt together with some water. This liquid is then called urine and it leaves the kidney in the ureter to go to the bladder. The blood, now with less urea and salt, enters the renal vein and is taken away from the kidneys. The urine contains more water, salts and urea than the blood, as is shown in the table below.

	Blood (%)	Urine (%)
Water	90–93	97.5
Salt	0.35	0.5
Urea	0.03	2.0
Other substances	7.1	none

Problems with the excretory system You may have heard of someone with 'kidney stones'. What are these? Tiny specks of solid material containing calcium can sometimes be built up into small 'stones' up to 25 mm in diameter. Stones smaller than 5 mm can pass into the ureter and cause pain as they travel down to the bladder. Drinking extra water – 5 litres a day – can help prevent their formation.

The role of the kidneys in excretion
Urea is made in the liver, but excreted by the kidneys. The diagram summarises the process.

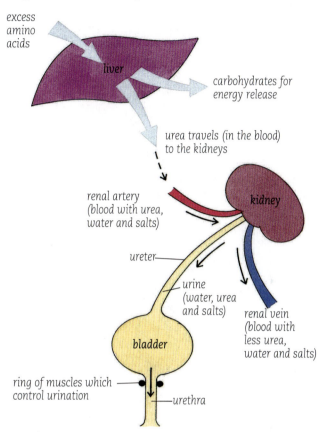

Excretion

This diagram illustrates how urea, salt, and water are excreted by the kidneys. On page 112 we shall look in more detail at the structure and functioning of the individual kidney tubules.

Questions
1. (a) Draw a labelled diagram of the male excretory system.
   (b) In what ways does it differ from the female excretory system?
2. Which of the following takes urine from the kidney to the bladder?
   (a) Ureter (b) Kidney tubule (c) Urethra (d) Renal vein
3. Which blood vessel: (a) brings blood to the kidneys (b) takes blood away from the kidneys?
   What is the blood like in (a)?
   What is the blood like in (b)?
   What are the reasons for the differences?
4. How do the amounts of salt and urea differ in the blood and the urine?

How do the kidneys work?

The kidneys work by blood being supplied to thousands of very small kidney tubules. Each tubule consists of a cup-shaped part and a long twisted tube. Below is a very simple diagram showing the main parts of a kidney tubule and how it works.

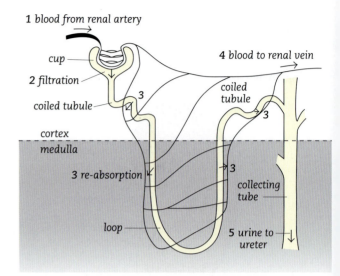

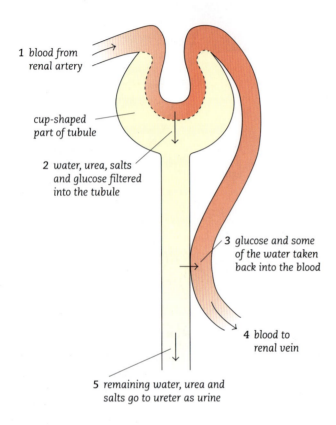

Kidney tubule

More detail of the coiling of the kidney tubule. The five stages of tubule functioning are labelled (see previous column)

Functioning of the tubule

1. **Blood comes to the tubule** in a branch of the renal artery. This blood contains water, urea, salts, glucose and protein molecules.
2. **Blood is filtered** through the cup of the tubule. All the water, urea, salts and glucose are forced into the cavity of the tubule. Blood cells and protein molecules are too big, so they stay in the blood.
3. **Useful substances are re-absorbed** from the tubule back into the blood. All of the glucose and some of the water is re-absorbed.
4. **Blood leaves the tubule** in a branch of the renal vein. This blood contains less water, urea and salts, but the same amount of glucose as the blood in the renal artery.
5. **Urine leaves for the ureter** The water, urea and salts left in the tubule is the urine. It collects and passes to the ureter.

Water balance or osmoregulation

In the kidneys, water passes from the blood into the tubules. But how much water is re-absorbed into the blood, and how much passes out in the urine?

1. If there is more water in the blood than usual, a lot of it will be lost in the urine and less will be re-absorbed into the blood. This is the case if the person has recently drunk a lot of water. Also, if it has been a cooler day than usual and the person has not been sweating very much, more water would need to be lost in the urine.

    In these cases our urine appears *lighter* in colour because it is more **diluted**.

2. If there is less water in the blood than usual, more of it will be absorbed back into the body, and less will be passed out in the urine. This is the case if the person has not been drinking much water. Also, if it has been a very hot day and the person has been sweating a lot to keep cool, there would be less water to get rid of in the urine.

    In these cases our urine appears darker in colour because it is more **concentrated**. In this case it is important to drink more water.

The amount of water in the blood is therefore kept constant and this is called **osmoregulation**. Osmoregulation is important because if the water content of the blood varies by more than a little bit, then this affects the functioning of the cells. This is why diarrhoea and dehydration can cause death – they reduce the amount of water in the blood.

Rehydration liquid must be given, especially to young children if they become dehydrated. The liquid replaces the water which the body needs, as well as sugar and salt.

Control of water loss

The lungs lose water depending upon the temperature and the amount of water vapour in the outside air.

The skin loses water depending upon the body temperature, the outside temperature and wind, and the amount of water vapour in the outside air.

The kidneys 'balance up' the amount of water lost day by day, so the amount in the blood remains constant (see below). The amount of water re-absorbed into the blood is under the control of a hormone which is produced in the pituitary gland (page 129). In the Caribbean it is important to drink a lot of water.

Cooler day or less physically active

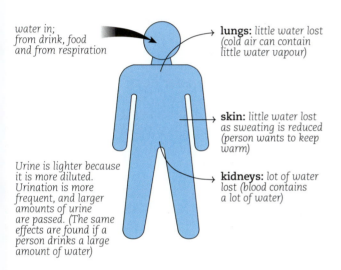

Hotter day or more physically active

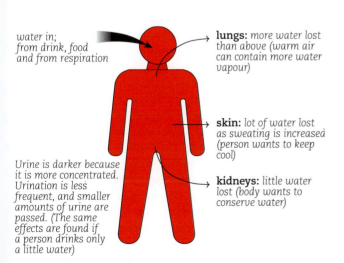

Kidney failure

If a person's kidneys are damaged, waste products such as urea and salts may not be filtered out of the blood. They could accumulate and cause poisoning. The person could use an **artificial kidney**. This machine filters and cleans the blood; a process known as **dialysis** (see below).

Blood is taken from an artery and pumped into the machine. The blood meets a layer of cellophane through which the unwanted urea and salts are able to pass. The blood cells and proteins stay in the blood. The concentration of the blood is adjusted by the machine and the clean blood is returned to the person in a vein. More kidney machines are needed in the Caribbean to treat people with these problems.

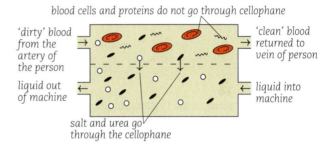

Dialysis – the principle behind the kidney machine

Diabetes

The amount of glucose in the blood is usually about 0.1% by volume. A person with diabetes has a much larger amount of blood glucose than a healthy person. When the blood goes to the kidneys this glucose passes into the tubules along with the water. But the diabetic cannot re-absorb sufficient glucose back into the blood: most of it will be excreted in the urine. Finding glucose in the urine is one test by which a person suffering from diabetes is identified.

Treatment of diabetes is to reduce the amount of glucose in the blood by diet and by giving insulin. The person tests their blood to see that the amount of glucose is kept low. Then, when the blood goes to the kidneys, the glucose can be re-absorbed and none passes out into the urine.

Questions

1. What do you understand by the following terms?
   (a) Filtration (b) Re-absorption (c) Dialysis
2. Why is the urine of a healthy person sometimes light and sometimes dark in colour?
3. What are the roles of (a) lungs (b) skin and (c) kidneys in controlling the amount of water lost by the skin? Which do you think is the most important organ in osmoregulation?

Why do temperatures change?

Activity | Can heat be transferred?

1. Get two tin cans the same size (about 200 cm³). Place about 100 cm³ of boiling water in one tin and an equal quantity of ice-cold water into the other.
2. Copy this table into your notebook.

Time in minutes	Temperature (°C) of Cold water	Hot water
0		
2		
4		
6		
8		
10		
12		
14		
16		
18		
20		
22		

3. Read the temperatures at the start of the experiment, to the nearest half degree Celsius if possible, and at 2-minute intervals.
4. Plot your results on a graph, temperature on the y–axis and time on the x–axis, to obtain two curves, one for the hot water and the other for the cold water.
5. Can you explain the shape of the curves? What do you think the room temperature is?

Given enough time, the water in both cans will become the temperature of the room. This means that heat from the boiling water will have been lost to the surroundings and heat from the surroundings will have been taken in by the cold water.

Activity | How is heat transferred?

1. Place a metal spoon in a beaker or saucepan of boiling water with its handle sticking out of the water. After two minutes, can you remove the spoon comfortably?
   The whole spoon becomes too hot to hold comfortably. The heat from the water must have travelled along the spoon from atom to atom inside the spoon. Heat transferred in this way is by the method of **conduction**.
2. Repeat the experiment, this time using spoons made of different materials including plastic and wood. Are these other spoons easy to hold after heating?

From the result of these experiments, you can conclude that metals are good conductors of heat while non-metals, such as wood and plastic, are poor conductors of heat. Therefore, it is not surprising that cooking utensils are usually made of metal. (See page 202.)

Metals readily absorb heat and pass it on by conduction to other materials in contact with the metal surface. Handles made of metal would also become hot and uncomfortable to hold. For this reason handles are often covered with wood, plastic, rubber or other materials, which are poor conductors of heat.

How does the conduction of heat take place?

Heat is a form of energy – energy is the motion or vibration of the atomic particles of matter. When matter gets hot, the atoms vibrate and emit electromagnetic radiation in the infrared (IR) regions of the electromagnetic spectrum (see page 320). In good conductors, atomic vibration (heat) easily passes from atom to atom. In poor conductors, atomic vibration (heat) is not easily passed on, probably because of a more rigid atomic structure.

How do liquids transfer heat?

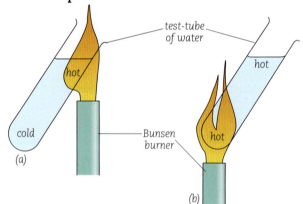

Heating water

A test-tube of water can be heated in these two different ways. In (a) the water will quickly boil on the surface of the test-tube while it remains cold at the bottom. In (b) the temperature of the water in the whole tube will rise steadily to 100 °C and then the water will boil.

The explanation is that heat is transferred from the glass container to the water by conduction. As the water gets hot, it gets less dense and more buoyant. Since water is not held together in a rigid structure like solids, the hot part of the water will rise because it is less dense. At the same time, the colder water sinks. If a drop of dye, a crystal of potassium manganate(VII) or some powdered chalk is placed in a tube of water being heated, the movement of the water in the tube can be seen. This type of movement is called **convection** and is the most important method of heat transfer in liquids and gases.

Ocean currents

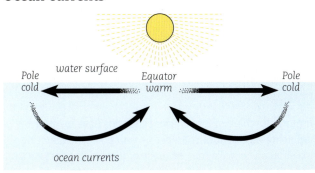

How ocean currents form

Ocean currents are brought about by convection in water. At the equator, the sun warms the surface water. The water in temperate and polar regions remains relatively cold. The warm water spreads northwards and southwards from the equator, forming a warm surface current. The cold water becomes covered and sinks, forming a deep cold current flowing back to the equator.

Land and sea breezes

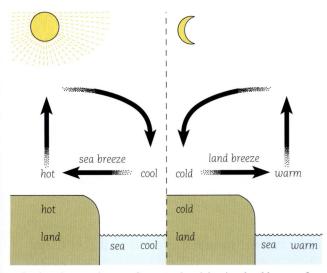

In the day-time sea breezes form   In the night-time land breezes form

Convection in air produces land and sea breezes. In day-time the land is heated by the sunshine. The air above the ground also becomes hot, and rises. The sea remains cool. Air from above the sea flows across the land to replace the warm land air which is rising. In this way a cool **sea breeze** is created.

At night-time **land breezes** form when air flows from the land to the sea. The land cools rapidly to below the sea temperature. Then the air above the sea is warmer than air above the land.

How is the sun's heat transferred to Earth?

The sun is about 1.5×10^8 km from the Earth. There is no connecting matter: therefore, heat cannot travel to the Earth by means of conduction or convection. The only other alternative is by **direct radiation** by infrared (IR) waves, which are always given off by hot bodies such as the sun. Infrared (IR) radiation is not visible to the human eye but has some of the properties of visible light. It can be focused by a concave mirror and a convex lens, and can be absorbed by various materials. The Earth's atmosphere, oceans and landmasses absorb IR radiation from the sun.

How do microwaves heat?

Examine the chart of the electromagnetic spectrum (page 320). Note the position of infrared waves and radio waves. The ultra-high frequency radio waves are also known as microwaves and have some of the properties of IR radiation. They are readily absorbed by water, causing a rise in temperature. For this reason they are used in microwave ovens. Substances such as glass or plastic are easily penetrated by microwaves without any effect, so if foods containing water are placed in glass or plastic containers they cook internally and externally at the same time. In this way foods can be cooked quickly and efficiently, losing much less of their nutritional value than by other cooking methods.

Microwave ovens today are fitted with safety devices making them impossible to operate if the safety procedures are not followed. This ensures that the person who is using the microwave oven cannot get burnt.

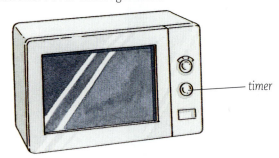

A microwave oven – a whole chicken will cook in about 45 minutes compared to more than an hour in an ordinary oven

Questions

1. What are the methods by which heat is transferred?
2. Explain the importance of convection currents in (a) water and (b) air.
3. How are sea and land breezes formed?
4. How would you identify a material as a good conductor?
5. Explain how microwaves are used for cooking.

Can we control heat transfer?

Activity | The vacuum flask
1. Set up an experiment with two vacuum flasks (also called thermos flasks), one with boiling water and one with ice water. Measure the temperature at the start, after 6 hours, 12 hours and 24 hours, or at other convenient intervals.
2. If the temperature of the water added to one flask was 100 °C and to the other flask 0 °C, how do you account for the different temperatures at the end of the experiment?

The greatest change from the original temperatures occurs after 24 hours. Heat has been lost from the flask which contained hot water, and heat has been gained by the flask which contained cold water. This happened very slowly because the flasks are specially made to reduce heat transfer, which may either be heat loss or heat gain.

> ### Question
> 1. The vacuum flask has the serious disadvantage that it is easily broken if dropped. The glass walls could be replaced by stainless steel. This kind of vacuum flask is shockproof, but it is not as good for keeping things hot. Why not?

A more durable but less efficient thermos flask is made by replacing the vacuum chamber by very poorly conducting materials. The walls may be made of plastic and enclose a layer of polystyrene foam, which is an insulator.

Why is polystyrene a good insulator?
The material itself is a poor conductor. When it is filled with air bubbles to form solid foam, it becomes a very light and even poorer conductor. The very tiny air bubbles are isolated from each other. Heat is not lost by convection because convection currents cannot occur between the air bubbles. Radiation heat loss is not a problem either, because radiation cannot penetrate the polystyrene layer.

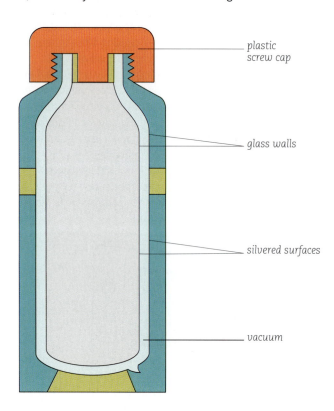

Polystyrene magnified × 200. The microscopic air cells make it a very poor conductor

How does the vacuum flask work?
The container is made of two walls of glass, with a vacuum between the walls. Glass is a poor conductor of heat. The stopper may be made either of cork or plastic, which are also poor conductors. This effectively reduces heat transfer by **conduction**. **Convection** is kept to a minimum because the inner wall is kept well away from air currents. The surfaces inside the vacuum are silvered. This reduces heat loss by **radiation**. Infrared (heat) rays are reflected back into the liquid inside the flask.

How is heat transfer applied in the kitchen?
The main sources of heat (fuels) for food preparation are wood, charcoal, oil (kerosene), natural gas and electricity. Each of these fuels may have advantages and disadvantages. Can you think of any?

Generally, a wood or charcoal fire is difficult to regulate. Gas and electric fires are very easy to regulate and much less energy will be wasted if they are used correctly.

The most efficient method of heat transfer for cooking is by conduction through metal pots. The heat source is either a flame or electrically heated surface. But most flames are well over 1 000 °C and thus metal pots in contact with flames can easily become so hot that the food chars and spoils.

Problems of heat regulation and distribution can be overcome by using water or oil in the pot. Usually, liquids cannot be heated to above their boiling points (see latent heat of vaporisation, next column). When water is used, the food is usually cooked at around 100 °C. However, a **pressure cooker** can be used to increase the boiling point of water to up to 121 °C. For most foods, cooking times are shorter at this temperature.

The use of oil in frying does produce high temperatures. The high temperatures produce the crisp drying and browning seen on potato chips and fried meat.

How can energy be saved in cooking?

Food prepared and not eaten is a waste of energy. Much energy may be saved by proper planning of the right quantities of food. Many different methods of preparing foods are necessary to produce the range of textures and flavours required. However, it is sensible to select the most energy efficient methods.

The use of an oven is particularly energy efficient. The oven is a closed well-insulated chamber, which traps heat from a source in its walls and distributes it by convection and radiation. Many pots can be heated at once and the rate of heating can be controlled.

Activity | How do you keep cool?
1. Let some tap water run onto both of your hands.
2. Keep one hand still and shake the other one about. Do you feel any difference between the two hands?
3. Place one hand in front of a fan or an open window in the breeze. How does it feel? How do you explain the sensation that you feel? Which hand dries faster, the one in front of the fan or the one held in still air?

Perhaps you think that your hand only feels cold and is not really cold. Try the next Activity and see if a thermometer can show a difference in temperature.

Activity | Cooling and temperature
1. Tie a small wet piece of tissue paper around the bulb of a thermometer. Hold it still, then read and record the temperature.
2. Shake the thermometer and paper for about a minute, then read and record the temperature. Is there a difference between the readings?

Shaking caused water to evaporate from the tissue paper quickly. Evaporation uses up energy so the temperature drops because heat energy is given up (heat is lost) by the thermometer bulb. Now can you explain why a wet hand feels cooler in moving air? (The way that humans use evaporation to keep cool is described on pages 118–19.)

Is energy required for evaporation?

Many substances can change their state if the conditions are right. For example, water can be solid (ice), liquid (water) or gas (steam). The metal mercury (used in thermometers) is a liquid at room temperature, but at very low temperatures it is solid and at high temperatures it is a gas.

Activity | Latent heat of vaporisation
1. Heat some water in a pan over an adjustable flame such as a Bunsen burner.
2. Bring the water to the boil and adjust the flame so that the water is boiling gently. Measure the temperature of the water.
3. Adjust the burner to increase the flow of gas several times, and read the temperature each time.
4. What is the temperature of the water when it is boiling gently? What is the temperature of the water when it is boiling vigorously?

Boiling water is the same temperature whether it is boiling vigorously or just simmering. When water boils at sea level it changes its state from water with a temperature of 100 °C to steam with a temperature of 100 °C. Energy transferred to boiling water at 100 °C does not produce a rise in temperature but rather a change in state, called a **phase change**, from liquid to gas. A certain amount of energy will change a certain amount of liquid at the boiling temperature to a gas at the same temperature, like this:

x g water at 100 °C + Energy → x g steam at 100 °C

The energy needed is called the **latent heat of vaporisation**.

How do trees keep cool?

When trees are in full sunlight, their leaves are a few degrees colder than the surrounding air temperature. This is because water is lost from the underside of the leaf by evaporation, so the temperature is lowered.

Questions

2. Explain how the materials and construction of vacuum flasks minimise heat transfer.
3. Explain why food will not cook faster with a high flame than with a low flame.
4. Discuss efficient and inefficient methods used in food preparation.
5. How would you show that energy is required to turn solid ice at 0 °C into liquid water at 0 °C?
6. How do trees keep cool?

How do living things control their temperature?

Poikilothermic animals

Poikilothermic animals cannot keep their body temperature constant. *Poikilo* means 'changing'. Their body temperature changes with the temperature of their surroundings, so if it is cold outside, their body temperature will also be low. If it is hot outside, their body temperature will also be high.

It is therefore better to speak of poikilothermic (changing temperature) animals rather than 'cold-blooded' ones.

Fish and amphibians As the water that fish live in warms up or cools down, so their body temperature goes up and down.

In countries with a cold winter, the water near the top of a pond or lake sometimes freezes. The fish will then sink to the bottom of the pond or lake until it gets warmer. The fish rest in the cold water until the ice melts and they then begin to warm up.

Amphibians, such as frogs and toads, are also poikilothermic. They tend to hide away in water or under stones in the day-time so as to keep away from the hot sun which would raise their body temperatures to dangerous levels, and also cause them to dry out. In cold conditions they may hibernate.

Reptiles Reptiles, such as lizards, tend to change their body temperatures with changes in their surroundings. They do not have any special structures (feathers or hairs) to help them to control their temperature. But they *behave* in certain ways so as to reduce the variation in their body temperatures to a small amount.

During the morning lizards lie out in the sun. They are 'warming up'. They become active and feed. At the hottest time of the day they will hide in the shade or under stones so that the sun will not overheat them. At night they will hide again so as not to become too cool.

The role of transpiration in plants

As water evaporates, it uses up the **latent heat of vaporisation** and so has a cooling effect (page 117) in addition to its role in the movement of materials (page 91).

In plants this evaporation is called **transpiration**. If you place a potted plant inside a plastic bag you can see the moisture that is lost. You will see drops of water from transpiration on the inside of the bag.

The leaves of a plant are partly covered with water-proof **cuticle** and water loss is restricted to the pores or **stomata** (page 70). If too much water is lost, the guard cells around the stomata collapse and close the pores, so stopping any more loss of water.

Homoiothermic animals

Homoiothermic animals can keep a constant body temperature. *Homoio* means 'the same'. Their body temperature stays almost the same (page 120). The animals have ways to control their temperature. Their body temperature is largely independent of the temperature in their surroundings, so if it is cold outside, their body temperature will not decrease, and if it is hot outside, their body temperature will not increase. Homoiothermic (same temperature) is a better description than 'warm-blooded'.

Birds Birds have feathers to help keep their body temperature constant. They have small **down feathers** close to their skin which hold a layer of warm air. The moisture evaporating from the skin builds up and so tends to reduce further evaporation. Birds are constantly eating. There is therefore a continuous release of energy that helps to keep them warm.

Birds cannot keep their body temperature as constant as mammals. In temperate regions birds may therefore have to migrate to warmer conditions during the winter. In very hot countries, the birds may migrate to cooler places during the summer.

Mammals Mammals have a heat-sensitive centre in the brain which controls body temperature (page 129), and triggers the action of hairs, sweat glands and the amount of blood circulating to the skin. Mammals also have fat deposits under the skin for insulation.

Humans are an unusual type of mammal. We do not have a thick covering of hair all over us but our covering of small fine hairs does help us to control our body temperature. We help the temperature control by wearing appropriate clothing (page 121).

The role of sweating

Loss of water by evaporation (in mammals) is called **sweating** or **perspiration**. The sweat evaporates from the sweat glands (page 109). The greater the evaporation of sweat the greater the cooling effect (page 117).

During exercise, when we get very hot, drops of perspiration will collect on the skin. This should be removed with a towel after exercise, because if it all evaporated quickly it might chill the body too much.

Perspiration is a natural process. **Anti-perspirants** which may block the sweat glands and upset their control mechanism are not recommended. Our skin can make the necessary adjustment to control the amount of sweating which is healthy for the body.

How the skin loses heat (becomes cooler)

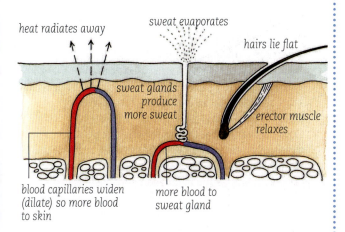

1. Blood vessels (capillaries) in the skin widen (**dilate**) so more blood is brought to the skin to be cooled. Heat is lost by radiation (page 115).
2. The amount of sweat production increases so the person is cooled by evaporation.
3. The hairs close to the skin lie flat, so little air is trapped and sweating occurs very easily.

How the skin retains heat (becomes warmer)

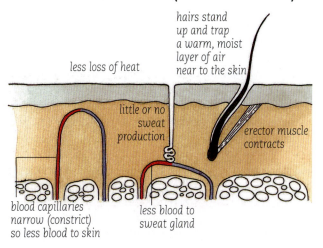

1. Blood vessels in the skin become narrow (**constrict**) so less blood is brought to the skin to be cooled. Therefore heat radiation does not occur.
2. The amount of sweat production decreases so less heat is lost by evaporation.
3. Hairs stand on end (goose pimples) so moist air is trapped near the skin and reduces sweating.

Advantages of homoiothermy

1. Chemical reactions in our body (including those involving enzymes) work best at a constant body temperature. If the temperature is too low, then the reactions will be slower. If the temperature is too high the enzymes may be damaged and no longer work properly (page 89).
2. In the early morning a homoiothermic animal does not take much time to 'warm up'. This means that it can be active immediately. It can start to chase its prey, or escape from its enemies, so it is more likely to survive. A poikilothermic animal takes time to warm up in the morning and is at danger from predators and also cannot quickly go looking for its prey.
3. A constant body temperature allows an animal to live in places where the temperature varies a lot during the course of the day. Because the animal is largely unaffected by the external temperature it is not restricted to places with a narrow temperature range. Poikilothermic animals can only live in places with a fairly constant temperature.
4. A constant body temperature also allows an animal, such as ourselves, to live in very hot and very cold areas of the world. In these areas we help to regulate our temperature by wearing appropriate clothing (page 121).

Questions

1. The graph below shows the body temperature of three animals during the day. Which of A, B and C are a fish, a lizard or a human? Give your reasons.

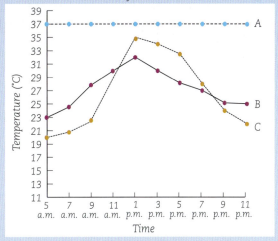

2. How are transpiration and sweating (a) similar and (b) different?

How do human temperatures vary?

Activity | Measuring body temperature

1. Heat about 200 cm³ of water to near boiling point in a beaker.
2. Insert the bulb of a laboratory thermometer (page 130) into the water, watch the temperature, and when the mercury is steady, record the result.
3. Remove the thermometer from the water and after 2 seconds read the temperature again. Record the result. Is there a difference between the two recordings? Can you explain why?
4. Examine this diagram of a clinical thermometer, or examine a real clinical thermometer. What are (a) the maximum and (b) the minimum temperatures on the scale? What do you think this thermometer is used for? Notice the kink or constriction in the tiny tube: what might its use be?

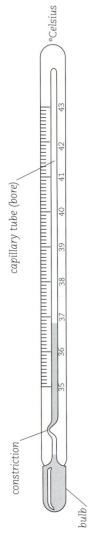

Thermometer

The range of temperature of the clinical thermometer is close to the human body temperature – it extends a little above and a little below normal temperature. The range is 35 to 43 °C. From the many divisions on the scale you can tell that the thermometer is designed to measure very accurately over this short temperature range.

In the Activity above you found that there was a slightly different reading for the temperatures with the **laboratory thermometer** in and out of the water. This is because when the thermometer was taken out of the water, the bulb with the mercury immediately began to cool. Therefore the mercury thread begins to **contract** (shorten), which in turn gives a lower temperature reading.

This is exactly what would happen if you were to measure human body temperature with a laboratory thermometer: you would get an inaccurate measurement. However, in **clinical thermometers** the position of the thread of mercury is fixed as soon as it reaches the temperature of the body. The constriction in the tube causes a break in the thread of mercury as soon as the bulb begins to cool down. When the thermometer is therefore withdrawn from the patient's mouth, the thread above the constriction remains intact, allowing an accurate reading to be taken.

Activity | Taking a temperature

1. Insert the clinical thermometer under your tongue and hold it in place for about 2 minutes to ensure that it reaches body temperature. (Normal human body temperature is 37 °C.) Shake it carefully but thoroughly afterwards to rejoin the mercury thread.
2. Does everyone have the same body temperature? Find the temperatures of all your classmates and plot the results on a graph. Make sure the thermometer is washed in antiseptic before someone else uses it. Arrange your results to get a **histogram** (see page 37). Discuss your findings in class. How much variation do you get? Is there any explanation for these variations?

Is temperature the same on all parts of the body?

In the Activity above, the mouth was used for the temperature measurements because in adults and older children it is the most convenient. However, in babies and small children especially where there is danger that the thermometer may be broken, rectal temperature or armpit temperatures are taken.

Activity | Temperature over the body

1. Use a clinical thermometer to measure the temperatures of various parts of your body. Measure the temperatures of some uncovered areas, for example lower arm, wrist, leg, and then some covered areas like upper arms, under sleeves and under socks. Make sure the thermometer is washed in antiseptic before someone else uses it.
2. Record your results. Is there a difference between the covered and uncovered areas?

The internal body temperature is 37 °C. Uncovered or exposed areas usually have a lower temperature, because they radiate and so lose heat. Clothes can restrict heat loss so that skin temperatures measured in covered areas may be higher than exposed areas. Of course, this is true when the air temperature is lower than the body temperature (as is usually the case in the Caribbean) and if the body is not exposed to direct sunlight.

The human body is able to cope with small changes from normal body temperature. If temperature falls, the rate at which food is burnt up, and therefore heat is produced, is increased. If it gets too hot, cooling by the sweat glands is increased (see pages 118–19).

Clothing

The tropics In the tropics, loosely hanging light coloured clothing, the traditional wear of many parts of Africa, offers good protection from the direct rays of the sun. Light colours absorb less of the infrared (heat) rays of the sun than the exposed (dark) skin. The loose fitting allows a free circulation of air over the skin underneath the clothes, so that moisture is continually being carried away by air.

Dark clothes absorb many more infrared rays and acquire a high surface temperature in the direct sunlight. But this causes increased convection currents and air flowing upwards under the loose robes increases the cooling effect.

Cold wintery climates In cold wintery climates, the most serious problem is heat loss. Clothing is therefore chosen to conserve heat. This is done by using layers of woollen clothing or garments made from furs or skins or padded with insulating materials. In this way, heat loss by conduction is kept to a minimum, air becomes trapped beneath the layers of clothing and heat loss by convection is also greatly reduced.

Among the different types of synthetic and natural materials available, the most popular materials for clothing are still natural fibres, for example cotton, wool and silk. They are used either in 'pure' form or mixed with synthetic fibres, for example cotton and Dacron mixtures.

Natural and synthetic fibres Natural fibres are found to be more comfortable than synthetic fibres for wearing next to the skin. The natural fibres that are used for clothing are of two main types: (1) of plant origin, for example cotton and linen and (2) of animal origin, for example wool, silk and leather. Natural fibres are able to absorb small quantities of water without getting wet. This allows loss of water from the skin and thus cooling on the skin without excessive drying. Most synthetic fibres do not absorb the moisture given off by the skin. They become wet on the surface instead, tending to stick to the skin and feeling uncomfortable.

Footwear Footwear is made from a wide variety of materials, both natural and synthetic. Open shoes or sandals are suitable for warm climates. A free flow of air through open footwear keeps the feet cool and dry. If your feet are damp with perspiration you are more susceptible to infections such as athlete's foot. In cooler climates enclosed footwear, such as a shoe or boot, is essential for keeping the foot warm and dry, since very little sweating takes place.

Activity | Does shape affect heat transfer?

Formulate a *hypothesis* (page 32), for example: that the shape of an object or body affects heat transfer. Next, design an experiment to test this hypothesis. For example:

1. Use equal volumes of hot water in differently shaped containers.
2. Read and record the temperature changes over time and plot cooling curves as on page 114.
3. Discuss your results and draw conclusions.

Alternatively, observe machinery around you which is concerned with heat transfer (losing, storing or acquiring heat), for example, a car radiator, solar collecting panel (for a solar water heater), hot water storage tank, vacuum flask, heat exchanger in a gas water heater, and so on.

What do heat storage systems have in common? What do heat exchangers have in common? Can you make a general statement about the shape of objects and thus their ability to transfer heat?

In modern societies, human body shape is usually not important for survival. Sometimes certain human shapes may have advantages over others, depending on the environment.

The shape of Inuit people (left) makes it easy for them to conserve heat in a very cold climate, while the shape of Masai people (right) makes it easy for them to lose heat in a very hot climate

Questions

1. What are the important features of the clinical thermometer?
2. How is the clinical thermometer used to measure human body temperature?
3. Why do natural fibres feel more comfortable next to the skin than synthetic fibres?
4. Why is loose fitting clothing best suited for hot conditions?
5. How might body shape be important in heat transfer?

What is the nervous system?

The nervous system is the main controlling system in the body.
1. It controls thinking, planning and learning. It allows us to interpret information from our senses (pages 125 and 262).
2. It controls our instinctive reactions which allow us to react to danger by reflex actions (page 124).
3. It controls the activities of all of our other systems. For example, it controls the movement of the limbs and the internal organs, and regulates our body temperature.

Parts of the nervous system
There are two main parts.
(a) The **central nervous system**, which includes the brain and the spinal cord (which runs down the body inside the backbone).
(b) The **peripheral nervous system**, which includes the cranial and spinal nerves, the sensory and motor nerve fibres, and the sense organs.

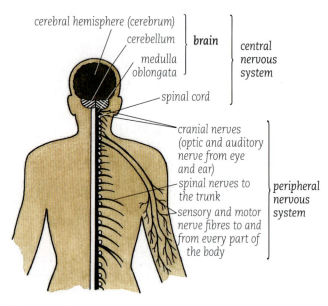

Nervous system

The nervous system links the **receptors** to the **effectors**.
- The receptors are sensory cells inside our sense organs.
- The effectors are various muscles and glands throughout our body.

The receptors are sensitive to changes called **stimuli** in the surroundings. The stimuli cause the setting up of tiny pulses of electricity which we call **nervous impulses**. Nervous impulses pass from the receptor along the nerves to the spinal cord and/or the brain. Corresponding nervous impulses are sent from the spinal cord and/or the brain to the effector, which responds. See the diagram in the next column.

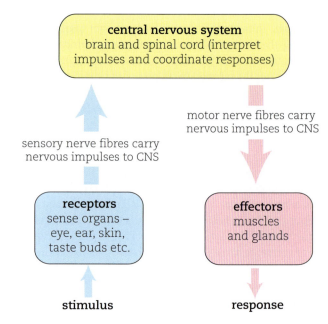

How the nervous system receives and responds to stimuli

Receptors
Different receptors receive different stimuli. For example, nervous impulses are set up in the eye by the stimulus of light. Other sensory cells are sensitive to stimuli which we call sound, smell, taste and an appreciation of touch, pain and heat.

The eye (page 279) The sensory cells in the eye are the **rods** and **cones** which are found in the **retina**. The rods are sensitive to the presence or absence of light, and so we see light and dark. The cones are sensitive to various colours and so we see in colour.

Nervous impulses are sent along the optic nerve to the sight centre in the brain (page 125). This interprets the upside-down image on the retina, so that we see the upright object in front of us.

The ear (page 269) The **semi-circular canals** have cells sensitive to changes in position, so they help us to balance.

The sensory cells in the **cochlea** respond to different frequencies of sound. Nervous impulses travel along the auditory nerve to the hearing centre of the brain (page 125) which interprets the impulses as sounds.

The nose When we draw in air through our noses it goes into a space called the **nasal cavity** (page 99). There are sensory cells in the lining of the cavity which are sensitive to particles in the air. Nervous impulses are sent from the sensory cells to the brain and so we can smell.

The tongue The sensory cells responsible for our sense of taste are found in the **taste buds** which are buried in pits on our tongue, see (a) on the facing page.

Chemicals in our food stimulate different sensory cells and impulses are sent to our brain. Different parts of our tongue are sensitive to different kinds of taste, see (b) below.

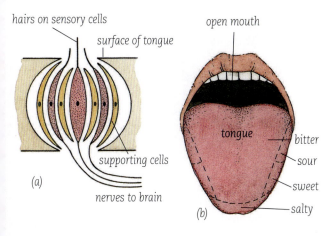

The tongue – (a) a taste bud (b) taste areas

The senses of taste and smell are interrelated. A large part of what we call 'taste', actually depends upon the vapours from our food which rise up into the nasal cavity and stimulate the cells sensitive to smell. We know this is true because when we have a blocked nose due to a cold, it is hard for us to 'taste' our food.

The skin There are special receptors in our skin which are sensitive to touch, pain, pressure, heat and cold as shown below.

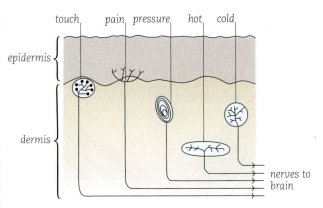

Sensory nerve endings in the skin

In different parts of the body there are different numbers of receptors. For example, there are more receptors sensitive to touch on our fingertips than on our elbows. When the receptors are stimulated, nervous impulses are sent along nerves to the brain.

Effectors

Sensory cells of the receptors send nervous impulses to the spinal cord, and/or brain, and appropriate impulses are sent to the effectors – muscles or glands – to make them respond.

Muscles Muscles respond to impulses from the spinal cord or brain by **contracting** (getting shorter) or **relaxing** (returning to their normal length). In our arms and legs the muscles occur in pairs, attached to the bones by tough inelastic straps called tendons. Here is a diagram of the pair of arm muscles.

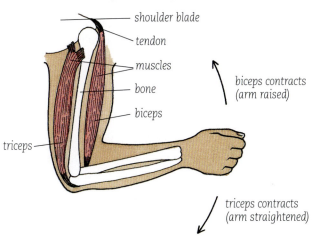

When the biceps contracts the arm is bent at the elbow and the arm is raised. The triceps relaxes. When the triceps contracts the arm is pulled straight. The biceps relaxes

Glands Glands respond to impulses from the nerves by secreting **juices**. For example, when we are eating food, our salivary glands are stimulated to secrete saliva, and the glands in our stomach (gastric glands) secrete gastric juice. These juices help to digest our food.

> ### Questions
>
> 1 Complete the following table.
>
> | Sense | Organ | Part responsible |
> |---|---|---|
> | Balance | Ear | Semi-circular canals in inner ear |
> | Hearing | | |
> | Sight | | |
> | Smell | | |
> | Taste | | |
> | Touch | | |
>
> 2 Give an example of (a) a receptor (b) sensory cells (c) a sense organ (d) a stimulus (e) an effector and (f) a muscle.

What are involuntary and voluntary actions?

Involuntary (reflex) actions

These actions are automatic: we do not have to think about them. Impulses from the receptors do not travel to the cerebrum of the brain. An example of an involuntary reflex action is the knee jerk.

Activity | The knee-jerk reflex

1. Work with a partner. The partner sits down and crosses his or her legs.
2. With the side of your hand gently hit your partner's knee just below the kneecap (see picture opposite). The lower part of the leg will spring up automatically.

Questions

1. What are (a) the receptor and (b) the effector in the knee-jerk reflex?
2. Describe how impulses pass from the receptor to the effector.

The knee-jerk reflex. The lower part of the leg springs up automatically when hit just below the kneecap.

A simple reflex arc

The diagram below shows that when a person's hand unintentionally touches a hot object (receptor, 1), the hand is automatically lifted away (effector, 5).

THE WAY IN WHICH THE NERVOUS IMPULSES PASS TO AND FROM THE SPINAL CORD ARE SHOWN. THIS IS A SIMPLE REFLEX ARC.

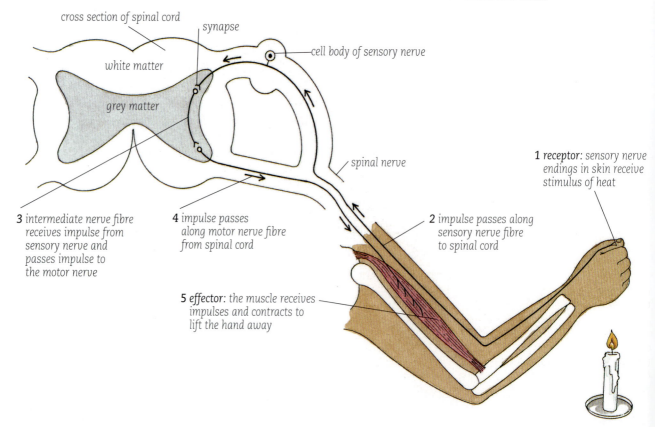

Reflex arc. Trace the pathway which links the receptor and effector

Voluntary actions

Voluntary actions are *not* automatic: they involve thought. Impulses from the receptors travel to the spinal cord and also to the cerebrum. An example is driving a car.

Look again at the simple reflex arc shown opposite. The **sensory nerve fibre, 2**, which brings the impulse to the spinal cord is in contact with an **intermediate nerve fibre, 3**. The place where the impulse passes from one nerve cell to the other is called a **synapse**. A corresponding impulse is then sent out along the **motor nerve fibre, 4**. This is a simple reflex and the hand is drawn away from the hot object.

Now what would happen if you pick up a hot object that is very valuable? Your automatic reflex would be to drop it. But instead, you would think about the consequences of breaking it and decide to try and put it down carefully.

What happens in this case? Impulses come in from the receptors and to the spinal cord in the same way. But then impulses also travel to the brain where a decision is taken. A new message is sent back to the spinal cord and out along motor nerve fibres *instructing* your hand to put the object down carefully. This is an example of **voluntary** action. Voluntary action involves thought, it does not happen automatically.

The brain The brain (below) controls voluntary actions and some involuntary ones. It consists of three parts, the large convoluted **cerebral hemispheres** (cerebrum), the smaller ridged **cerebellum** and the stalk-like **medulla oblongata**. The cerebrum is mostly concerned with thinking and with voluntary actions, and the cerebellum and medulla oblongata with involuntary actions.

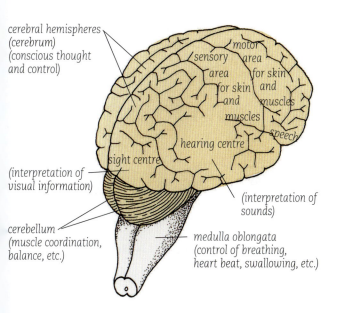

Parts of the brain

How the nervous system controls our actions

Involuntary (reflex) actions Involuntary actions are important in helping us to keep safe. They are very quick automatic reactions which make us move quickly away from hot objects or other things that may hurt us. The coordination of our movement in walking and running are also examples of involuntary actions.

The **instinctive** reactions of animals, for example the mating and nest-building activities of birds, are involuntary actions. They do not have to be learned.

Voluntary actions Voluntary actions involve thought by some area of the cerebrum of the brain. Impulses coming in from sensory cells are judged against the experiences stored in the brain, and appropriate action is undertaken. We can **learn** from our experiences. Our cerebrum is the part of our brain which allows us to read, understand and remember. It stores our memories for a later time.

Driving a car is a voluntary action. The person has to think carefully about each reaction. However, after a while things become easier: we learn *how* to drive.

The pathways along which impulses travel have become familiar and the responses happen more easily. Our voluntary activities then seem to involve less conscious thought. These actions are then known as **conditioned reflex** actions (see also page 252).

The autonomic nervous system

Activities which go on inside our bodies of which we are not aware, are under the control of another set of nerves belonging to the **autonomic nervous system**.

Some examples of such actions are the movements of the gut, the heartbeat, the rate of breathing, and the size of the pupil in the eye. These actions are mainly under the control of the medulla oblongata and spinal cord.

The endocrine system

The actions we have described so far all involve electrical impulses being sent along nerves. But activities can also be controlled by chemicals released into the blood, as we shall see on pages 126–7.

Questions

3 (a) Describe a simple reflex arc (involuntary action).
  (b) How is a voluntary action different from this?
4 Describe three different actions which are under the control of the three different parts of the brain. Which ones are voluntary and which are involuntary?

What is the endocrine system?

The coordinating activities that we have looked at so far, have been under the control of nervous impulses (pages 122–5). But there is another way in which some of the activities inside our bodies are controlled. This system (the **endocrine system**) relies upon chemicals or secretions called **hormones** which are produced by special glands. These glands do not have tubes or ducts, but pass their juices (secretions) directly into the bloodstream.

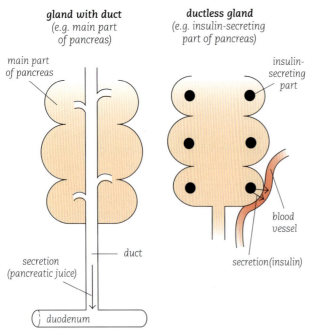

Left: gland with duct – secretion passes along duct into part of the body. Right: ductless gland – secretion passes directly into the blood

The hormones pass into the blood and are carried around the body. Each one affects a particular organ, causing it to slow down or speed up its activity. The ductless glands (also called **endocrine glands**) and the hormones which they produce are shown in the table below and the diagram alongside.

Endocrine gland	Hormones produced
Pituitary	Hormones controlling: growth (growth hormone), thyroid gland, amount of water in urine, reproductive systems
Thyroid	Thyroxine
Pancreas	Insulin
Adrenal	Adrenaline
Ovary	Oestrogen and progesterone
Testes	Testosterone

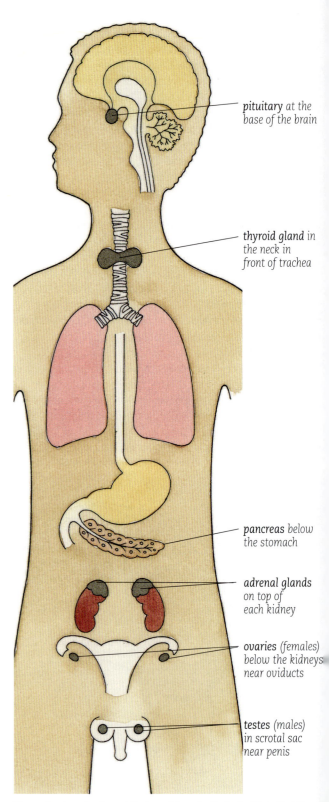

The endocrine glands and their positions in the body. Note that the alternatives for female and male bodies have been shown

How do the endocrine glands affect development and everyday activities?

Pituitary gland This is also called 'the master gland' because it secretes many hormones. These hormones influence growth and also control other glands in the body. Some of the pituitary hormones are:
1. **Growth hormone** If too much growth hormone is produced, the person will be a **giant**, but if too little is produced the person will be a **dwarf**.
2. **Hormone controlling the thyroid gland** When this is secreted the thyroid produces thyroxine (see below).
3. **Hormone controlling the amount of water in the urine** This affects the kidneys. The hormone controls conserving water on hotter days, and losing extra water in the urine on cooler days (page 113).
4. **Hormones affecting the reproductive system**
   (a) They cause ovaries to produce eggs and testes to produce sperm.
   (b) They make the uterus contract to expel the baby at birth, and they make the breasts produce milk.
   (c) They make ovaries and testes produce their own hormones.

Thyroid gland The thyroid gland produces thyroxine, which controls the rate of chemical reactions in the body. If too little thyroxine is produced during childhood, the person will be a dwarf and will also be mentally disabled. People who have this disability are called **cretins**.

In adults, too little thyroxine causes body processes to slow down. Less energy is needed from food, so the people get fat. They will also feel tired and worn out and their faces may become swollen and puffy. This is called **myxoedema**. Too much thyroxine causes body processes to speed up. More energy is needed so the people become thin. They become excessively active. Their eyes may bulge. Both conditions can be treated by a doctor.

Pancreas Produces insulin, which controls the amount of glucose in the blood. Insulin makes the cells take up glucose from the blood to use for the release of energy. It also makes the liver absorb and store the rest as glycogen. Insulin therefore lowers the glucose level in the blood (page 129).

If too little insulin is produced, glucose accumulates in the blood and some of this is excreted in the urine (page 113). The person suffers from **diabetes**.

The person will urinate more often and will be quite thirsty. Because glucose is not being taken up and used by the cells, the person will be tired and weak. There may also be a decrease in mass as fat and protein (muscles) are burned to provide energy.

Diabetics should eat a diet low in carbohydrates. Daily injections of insulin may also be needed.

Adrenal glands These glands produce adrenaline. Adrenaline is the 'emergency hormone' which helps us to prepare for 'fight or flight'.

In a physically dangerous situation extra adrenaline is secreted into the blood which affects various organs. The breathing rate increases and the heart beats more quickly. Extra glucose is released into the blood and the blood vessels to the gut become narrower. Extra blood containing oxygen and food is pumped to the muscles. These can then act more effectively to allow the person to be more physically active.

These effects may be felt if, for example, you are in danger of being run over by a bus, or if you are very angry with someone. The 'sinking feeling' in the gut before an examination is also the result of adrenaline.

Adrenaline is also released if a person is continually anxious or under stress. The long-term effects on the body can harm the circulatory system (page 97).

Ovaries (females) At puberty the ovaries are stimulated by a hormone from the pituitary, and start to produce female sex hormones (oestrogen and progesterone) (pages 48–9).

Hormones from pituitary	Hormones from ovary
1 Cause development and release of egg in the ovary	**oestrogen** which causes repair of uterus wall and development of breasts
2 Stimulate ovary to produce	**progesterone** which causes uterus wall to become thick and full of blood

Testes (males) At puberty the testes are stimulated by a hormone from the pituitary and start to produce the male hormone **testosterone**. This in turn causes the development of secondary sexual characteristics (page 49).

Testosterone and related chemicals have a body building effect and might be taken (usually illegally) by both male and female athletes (page 253). They are taken as tablets, by injection or implanted in muscle tissue. Both men and women become stronger and more muscular, but they also face health problems.

Questions

1. Which endocrine glands produce (a) thyroxine (b) testosterone (c) insulin (d) growth hormone and (e) adrenaline?
2. What are the results of (a) over-secretion and (b) under-secretion of (i) growth hormone and (ii) thyroxine?
3. What are the results of adrenaline secretion?
4. The nervous system and the endocrine system control activities of the body. List three ways in which the systems are different.

How are life processes interrelated?

Living organisms carry out various activities (pages 6–7). These activities take place in individual cells which have to receive raw materials, process them and get rid of waste products. The various systems of the organism are interrelated so that the cells, and the organism as a whole, can carry on their activities.

In plants, which are usually less complex than animals, the interrelation between processes is simple. For example, *photosynthesis* and *respiration* are interrelated by the gases which they use and produce (pages 101 and 108). If sufficient food is built up by the plant it can *grow* and *reproduce*. Plants are also sensitive to important stimuli in their environment and *respond* by growing, which is itself under the control of plant hormones.

In animals, the cells are surrounded by tissue fluid (below). Let us see how the systems are related.

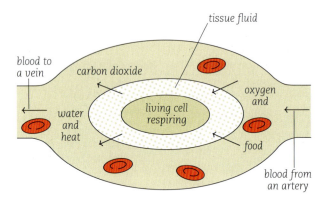

A living animal cell

Circulation Blood brings oxygen and food to the cells, and takes away carbon dioxide and waste products.
Digestion Food is broken down in the digestive system and brought to the cells.
Respiration Respiration occurs in the cells by using food and oxygen and releasing carbon dioxide, water and heat. The oxygen is taken in by the lungs, and the carbon dioxide, water and heat are excreted by the lungs. The energy released is used for all other activities.
Excretion Waste products such as carbon dioxide, urea and heat are excreted.
Growth and reproduction The whole organism grows as a result of feeding, and may reproduce.
Response The nervous and endocrine systems allow the animal to respond to external stimuli and also to keep its internal conditions constant.

The lymphatic system
The tissue fluid which surrounds each living cell in our bodies comes from blood in capillaries from an artery. Most of the tissue fluid returns to other capillaries which join to make a vein. But some of the tissue fluid enters **lymph vessels** which are part of the **lymphatic system**.

The lymphatic system is mostly made up of the tube-shaped lymph vessels which contain valves, and of swellings called **lymph glands**. It is in the lymph glands that the white blood cells called **lymphocytes** (page 92) are made. Lymphocytes are important because they produce antibodies and antitoxins which are a defence against germs (page 161).

When we have an infection (for example, see page 53) we may notice that our lymph glands, especially in the groin and the armpits, become swollen and ache. This is because they are producing extra lymphocytes, and removing any dead germs and lymphocytes from the body.

Another part of the lymphatic system are the **lacteals** (page 84) which are finger-like projections found in the villi of the small intestine. When fats have been digested in the gut to make fatty acids and glycerol, these are absorbed into the villi. The fatty acids enter the blood capillaries and the lacteals, while glycerol only enters the lacteals. These substances are taken to the liver.

Homeostasis
Living cells can only work properly within fairly narrow ranges of pH, temperature and concentration of different substances. They are bathed with tissue fluid which comes from the blood. There must therefore be mechanisms in the body to keep the pH, temperature and concentration of the blood fairly constant. This is one aspect of **homeostasis** or 'steady state' within the body.

Feedback control or **feedback mechanisms** are the processes by which the body brings about the steady state and keeps conditions in the body within narrow limits.

The role of the liver in homeostasis
When we digest our food, the end products of digestion are absorbed into the villi in the small intestine (page 84), and pass to the liver in the hepatic portal vein.

If all this food passed on into the bloodstream at once, there would be a very high concentration which could upset the functioning of the cells. On the other hand, if we did not eat for several hours, the food levels could become very low. To avoid these large changes, the liver controls the amount of foodstuffs passed on, and stores some of them for later use (see summary in the next column).

Blood is brought to the liver in the hepatic artery (from the aorta) and the hepatic portal vein (from the digestive system). Blood leaves the liver in the hepatic vein.

The liver adjusts the amounts of foodstuffs in the blood as follows:

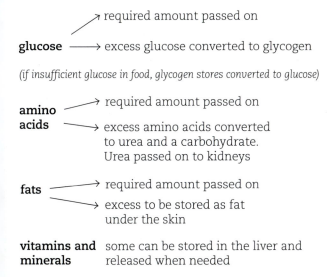

Liver control

The role of the endocrine system in homeostasis

Let us take as an example the control of the amount of glucose in the blood (below).

Control The pancreas secretes insulin which converts glucose to glycogen. Other hormones, for example adrenaline, convert glycogen to glucose.

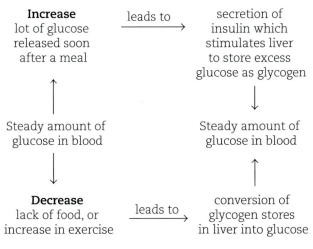

Endocrine system control

Another example of control by the endocrine system is water balance in the body. This is controlled by the pituitary gland which is sensitive to the amount of water in the blood. If there is too much water, then a lot of water-balance hormone is secreted which causes more water to be lost by the kidneys. But, if there is too little water then no hormone is secreted and water is conserved by the kidneys (page 113).

The role of the nervous system in homeostasis

Let us take as an example the control of the amount of carbon dioxide in the blood (below).

Control A 'breathing centre' in the medulla oblongata (page 125) senses the amount of carbon dioxide in the blood and sends messages to the respiratory system to increase or decrease the breathing rate.

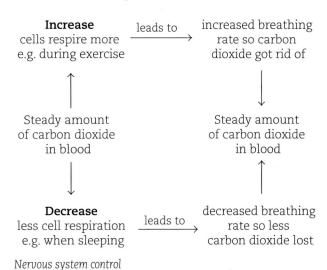

Nervous system control

Another example of control by the nervous system is the control of the temperature of the blood. A heat-sensitive centre in the medulla oblongata senses the temperature of the blood. If this is too high (because it is a very hot or humid day or because the person has been exercising or has a fever) then messages are sent to the skin (page 119) to lose heat and to the liver to cut down on activities releasing heat.

If, however, the temperature of the blood is too low (as it is a cool day, the wind is blowing or the person has not been exercising) then messages are sent to the skin to conserve heat, and to the liver to increase the production of heat.

> ### Questions
>
> 1. How are photosynthesis and respiration interrelated?
> 2. How do your body cells stay alive?
> 3. Give two reasons why the lymphatic system is important.
> 4. Give three reasons why the liver is important.
> 5. How is (a) the temperature and (b) the water concentration of the blood kept constant? Illustrate your answers with diagrams.

How do we measure and regulate temperature?

Activity | How do we measure temperature?
1. Get three bowls or basins that can hold 4 to 5 dm³ (litres) of water.
2. Place 2 dm³ of ice-cold water (from the fridge) in the first, 2 dm³ of cold tap water in the second and 2 dm³ of warm tap water (about 40 °C) in the third.
3. Put your left hand in the first bowl (cold water) and your right hand in the third bowl (warm water) for about 1 minute. Now place both hands in the middle bowl (tap water). How does your left hand feel? How does your right hand feel?

From this Activity it should be clear that your judgement of temperature – hot or cold – is not very accurate. Often it is only relative to what you have become accustomed to. Scientists living in different parts of the world need an accurate reference point or independent scale by which they can measure temperature. A thermometer is used to measure temperature and scientists have universally adopted the **Celsius** scale. Temperature is measured in **degrees Celsius (°C)**.

How do thermometers work?
The most common type of thermometer is the **mercury-in-glass thermometer** such as the laboratory thermometer (see below). There are other types which use other liquids, for example the coloured **alcohol thermometer**, and there are also **electronic thermometers**. The liquid-in-glass type works by using the properties of expansion and contraction of matter.

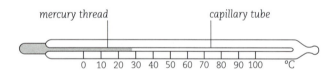

The mercury-in-glass thermometer

Activity | Capillary tubes
1. Get a set of test-tubes of the same size (volume). Close them with close-fitting corks or rubber bungs, each fitted with capillary tubes of different capillary diameters.
2. Fill the test-tubes with coloured water and carefully insert the rubber bungs, making sure that there is no air in the tubes, as on the right.
   You must also be very careful to exclude all air bubbles from the capillary tubes, as these will dramatically distort your results.

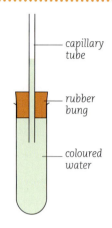

3. Place the test-tubes in a cold-water bath for a few minutes, then mark the height of the water in each capillary tube.
4. Transfer the tubes to a warm water bath and allow the water in the tubes to adjust for 5 minutes. Mark the final positions of the water in the capillary tubes.
5. Measure the distances between the two marks in the capillary tubes and record your results.

Your results should clearly show that the smaller the inner diameter of the capillary tube, the more the height of the liquid increases when heated. How would altering the size of the test-tubes affect the results? What would happen if hotter water were used in the water bath? The principles demonstrated in this Activity are applied in the construction of all liquid-in-glass thermometers.

Making a thermometer scale The scale on a thermometer is made from two fixed points. These are the melting point of ice and the steam point (the temperature at which steam is produced from water). These two temperatures are the same all over the world at sea level (760 mm Hg): the ice point gives the 0 °C position while the steam point gives the 100 °C position. The distance between the two fixed points is divided into a hundred equal parts to obtain the Celsius scale.

Are there thermometers for special purposes?
By varying the length and inner diameter of the capillary tubes in a thermometer the range and division of the scale may be varied. The clinical thermometer has already been described (page 120). Other types of thermometer include the **wet bulb thermometer**, used for measuring humidity (the use of this is described on page 135), and the **maximum** and **minimum thermometers**.

In the *minimum thermometer*, alcohol is used. Inside the capillary tube is a small loose-fitting glass index. When the temperature is low, the alcohol contracts, so the level falls, pulling the glass index with it to the same height as the surface. If the temperature increases again the alcohol expands and flows past the glass index so the minimum temperature is still recorded.

In the *maximum thermometer*, the liquid used is mercury and the capillary tube has a constriction like the one in the clinical thermometer. When the temperature falls, the mercury thread contracts and breaks below the constriction.

The thread above the constriction remains intact, recording the maximum temperature reached.

How can temperature be regulated?

Some home appliances, for example electric irons, hot water storage tanks, and ovens, have knobs for temperature regulation. How do they work?

The electric iron The electric iron uses a bimetallic strip (see (a) below) made up of two different metals bonded together, one of which expands much more than the other.

In a **thermostat** (see (b) below) the current going to the heating element passes through the bimetallic strip which acquires some of the heat. When the strip becomes hot it bends away from the element so that the circuit is broken and the strip is no longer heated. When the strip cools down it straightens out again, makes electrical contact in the circuit and repeats the cycle. By turning the knob, contact with the metal strip may be increased or decreased, thus varying the maximum temperature that may be attained. The bimetallic strip is also used in other household appliances which are heated electrically, for example, the electric oven and electric water heater.

How does the gas thermostat work?

The gas thermostat also makes use of the difference in expansion between metals, like the bimetallic strip. However, instead of using two metals, it uses a brass tube and an **invar** rod. Invar is an alloy of steel and nickel and it has an exceptionally low expansion rate. When the oven is lit, gas flows through the valve head. As the oven heats, the brass tube expands, reducing or cutting off completely the flow of gas through the valve head. If the temperature of the oven falls, the brass tube contracts, increasing gas flow again.

The gas thermostat is normally located at some convenient point in the oven. The thermostat also has a control knob to regulate the temperature.

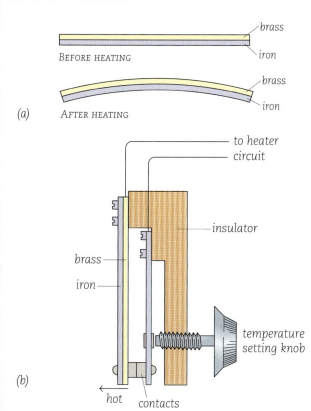

(a) A bimetallic strip (b) A thermostat as used in, for example, an electric iron

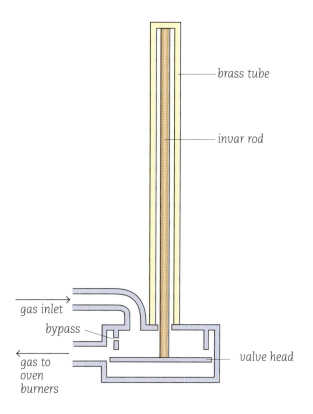

The gas thermostat

Questions

1. Why are instruments needed for measuring temperature?
2. Explain how the bimetallic strip is used to regulate temperature.
3. How is the scale on a thermometer made?
4. Explain how the gas thermostat works.
5. Could water be used in a thermometer instead of alcohol?

How do we use light to see?

Activity | Are there different kinds of light?
1. Look around you and list some objects which are different colours in daylight.
2. Look at these same objects through clear celluloid or acetate filters. Try to use red, green, blue and yellow filters. A filter allows through light which is the same colour as the filter. Do the objects appear to change colour?
3. Record your results in a table like the one below.

Name of object Colour	(a)	(b)	(c)	(d)
In daylight				
Through blue filter				
Through green filter				
Through red filter				

4. Find some pieces of coloured card. Look at the card in daylight. Then look at them again under street lighting. Street lighting is usually white (mercury vapour type) or yellow (sodium type). Do the colours appear to be the same?

Why are objects visible?
You see objects by the light they appear to give off. This light enters your eyes and forms an image of the object on your retina (page 278). The light from the objects you see originates somewhere else, unless the object is **luminous** (for example, a lamp). Most objects viewed in daylight **reflect** the light from the sun and that is how they are seen.

In the Activity above, light reflected by the object passes through the coloured filter you look through. An object looks red because most of the light it reflects is red. When viewed through a red filter the object continues to appear red. But when viewed through a green filter the object will appear dark or even black.

In the second part of the Activity the **sources** of light are different. The colours you see in daylight are not seen under street lighting, for example, objects which appear blue in daylight, appear *black* under sodium vapour lighting. The object appears black, which means that light is being **absorbed**. There is no blue light from the source to be reflected to the eye.

Does sunlight contain different colours?
From the previous Activity you can deduce that sunlight contains light of different colours. You may also demonstrate this directly by using a glass prism (as Sir Isaac Newton did in the 17th century). Or, observe the colours of the rainbow produced in sunlight.

How is light produced?
You must be familiar with the process of burning. Striking a match sets off chemical reactions in the match head. Very high temperatures are produced and the hot gases resulting from the chemical reactions produce light. When you light a candle or a gas burner the result is light and heat. In these examples, there is a flame. However, if you place a piece of wire in a hot gas flame, it will glow and continue to do so for a short time *after* you have removed it from the flame. Burning is therefore not the only way to produce light.

The electric filament or incandescent lamp This makes use of electricity to produce light. The incandescent lamp has a filament made from tungsten, a metal with a melting point of 3 400 °C. When an electric current at a suitable voltage is passed through the filament, it gets hot, rising to a temperature of 2 000 to 3 000 °C. (See page 320.) In this range, not only is **infrared radiation** (heat) emitted, but so is a great deal of visible light. The glass envelope (bulb) usually contains the inert gas argon, which prevents the tungsten in the filament from evaporating and blackening the glass bulb.

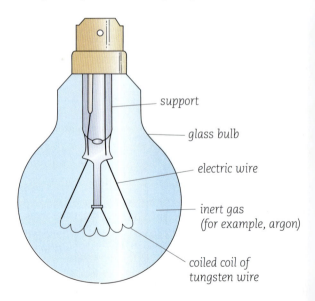

An electric filament lamp

The fluorescent lamp Another popular and efficient light source is the fluorescent lamp. This consists of a glass tube containing argon and a droplet of mercury. It also has electrodes made from tungsten. When the lamp is connected and switched on, the mercury vaporises and electricity flows through the argon and mercury gas. Only a small amount of visible light is produced. Most of the light is **ultraviolet (UV) light**. This UV light is absorbed by a thin layer of phosphor that coats the tube and this produces visible light (a process called **fluorescence**). Lamps that fluoresce to produce white light or other colours are made by adding certain substances to the phosphor. Far more of the electrical energy used by the fluorescent lamp is converted into light than is the case with the incandescent lamp. Some fluorescent lamps produce more than four times as much light energy as incandescent lamps of the same power.

Fluorescent lamps are a very economical form of lighting and are used extensively in offices, shops and homes. Because the lighting is spread along a tube it has a 'soft' effect, i.e. no harsh (distinct) shadows are produced, and it is not dazzling to look at. On the other hand, the incandescent lamp has the advantage of being able to produce a spot of strong light, which is very useful for illuminating specific areas, for example sports grounds at night. Powers of several thousand watts are needed for illuminating airfields. At the other extreme, incandescent lamps can be made smaller than a pea. Fluorescent lamps require high operating voltage to work properly, whereas incandescent lamps can be made to operate over a range of voltages.

How do our eyes respond to lighting?

In the workplace, school, office or home adequate lighting is necessary if you want to achieve maximum efficiency. The positioning of light sources for proper illumination is therefore very important. Fluorescent lighting and incandescent lighting both have advantages and disadvantages and this must be considered in relation to the specific area you wish to illuminate.

The softening of shadows has been mentioned as an advantage of fluorescent lighting. However, in some instances it could be a disadvantage. When we see objects, we use not only illumination but also shadows to identify objects in space. If the shadows have been softened by fluorescent lighting, we may not see so clearly.

How do our eyes respond to bright light? You should never look directly at the sun while your eyes are unprotected because it could blind you. If you try to look at the sun for too long, the image of the solar disc will be burnt onto the retina and permanent blindness may result.

Light sources such as the electric arc, used in welding, and the burning magnesium ribbon, produce ultraviolet light, which can damage or destroy the light-sensitive cells of the retina and may eventually lead to blindness. Industrial workers who are exposed to such intense sources of light must wear proper protective shields over their eyes.

Ordinary sunglasses or 'shades' do not give us protection from the dangers of ultraviolet light.

How do our eyes respond to low-level lighting? The amount of light entering the eye (through the pupil) is controlled by the iris diaphragm (page 280). In normal light, the pupil is fairly small. With this narrowed pupil, the light is focused sharply onto the retina and objects are seen clearly. This is because most of the rays entering the eye are parallel. In low-intensity light, the pupils are widened so that the eye can let in more light. At the same time, light from widely different angles enters the eye and a sharp focus is not easily achieved. The efforts by the eye muscles surrounding the lens to find a focus can cause eyestrain, discomfort, fatigue and impairment of vision. The result is that objects can be misjudged for shape, size and position and serious accidents could occur, for example while driving at night.

BRIGHT LIGHT — pupil smaller
DIM LIGHT — pupil dilates

How the pupil responds to variation in light intensity

Questions

1. Explain how a red or yellow object could be made to appear black.
2. Make a list of some luminous objects. (Objects which produce their own light.)
3. How does (a) too intense lighting and (b) too dim lighting affect our vision?
4. Explain how electrical energy is used to produce light.

Why is ventilation important?

If you heat a pan of water over a steady flame, it will boil and continue to boil until all the water disappears. Wet clothes placed on a line in the morning will be dry before the end of the day. In each case, water is present at the beginning but later disappears. Where does it go?

Activity | Does air contain water vapour?
Pour some ice-cold water (from the fridge) into a clean, dry container such as a clean drinking glass, beaker or tin can. Observe the outer surface of the container. Does it become cloudy or lose its gloss? Now touch the surface. Is it wet? Where do you think the moisture comes from?

The air around us holds a great deal of water vapour. The higher the air temperature, the greater the capacity for holding water. Conversely, the lower the temperature, the lower the capacity for holding water vapour. The actual amount of water in the air depends on the weather. When air can no longer hold water the vapour condenses, forming tiny droplets, which may remain suspended as mist or as clouds. If these droplets join together, they form water drops.

Can the amount of water vapour in air vary?
Air can be very dry – containing very little water vapour, or be very moist – containing a lot of water vapour. When air contains the maximum amount of water vapour it can hold, it is called **saturated** air. The amount of water vapour in air is measured by its **relative humidity**.

Relative humidity

$$\text{Relative humidity} = \frac{\text{mass of water vapour in a certain volume of air}}{\text{mass of water vapour in the same volume of saturated air}}$$

Does saturation affect drying?
If a wet strip of cloth is left on a watch glass on a laboratory bench it will eventually become dry. But if a similar strip of cloth is covered by a bell jar, it will remain wet. The small quantity of air in the bell jar quickly becomes saturated with water vapour, whereas the entire laboratory and the air currents from outside are able to take up the water from the exposed cloth strip.

Activity | How does speed of drying vary?
1. Design an experiment based on the principle outlined above. Use four jam jars and equal sized pieces of wet cloth. Make all the conditions in the jars equal except for the *extent* to which each is covered.
2. Observe and record your results. Does the speed of drying relate to how well the jar is covered? Discuss your results in class.

On a bright sunny day, the warm air is able to take up much more water vapour than on a colder day. This is the best kind of day for drying clothes on a line. How could you investigate the effect of temperature on drying?

If the day is windy as well as sunny, clothes will dry even faster. This is because air from around the clothes, which is laden with moisture, is quickly removed. The clothes will therefore lose water faster.

The same thing happens to plants. The temperature, wind and relative humidity all affect the rate at which water is lost from the leaves (see page 91).

After a heavy shower of rain, the air is saturated with water vapour and very little drying can take place. On the very humid days that we often have in the Caribbean we feel 'hot and sticky'. This is because very little perspiration is able to evaporate from the surface of the skin and cool us (see page 119).

Is there an ideal humidity?
For comfort, a balance must be found between temperature and relative humidity. A high temperature may be comfortable if there is low humidity, because easy evaporation of perspiration from the skin will have a pleasant cooling effect. A lower temperature could be comfortable with a higher relative humidity.

How can we measure humidity?
One method of measuring relative humidity is by using the temperatures obtained from a wet and dry bulb **hygrometer** (see the next column).

This hygrometer consists of two glass thermometers that are mounted side by side. One thermometer has the bulb covered with wet muslin. The water from the wet muslin evaporates, producing a cooling effect on the thermometer and causing the temperature reading to fall relative to the other one. Using the two temperatures, relative humidity is estimated from **psychrometric tables**.

From the wet and dry bulb temperatures it is also possible to calculate the degree of comfort or discomfort, measured by the **Temperature Humidity Index (THI)**. This is based on experiments which have determined the exact temperature and humidity conditions that affect the comfort or discomfort of people.

Part 1

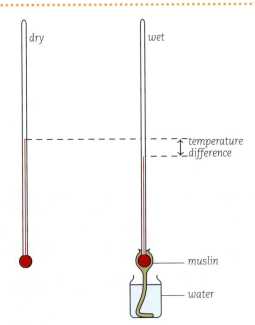

Wet and dry bulb hygrometer

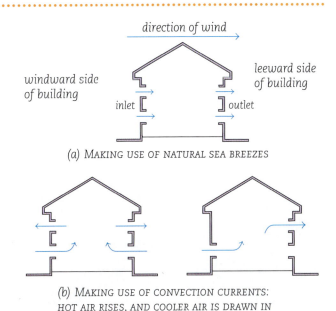

(a) MAKING USE OF NATURAL SEA BREEZES

(b) MAKING USE OF CONVECTION CURRENTS: HOT AIR RISES, AND COOLER AIR IS DRAWN IN

Methods of natural ventilation

Why is ventilation needed?

When you open the windows to let the breeze blow through, or use a fan, you are to some extent trying to control the humidity.

The air around your body has probably become full of water vapour, changing the temperature humidity index so that you feel uncomfortable. Ventilation makes the surroundings more comfortable by renewing the air.

Air can become unpleasant not only because of high humidity but also because of the presence of fumes, for example, cigarette smoke, or because of stale smells or airborne particles, for example, dust and germs. The air we breathe must also be renewed because of the reduced oxygen content and increased carbon dioxide content. In schools, churches and other places where people congregate, indoor conditions can rapidly change from pleasant to intolerable in the absence of satisfactory ventilation.

In the Caribbean, the islands are usually swept by sea breezes and in the past the positioning and construction of houses used to take full advantage of this.

Today, the older types of wooden windows and wooden shutters which were built by craftsmen have been replaced by glass windows. Glass windows admit light, even direct sunlight, so they not only increase the need for ventilation, but they also restrict the flow of air through a building.

In our cities, overcrowding and intense competition for limited space restrict airflow through buildings. The air is often dust laden and polluted by the exhaust of motor vehicles, so that the 'natural' untreated air may be very unpleasant.

Even when air is satisfactory in terms of oxygen content, carbon dioxide content and temperature humidity index, feelings of stress, discomfort and fatigue may still arise.

This fact led to the discovery of the influence of charged air particles called negative ions and positive ions. An excess of positive ions in the air is uncomfortable, while an abundance of negative ions is beneficial. Negative ions are produced by vegetation and by the sea, and are abundant in the countryside. Positive ions are abundant in the cities.

Questions

1. How does the use of a fan increase the rate of heat loss from the skin? Is it mainly by (a) convection (b) conduction (c) evaporation or (d) radiation?
2. How serious is the problem of air pollution in your country? What could be done to improve the situation?
3. Feeling hot or cold does not depend on temperature only. Discuss.

Why is ventilation important?

Activity | How can ventilation be assisted?

1. Visit an old-style house, for example a plantation house and look for features which assist natural ventilation. Look at the style and number of windows, the spacious layout, covered verandas and balconies. Also consider the use of plants for shading. Look at the high sloping roof, and the air vents which assist convection.
2. Compare these features with a modern building of similar size. The photographs below may help you.

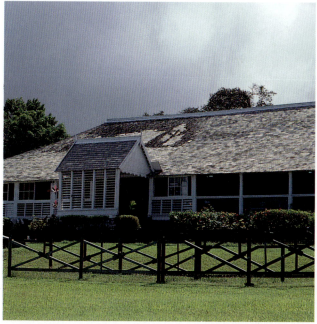

An old building. Which features assist natural ventilation?

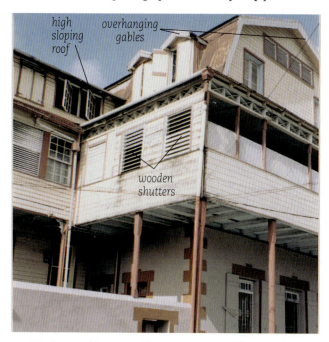

An older house with a narrow verandah. The building does not have a profusion of plants but it does have some features which assist natural ventilation. Describe how each of the labelled features are important

A block of flats. Put the labels into two groups: those that assist and those that discourage ventilation

Air-conditioning units. These are needed especially on modern houses which have unshaded louvre windows which allow a lot of heat to enter

Part 2

What are air conditioners?

If the air brought in from outside is already humid, dust-laden or otherwise polluted, then neither the building's design features nor mechanical assistance (for example, fans), will be able to create a suitable, comfortable, environment. Therefore air conditioners filter the air and alter temperature and humidity. Refrigerant leaving the compressor flows into the outer air coil. There it loses heat to the outside air. The refrigerant then flows to the expansion valve, which is on the outside of the building. From the expansion valve the refrigerant expands and flows into the air coil inside the room. The expanded refrigerant is cold, so the warm air inside the room cools down as it warms up the coil. The refrigerant flows back to the compressor and the cycle is repeated.

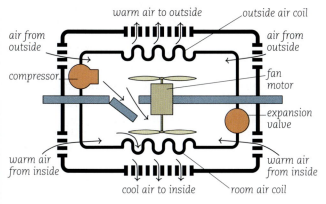

This is how an air conditioner works

Do fans help?

In modern buildings, natural airflow may be restricted, and artificial ventilation is therefore a design feature which must be incorporated. It is possible to use fans like the one shown below, to create a draught by drawing air around a room.

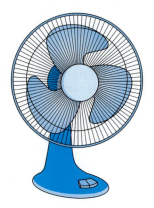

A fan can be placed on a table to help keep a room cool

A vaned unit, which is turned by the wind, uses airflow around it to remove air from the building. Electric fans are used extensively for assisting ventilation. They must be either **blower fans** that force air into a building, or **extractor fans** that take air out of the building, thus speeding up the flow through windows and openings.

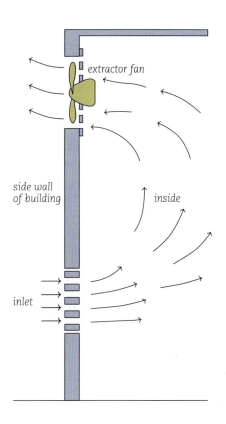

Artificial ventilation using an extractor fan

Questions

1. How would you show that air contains water vapour?
2. What is the relationship between temperature and the ability of air to hold water vapour?
3. Explain how air can become stale.
4. Explain how convection currents in air assist ventilation.
5. What design features in buildings can assist natural ventilation?
6. Why are large modern buildings almost always dependent on mechanical means of ventilation?
7. Explain why fans are used in overcrowded offices.

Why is water important for life?

1. *Living things are mostly water* Inside each living cell of all plants and animals is a cell membrane which encloses the **protoplasm**. Protoplasm (page 10) consists of the nucleus and cytoplasm and its contents, and is 70% water.

    We lose nearly 2 litres of water each day as moisture from the lungs, and in urine, sweat and faeces (page 108). We need to replace this in our food and drink. We have to be sure that the water we drink is free from impurities (pages 140 and 146–7) and organisms that can cause infections (page 156).

    We can only live for a few days without water, although we could live on water and no food for several weeks. If humans, especially babies, lose a lot of water because of diarrhoea they become **dehydrated** and may die. They should drink rehydration fluid – sweetened, slightly salty water – to replace their body fluids.

    Some animals, such as desert rats and termites, are specially adapted to conserve water, and get all the water they need from their food. Desert plants, such as cacti, have swollen water-retaining tissues so they can survive drought.

2. *Water takes part in chemical reactions* Water molecules take part in the breakdown of food (page 88) during digestion. The digestive juices, including saliva, are mainly water. Water is a very important **solvent** (a liquid in which other substances dissolve). For example, when we digest our food (page 84) some of the particles are broken down and go into solution before they can be absorbed.

    Water is also used by plants for the breakdown of food stores (page 72) and in photosynthesis (page 67).

3. *Water is necessary for transport* In animals, blood (page 92) is mainly made up of water. There is a layer of moisture in the alveoli of our lungs (page 100) in which the respiratory gases dissolve and so can be transferred between our body and the outside air.

    In plants, water is important in the transfer of mineral salts in the xylem and food materials in the phloem (pages 90–1). Moisture in the leaf cells is also important for the transfer of gases (page 70).

4. *Water is a habitat for organisms* Aquatic habitats such as the sea and freshwater (pages 144–5 and 306–7) are important for many organisms that could not survive on dry land.

5. *Water is necessary for support* Water is needed to keep plants upright, especially plants without much woody tissue. The individual cells of the stem become swollen (turgid) with water and so the plant stands upright. If there is not enough water the plant droops and becomes **wilted**.

How can we make good use of water?

1. *By building dams and reservoirs* These are built to hold water from a river. As the water is needed it can be purified (page 140) and then used.
2. *By storing water* We can catch rainwater in bowls or buckets and use it for watering our plants. Gutters, downpipes and a storage tank are also useful so that water can be stored when it is plentiful and can be used later on when there are water restrictions.
3. *By conserving water* Use water carefully, for example showers usually use less water than baths, and use a cup of water for cleaning your teeth instead of running the tap. You can also keep water from one use, such as washing up, for cleaning the car, and use only as much water as is essential. We can also use water from washing ourselves for watering plants. However, water which has had detergents (pages 206–7) added to it may contain chemicals which are harmful to plants.
4. *By watering plants effectively* It is best to water plants in the early morning or the late evening. At other hotter times of the day a lot of the water will evaporate quickly and will not go into the soil to supply water to the plants.

    A plant takes up water through its roots; it is therefore best to put the water into the soil and not onto the leaves. In some irrigation systems water is taken underground in pipes directly to the roots.
5. *By planting trees* Trees give shade and create cool and humid conditions. In the shade, less moisture is lost from the small plants on the ground, and any animals that live there will also benefit from the coolness and moisture. The leaves of the tree will transpire water into the atmosphere, and this is an important part of the water cycle.

Activity | How much water do you use?
Do this Activity at home.
1. Find out the volume of water you use to (a) have a bath (b) take a shower and (c) wash with a basin of water. How do the amounts compare?
2. (a) Find out what volume of water is used each time you flush the toilet. (b) How often, on average, is the toilet flushed each day? (c) If you were to put a brick into the cistern, which reduced the volume of the cistern by a quarter, how much water would you save in a week?

The water cycle
Water, like many other substances (page 247), circulates or **recycles** in nature. The water cycle shows how we obtain a continuous supply of fresh water.

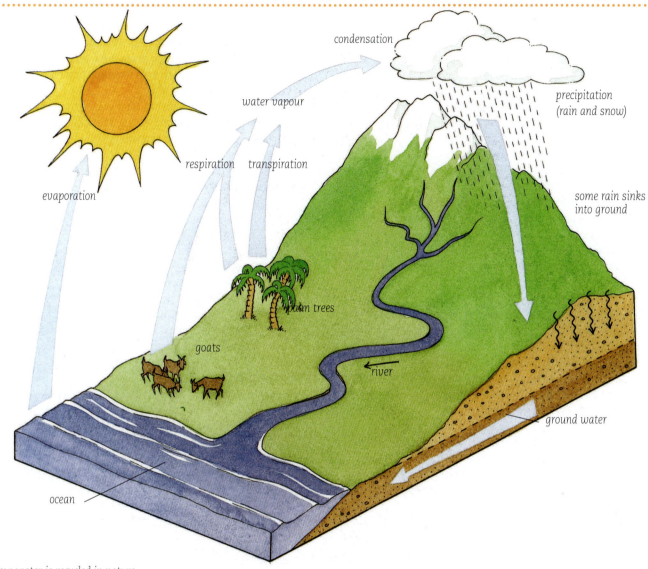

How water is recycled in nature

Questions

1. With reference to the water cycle above describe (a) three processes by which water is lost from living things into the atmosphere (b) one process by which water is lost from non-living things into the atmosphere and (c) how clouds (page 310) are involved in recycling water back to the land.
2. What will be the effect on the water cycle if we (a) cut down large forests and (b) build large dams and reservoirs?
3. Describe two examples each that show why water is important for life processes in (a) animals and (b) plants.
4. If a baby has diarrhoea it may lose a lot of water.
   (a) Why is this dangerous?
   (b) What can be done about it?
5. How can we best use water in our homes?
6. In the middle of the day some potted *Coleus* plants were out in the sun. They were droopy and wilted. Describe *exactly* what you would do to restore the plants to their upright position with the *minimum* amount of water.

How do we obtain the water we need?

How is impure water made safe for use?
Life without water is virtually impossible. Some of you may have lived through a period after a hurricane has struck your country. One of the urgent needs in such a situation is a safe supply of water. If you think about the number of times you use water in a day, you will see how important water is. (You could start by checking the number of times you turn a water tap on each day.) Water is also important to all living things – see pages 138–9.

Activity | Water sources
There may be a number of different sources of water in your country. These could include streams, rivers, wells and reservoirs.
1 Make a list of these different sources.
2 Which is the most important source of water in your country? Why?

The countries of the Caribbean vary greatly in the nature and extent of their water sources. For example, in Barbados and Antigua, there are few (if any) rivers. By contrast, Dominica and Guyana are blessed with an abundance of rivers. The annual rainfall figure is a key piece of information about the amount of water that might be available. If there is high annual rainfall, it should be possible to store some of the rainwater so that it can be used as a water supply.

Water that has been stored or is found in a river, stream or underground well is likely to contain many things (often called impurities) other than water. These may include soil, bacteria, traces of fertiliser, detergent and so on. Before we make use of a water source we need to know what it contains.

3 Starting with the water sources you have listed, make a list of the types of impurity you might expect to find in each source. Discuss this with the class.

Activity | Impurities in water
It is now possible to test water samples for almost any impurity. For example, you may want to know which salts are present. You can test for these and for micro-organisms.

1 Take a small sample of water from each of the water samples provided for the class. Test each sample for salts – your teacher will demonstrate the methods.
2 Add a few drops of each water sample to a dish containing nutrient agar.
3 Examine each of the dishes after two to three days. Which of the water samples seems to contain most micro-organisms? How can you tell?

You should find that rainwater is fairly clean, containing few impurities. This would not be true if you collected rainwater in London or New York, for example. Did you make sure that the container in which you collected the rainwater was clean? If not, impurities may have been introduced.

In a number of Caribbean countries, rainwater storage tanks are a common sight. People use them to store water for gardens, for washing and for cleaning. You have to be careful not to let the surface of the water lie exposed, or it will be a good breeding ground for mosquitoes. For many of us, our water supply comes from a tap. The water is purified and distributed by a water authority, for example.

Water purification
We all need a good, reliable source of reasonably clean water. This water will often be purified at a waterworks. Although the process is more complicated than the way it is explained here, there are four main stages in purifying water. See the diagram below.
1 Impure water is filtered, to remove large solid particles.
2 The filtered water is then passed through a sedimentation tank. Chemicals are added and smaller solid particles stick together and sink to the bottom.
3 The water is then passed through a fine filter to remove the smallest solid particles.
4 Chlorine gas is added. This acts to kill off any bacteria.

Water supplied to the home has to be reasonably clean and free of bacteria, or it would be a health hazard. Pure water is very important in hospitals and science research laboratories. It can also be important in many industries.

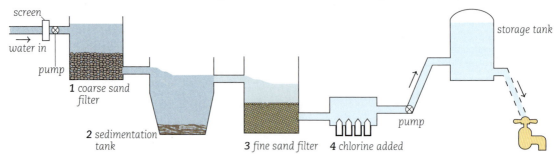

Stages of water purification

> ### Questions
> 1 Why is a supply of clean water important for human life?
> 2 Try to visit a place where water is purified for use by humans. Write a brief description of the processes you see.
> 3 'Water supplied to your home is likely to be clean but not pure.' Do you think this statement is likely to be true?

You have listed some of the impurities that may be present in water obtained from different sources. The process of water purification should be able to remove some, if not all, of these impurities. On a small scale, we can also try other methods of purification, such as boiling and distillation.

In many homes across the Caribbean, there is not yet a reliable supply of fresh water. If water is obtained directly from a stream or river, it will be important to boil it before use for drinking, for example. This helps to destroy bacteria.

In some countries, such as Antigua, there are relatively few water sources. Since there is a great deal of water in the sea, it should be possible to get fresh water from this source. Sea water contains salts, such as sodium chloride, and these have to be removed. A **desalination plant** is used for this purpose. There are desalination plants in Antigua, Barbados and Trinidad. The taste of the water is different from that obtained from a regular water supply.

A desalination plant in Barbados

Why should we conserve water?

In many countries of the world, water is becoming scarce. The control of water supplies will become an increasingly critical issue for governments as populations increase in many countries. In the Caribbean, there is little or no direct competition between countries for water resources. However, every government wants to ensure that its people have adequate water supplies. Water security has now become a more important issue in the Caribbean.

We have to think about ways in which we can conserve water, since it is such an important commodity. What water conservation strategy would you propose for your school?

Activity | Discussion
1 Divide the class into small groups. Discuss ways in which water is used and how it can be conserved.
2 List the methods of water conservation suggested.

How can we conserve water in the home?

Your home is the place where you use most water. Can you think of ways in which water could be used more efficiently at home? For example, could you make more use of rainwater? Could waste water from the kitchen be used for watering plants and flowers in the garden?

Activity | Do we waste water?
1 Adjust a water tap in the laboratory until it is dripping very slowly, one drop every few seconds.
2 Collect the water drops in a measuring cylinder over a given period of time – perhaps 2 to 3 minutes.
3 How much water would be collected in (a) 5 minutes and (b) 10 minutes?
4 Now work out how much water would be collected in (a) one day (b) one week and (c) one month. A dripping tap can waste a great deal of water.
5 Make a list of the different uses of water in your home. Try to work out which requires most water and which requires least. Try to estimate roughly how much water is used each day in your home. Do you think any water is wasted? If so, how could it be used more efficiently?

Conservation of water and conservation of energy are closely linked. Water purification requires energy and energy is needed for pumping water from a waterworks to homes and industries. Energy costs money to provide and has to be paid for.

Activity | Discussion
1 In small groups, discuss and then list ways in which water can be saved in the home. Make sure that there is a thorough discussion before the list is drawn up.
2 Each group should put their suggestions to the whole class, who can then discuss all the methods proposed.

> ### Questions
> 4 What is the source of the water supply for (a) your home (b) your school and (c) the nearest hospital?
> 5 How does the water authority charge people for water use in the home? Is there a water metering system, for example?
> 6 How can water be conserved in (a) your school science laboratory and (b) the school generally?

Would you rather be a fish?

What do you think it would be like to live under water? What problems would you have?

1 **How would you breathe?** Humans need oxygen from the air to breathe. If you go under water you have to take a gulp of fresh air before you go down. How long could you hold your breath under water?

If you want to stay under water for longer than you can hold your breath, you will have to use a snorkel or take cylinders of air and breathing masks with you. You also need to take certain precautions (pages 256–7).

How do aquatic organisms manage? The very small ones, and plants, exchange gases through their surfaces (page 287), while larger ones, such as fish, have gills.

2 **How would you stay under water?** When we lie stretched out in sea water, we float. If fish also floated on the surface they would be easily caught by their enemies such as sea birds and fishermen.

Bony fish have a **swim bladder** near their backbones which is like a balloon containing air. When air is expelled from the bladder, the fish sinks in the water. When air is let into the bladder, the fish becomes more buoyant and rises towards the surface.

3 **How would you move?** If you tried walking under water you would find a lot of resistance from the water. We are also not very good swimmers, the world record for humans for a short distance is less than 10 km/hour, while a sailfish can swim at over a 100 km/hour.

Fish are **streamlined**: narrow at the front, covered with backward-pointing scales and slippery mucus, and with few appendages to break up the movement through the water. They also have a very muscular tail, like this.

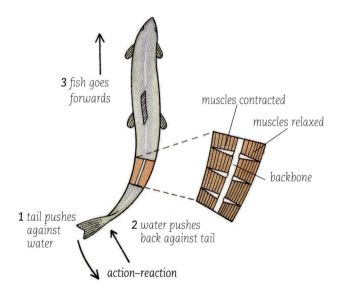

How a fish moves

4 **Could you live in freshwater or sea water?** How are freshwater and sea water different? The main physical difference is that it is easier to float in sea water than in freshwater. The main chemical difference is that sea water contains a lot of salt (about 3.5 g per 100 cm^3 of water, or 3.5% salt), while freshwater contains almost none.

If a fish lives in freshwater, its body fluids contain more salts (i.e. are more concentrated solutions) than the water. Because of osmosis (page 69), water will tend to enter the fish from the freshwater outside. The fish needs a way to get rid of excess water (see (a) below).

If a fish lives in sea water, its body fluids have less salt (i.e. are less concentrated) than the water. It therefore tends to lose water by osmosis into the sea water. It must conserve water (see (b) below).

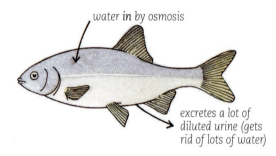

(a) Water balance in a freshwater fish

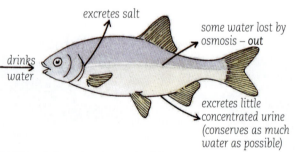

(b) Water balance in a sea-water fish

Activity | Freshwater and sea-water plants

1 Strip off a thin outside layer of pondweed (from freshwater) and seaweed (from sea water).
2 Mount some of the pondweed strip in fresh water and some of the seaweed strip in sea water. Examine the cells under a microscope.
   - How do the cells look in the liquids in which they are usually found? Explain your results.
3 Mount some of the pondweed strip in sea water and some of the seaweed strip in fresh water. Examine the cells under a microscope.
   - How do the cells look in the liquids in which they are *not* usually found? Explain your results.

Activity | How are fish adapted for life in water?

1. Observe a living fish in an aquarium (below).

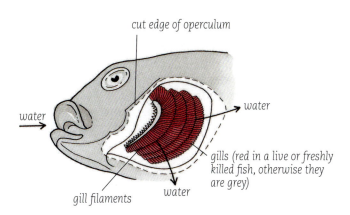

Operculum removed to show the gills

Watch how it swims

(i) Its body is **streamlined**. Notice how the pointed head goes through the water first and the scales all point backwards. This reduces the resistance to the water. The streamlining is similar to that of a bird or aeroplane (pages 298–9).

(ii) Its body is muscular. Muscles on either side of the backbone contract and relax alternately. As the tail pushes against the water, the water pushes back and drives the fish forward in the water. Look at the picture in the first column on page 142.

(iii) The **fins** help in swimming. The **tail** fin pushes the fish through the water. The **dorsal** and **ventral** fins keep the fish upright and on a straight course. The **pectoral** and **pelvic** fins, together with the swim bladder, control the depth at which the fish swims, and help it to stop.

Watch how it breathes

Inspiration – The operculum (part that covers the gills) is flat to the body, and the mouth is open. The floor of the mouth cavity is lowered. Water enters the mouth cavity.

Expiration – The operculum is raised away from the gills, the mouth is shut. The floor of the mouth cavity is raised so that water is forced out over the gills (see pictures on page 100).

2. Observe a dead fish where the operculum has been cut away (top of next column). Look at the gills, made up of finger-like gill filaments.
   - What colour are they? Why?
   - Why do you think a fish dies out of water?
   Try to estimate the surface area to volume ratio (page 286) of the gill filaments.
   - Why is a high surface area to volume ratio important?

3. Look at the diagram in the next column which shows how blood circulates to and from the gills.
   - Describe in your own words how the oxygen that is dissolved in the water enters the blood.

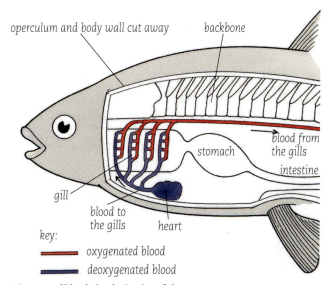

Diagram of blood circulation in a fish

Questions

1. What problems are there with living under water? How have fish solved these problems?
2. How is (a) a freshwater fish adapted to life in freshwater? (b) a sea-water fish adapted to life in sea water?
3. What do you think would happen to:
   (a) a freshwater fish put into sea water?
   (b) a sea-water fish put into freshwater?
   (c) Why should we not do this experiment?
4. Fish and other organisms living in estuaries and mangrove swamps (page 306) are adapted to live in a wide range of salinity (0.0 to 3.6% salt). Why do you think this is important?
5. How do the body (shape and muscles), fins and swim bladder of a fish help it to swim?
6. How does a fish obtain oxygen from the water?

Where do fish fit into food chains?

Food chains show which organisms eat other organisms. At the beginning of the food chains are green plants, **producers**, which produce food by photosynthesis (page 66).

Consumers are animals.
- Animals which only eat producers are called **first order** or **primary consumers (herbivores)**.
- Animals which eat the first order consumers are **second order** or **secondary consumers**, and the animals which eat them are **third order** or **tertiary consumers**.
- Second and third order or tertiary consumers which eat only animal food are **carnivores**.
- Consumers which eat both plants and animals are called **omnivores**.

Organisms which eat other *living* things are **parasites** (page 148). Organisms which eat decaying organisms are called **scavengers**. Organisms causing decay are called **decomposers**. The complex feeding relationships in a community make up a **food web** (see also pages 244–5).

Food chains in freshwater

The producers are the green pondweeds such as *Elodea* (Canadian pondweed) and microscopic green algae (plankton).

First order consumers are animals such as mosquito larvae and pupae which feed on the plankton, and water snails and tadpoles which eat pondweed.

Second order consumers such as young, newly hatched fish (also called fry) and small fish such as guppies, eat the primary consumers. They in turn are eaten by third order consumers – large fish such as pike, or birds like the kingfisher.

Food chains in sea water

The basis of the food chain is plankton. Plankton consists of different kinds of organisms. The plant-plankton are the producers. They are microscopic algae which carry out photosynthesis (picture (a) below). The animal-plankton are the first order consumers (picture (b) below) and they eat the producers. They are mostly small aquatic larvae and adult crustaceans.

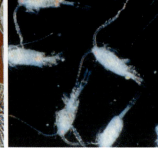

Plant-plankton – Microscopic diatoms which carry out photosynthesis. Approximately × 50

Animal-plankton, e.g. copeopods eat the plant-plankton. Approximately × 30

Small fish such as sardines and the young fish fry of larger fish, eat the plankton and are the second order consumers. Larger fish such as grouper, snapper, king fish and flying fish, the third order consumers, then eat the second order consumers. They in turn may be caught and eaten by people, who are the fourth order consumers.

microscopic plankton (producers) — mosquito larvae and pupae (first order consumers) — guppies (second order consumers) — pike (third order consumer)

A food chain in freshwater

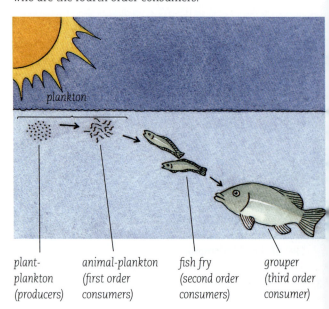

plant-plankton (producers) — animal-plankton (first order consumers) — fish fry (second order consumers) — grouper (third order consumer)

A food chain in sea water

Activity | Catching and eating fish

Carry out a survey to find:
1. Which fish that we eat are most commonly caught in (a) freshwater and (b) sea water?
2. What methods are used for catching the different kinds of fish?
3. Which fish are (a) most popular and (b) least popular with the people who buy them?
4. How are three kinds of fish dishes prepared?
5. Which is your favourite fish dish?

Methods of catching fish

Spear-fishing is used for catching barracuda. It is a highly skilled method and cannot be used for catching large numbers of fish.

Fishing rods and lines are baited with food or other items which are attractive to the fish you wish to catch. With a fishing rod, you can catch only one fish at a time. This method is not used for commercial fishing, but is rather seen as a sport or hobby.

Fish traps have baits and one-way trap doors so that fish which enter cannot get out again. The traps have to be emptied regularly, because if fish die in them they would leave a smell that would deter other fish from entering.

Fish nets can range from very small handmade nets used in local streams, to large *trawling* nets which are pulled along behind a boat (trawler) to catch fish in the sea.

The basic principle behind choosing a net depends upon the size of the holes. If the holes are too big, then the fish that are 'caught' will be able to swim out again! (This is also the reason why fishing nets have to be mended.) But if the holes are too small, *all* small fish will be caught. This will include young fish fry which are too small to eat. If the fry are caught, then there will be less fish in the next generation. Catching and removing too many fish is called **over-fishing**.

Where do we catch fish?

Freshwater streams, rivers and lakes We can use fishing rods and lines, fish traps and nets to catch fish in these areas. **Fish farms** have also been set up where immature fish, such as *Tilapia*, are reared in special enclosures. They feed on mosquito larvae and organic wastes, and every month or so the mature fish are caught and sold.

Shallow sea water To fish here you do not need an expensive boat, and it is relatively easy to transport the fish to the buyers. But in shallow water there may be too much fishing and so there is more competition.

Deep-sea fishing People who fish further out to sea use trawling nets. They have more chance of getting a big haul, as there is less competition. But they will have to buy bigger, more powerful boats for the journey and to withstand any storms. They may also need refrigerators and ice to store the fish for the longer journey back to shore.

The fishing industry

Fish is a good source of protein. In the Caribbean it provides a large proportion of the protein eaten, as well as supplying the tourist trade. There are, however, some problems in maintaining a regular supply of fish.

Marketing The cost of getting fish to the customer can be high. It first has to be transported to land, and then kept fresh until it is sold. The people who sell the fish to the customer also take their profit. Fish needs to be packed in ice, or salted to keep it fresh. It may also be gutted (that is, have its gut removed from the ventral surface) to keep it fresh for longer.

Consumer choice People have traditions about which fish they like, and which they dislike. If a new fish is caught, we may have to influence people's taste for it. Using different types of fish and making new fish dishes in school could be a way to encourage the use of a wider range of fish.

Over-fishing If there is too much fishing in one area of the sea or in freshwater, or if the nets used have only small holes, then very young fish will be caught. This over-fishing will gradually cause the fish population in that area to go down. To help prevent this, several countries have quotas (allowed numbers) of fish which may be caught in a certain period of time. Fishermen are also encouraged to return small fish to the sea.

Pollution Fresh- and sea water can become polluted (page 147). The chemical wastes could poison the fish. The organic wastes decay and use up oxygen from the water so that there is not enough oxygen for the fish to breathe. Oil spilt onto water can also damage the gills and cause death.

Questions

1. Describe the food chain shown in the freshwater diagram. (on page 144)
2. Where could humans fit into this food chain?
3. 'We depend on plankton for the food we get from the sea.' What is meant by this statement?
4. Describe the meaning of each of the following terms and give a named example from an aquatic habitat:
   (a) producer (b) first order consumer (c) herbivore (d) second order consumer (e) carnivore (f) omnivore (g) decomposer.
5. 'Fish occupy a central position in the food chain from microscopic plants to humans.' What is meant by this statement?
6. What is meant by the following terms?
   (a) Fish fry (b) Fish farming (c) Over-fishing
7. What are some problems of the fishing industry? How could they be overcome?

Should we try to keep water clean?

What happens to life when water is very dirty?

When you use water from a tap, or rainwater which has been kept in a clean container, you know that you are using water which is safe for human beings and other animals. At least you should be able to say this. Not everyone is so lucky. For example, a World Health Organisation report on the water supply of Karachi in Pakistan suggested that it was unfit for human consumption.

We know that water which is to be drunk by humans has to be clean, even if it still contains some dissolved salts. But water used for other purposes need not be clean or pure. For example, water used for irrigation on farms need not be pure. Water used for washing crockery can be used for watering plants. But, highly polluted water cannot be used for plants and animals as it may poison them.

Activity | The effect of different water samples on mosquito larvae

1. You will need a supply of mosquito larvae. Leave a jam jar with some rotting plant stems in water in a safe place outside. Female mosquitoes will lay eggs in this stagnant water. There should be many larvae by the end of a week.
2. You will then need three different samples of water. The first should be a sample of polluted water which your teacher will provide. Prepare the other samples as follows. Boil some tap water for about 5 minutes and then allow it to cool. Divide the water into two equal portions. Use a tube to blow air into one portion and leave the other as it is.
3. Place equal volumes of each water sample into separate measuring cylinders.
4. Place about 10 mosquito larvae in each sample. You could use a dropping pipette to do this.
5. For each sample, count the number of times the larvae visit the surface to obtain air over a fixed period (5 minutes may be long enough). Record your results for each sample.
6. Leave the water samples for a few days, and observe what happens to the larvae.

At the end of your Activity, kill the larvae so they do not complete their life cycle to become mosquitoes.

Questions

1. Do the larvae behave in the same way in each water sample?
2. Why are the polluted sample and the sample which is boiled and not aerated, similar?

Water which contains very little oxygen can be dangerous to life. Unfortunately, there are many examples of water which is so polluted that very little can live in it. The photograph below shows wastes being dumped from a pipe into the sea. It is important that countries should be more aware of the problems, and try to keep their water supplies free of pollution.

Waste water being discharged into the sea

Pollution usually results from industrial wastes. One river in the United States was so heavily polluted with oil that it actually caught fire! Excess amounts of some metals can be poisonous. For example, **mercury** is extremely poisonous. Mercury compounds in water caused the deaths of a number of people in Minamata in Japan. Fish ate the mercury compounds and people who ate the fish had too much mercury in their systems and died.

It is very important to know what is contained in any water supply. For example, you might be very doubtful about using water obtained from a harbour in which many ships are found. It is important that the sea should not be used as a dump for untreated industrial and human waste products. The sea cannot cope with everything which we put into it!

Pollution in a harbour. What are the possible dangers of this?

> **Questions**
>
> 3 Can you explain why pollutants reduce the amount of oxygen in water?
> 4 How would you try to reduce the amount of waste products from industry to a low level?

What happens when lakes, rivers and seas become polluted?

Rain is an important part of the water cycle. Look back to the water cycle on page 139 to check on this.

But, rain can be polluted. We burn all sorts of things on the surface of the Earth and the products of burning often end up in the atmosphere. We burn fuel in cars, for example, and the exhaust gases end up in the atmosphere (page 193).

Acid rain A number of industrialised countries have become very concerned about this. Acid rain is rain which contains substances such as sulphuric acid.

Oil and coal often contain sulphur. When oil or coal is used as a fuel for a power station, sulphur dioxide may be formed. This is an *oxide* of sulphur. In the atmosphere, sulphuric acid may be formed from this oxide. If there is a large amount of acid in the rain this can have disastrous consequences. For example, forests and lakes in Canada, the United States and Scandinavia have been damaged by acid rain.

Soil chemicals We use all sorts of substances on the land, such as fertilisers, herbicides and pesticides. These may find their way into water supplies as rain washes them down through the soil. It is possible to find quite high concentrations of nitrates and phosphates in water where large amounts of fertiliser have been used by farmers. Chemicals used to control pests can be very dangerous indeed (page 77).

Farmers and others need to encourage the growth of their crops. They also need to discourage the growth of pests and weeds which will affect their crops. But the use of chemicals has to be controlled; farmers need to understand what they are doing with chemicals and the effects they may have on the environment.

Industrial wastes In the same way, those working in industry have to make sure that their processes work without having a bad effect on their environment. In both Jamaica and Guyana there are problems with 'red mud' which is a waste from the bauxite industries. The chemicals in the mud may damage the environment.

Activity | Water pollution issues

1 The class should be divided into groups of four or five.
2 Each group should discuss one issue relating to the control of water pollution for about 30 minutes and record its recommendations. Examples of possible issues are:
   (a) Should a government attempt to control the use of pesticides and herbicides? If yes, how? (See also page 155.)
   (b) Is the use of artificial fertilisers always a good thing?
   (c) What efforts are made by industries in your country to take care of the water they use and then release into the waterways?

When you discuss your issue, you will probably find that there are no simple answers. Remember it is not just a question of scientific principles and how they are used. There may be questions about costs, political control, relations with other countries and so on.

You are not expected to work out the costs of particular processes or precautions but should be aware of their importance.

What is being done about water resources in your country?

Find out what you can do about the use and conservation of water resources in your country. One way to tackle this may be to work in small groups and concentrate on one or two topics as the basis for a short project. Think of who you might ask for help, and what resources you can find from the library or local businesses and industries. Some issues which might be useful as a start are:

(a) What is the water supply authority doing about the conservation of water? This might include maintenance of reservoirs, water piping systems, etc.
(b) Are there any regulations in your country about the pollution of water by industries? How are these regulations enforced?
(c) Is there a government or local government department who is responsible for water analysis? What happens at this department?
(d) Are there any regulations about the dumping of industrial and human waste in the sea? Does sewage have any effect on the environment close to the shore?
(e) Would you expect an oil refinery to have a bad effect on its local environment? If there is an oil refinery close by, visit it to find out what precautions are taken to prevent waste products from polluting the environment.
(f) Are there areas of water in your country which have been so polluted that they cannot be used any longer? Is it possible for this water to be made clean again?
(g) Is rainwater used as a supply of water for homes? What are the advantages of households collecting rainwater directly rather than relying on piped water or other sources?

What are parasites?

A **parasite** is an organism that lives and feeds on or inside another living organism, which is called its **host**. The parasite usually harms the host in some way, and may even kill it.

Parasites live with their hosts. They often have adaptations such as claws or suckers to hold on with, or they live inside the host's cells.

- If they are attached on the outside they are called **ectoparasites**, for example head lice and love vine (dodder).
- If they live inside the host they are called **endoparasites**, for example hookworms, malarial parasites, and viruses and bacteria.

Problems faced by parasites

Parasites get their food from their hosts. Animal ectoparasites, like lice, fleas and bedbugs have biting mouthparts and suck blood from their hosts. The love vine, a plant ectoparasite, is made up of yellowish-orange threads with suckers which grow into the host and take away some of its food.

Endoparasites, such as hookworms and tapeworms, usually live in the alimentary canal and are surrounded by the food that the host eats. Germs such as viruses and bacteria live inside the host's cells and feed on them.

A big problem for some endoparasites is how their young can find a new host. Parasites usually produce a large number of eggs or spores, at least some of which grow and spread successfully. Some parasites have to use another organism, which is called a **vector**, to spread. For example, the female *Anopheles* mosquito is the vector which transfers the malarial parasite from one human to another.

How can we control parasites?

1 Kill a stage in their life cycle The malarial parasite is injected into a human by the bite of an infected mosquito. It reproduces in the red cells in the blood (page 151). Quinine and related synthetic drugs can kill it.

Worms which live in the alimentary canal can also be killed with drugs. They are then digested or passed out of the body.

2 Stop eggs and spores from being transferred The eggs and spores of many worms and other parasites pass out in the faeces of an infected person and can get onto food or into drinking water. Proper disposal of faeces will stop the parasites from spreading (page 152).

3 Kill their vector The blood fluke which causes schistosomiasis (bilharzia) is spread by a water snail (page 152). If this vector is killed, the parasite cannot spread. Killing mosquitoes (page 154) also helps to control malaria.

What harm can parasites do?

Parasites
- attack crop plants, causing poor growth and reducing fruit or seed production
- attack animals reared for food, causing poor growth and reducing meat and milk production
- attack humans, causing ill health or death and reducing the amount of work that can be done

Parasites of plants

A common ectoparasite of plants is love vine (dodder) which depends upon its host for food. It puts suckers inside the host to take away some of its food (below).

Twining parasitic stems of love vine

Many of the endoparasites of plants are viruses and fungi. Rusts and smuts are parasitic fungi which attack, for example, maize, rice and other cereals. Rusts cause red or brown spots or stripes on the leaves. This reduces photosynthesis, and the leaves may die. Many spores are produced, which are shot out to infect more plants.

The spores of the smut fungus settle on a plant and fungal threads grow in the leaves, stems and cobs. They cause maize grains to swell to a large size. The plant is killed, and the fungus makes black spore cases which open and spread more spores.

Parasites of animals

Blood-sucking ectoparasites such as lice, ticks and fleas are found on many animals which we breed for food. They are a nuisance to the animals and can also spread bacterial and viral diseases.

The main endoparasites are various roundworms and tapeworms which are picked up from infected faeces. These worms live in the alimentary canal or blood system where they take food which was intended for the animal (pages 150 and 152). The animal does not grow so well, and may become weak and die.

The head louse: an ectoparasite

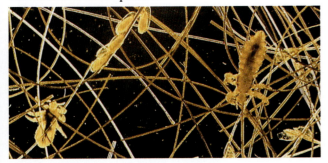

Head lice and their eggs (nits) on human hair

The head louse (see photo above) lives amongst the hairs on the human head. It has a flattened shape so it can lie close to the skin. It also has claws to hold onto the hairs so it is not combed out.

It has biting mouth parts to bite into the skin and suck blood from small blood vessels. The bites cause itching, and infections might be spread. Lice remove some blood, but are more of a nuisance than really harmful.

Life cycle The female lays eggs, also called nits, which are firmly attached to hairs so they are not easily removed. They are pale and shiny. In one to three weeks small nymphs hatch out and begin feeding on blood and any material between the hairs. They grow and moult. After about three weeks they are adult and can mate and produce more eggs.

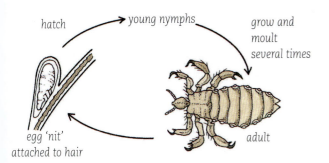

Life cycle of head louse

Head lice may be especially common in primary schools and may spread if the children share combs or brushes. For treatment use a special shampoo or cream. The first time it is used the adults will probably be killed, but it will have to be used another one or two times to kill the nits. The nits can then be combed out with a fine comb. Any affected hats or clothing should be disinfected. Children should not share combs and brushes, as it is easy to become re-infected.

The flea: an ectoparasite The flea (below) is flattened along the sides and stays close to the skin. The claws on its legs hold onto hairs. Its legs are long and strong so it jumps from one host to another. It lays eggs that hatch into nymphs.

It bites and sucks blood. The bites are painful and can also spread infection. The flea is a vector which picks up bacteria from rats which can cause the diseases of plague, typhus fever and leptospirosis (page 151).

Flea (enlarged)   *Bedbug (enlarged)*

The bedbug: an ectoparasite The bedbug (above) has a flat oval abdomen and elongated piercing mouth parts. During the day bedbugs hide away from the light in cracks in furniture and walls. At night they come out to feed on human blood. They pierce the skin and inject saliva – which produces pain and disturbs sleep. Young and adults suck blood. The young (nymphs) look like small adults.

Cleanliness discourages bedbugs. Cracks should also be filled in. Bedbugs can be caught and killed, or insect sprays can be used – these need to be repeated after 10 days to kill newly hatched nymphs.

Questions

1. What is (a) an ectoparasite and (b) an endoparasite?
2. Imagine that you are (a) an ectoparasite and then (b) an endoparasite. In each case describe the problems you would have, and how you would overcome them.
3. In what ways can parasites cause us to lose money (that is, cause economic loss)?
4. List three methods by which we control parasites.
5. Describe the life cycle of a named ectoparasite. How is it adapted for its way of life?
6. Which are (a) the most widespread and (b) the most dangerous parasites in your country? To find the answer you will need to talk to farmers, and to visit a local hospital or clinic. Group the various parasites as either ectoparasites or endoparasites and, for each parasite, describe one way in which it could be controlled.

Which parasites might we find in humans?

Endoparasitic worms

Hookworm

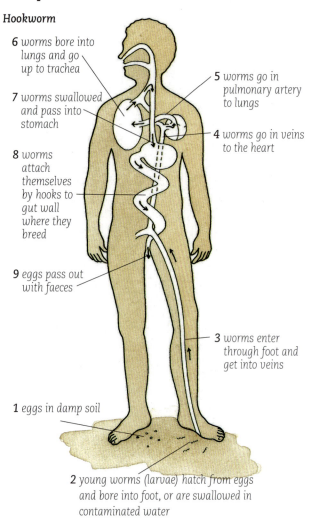

6 worms bore into lungs and go up to trachea
7 worms swallowed and pass into stomach
8 worms attach themselves by hooks to gut wall where they breed
9 eggs pass out with faeces
5 worms go in pulmonary artery to lungs
4 worms go in veins to the heart
3 worms enter through foot and get into veins
1 eggs in damp soil
2 young worms (larvae) hatch from eggs and bore into foot, or are swallowed in contaminated water

The life cycle of a hookworm (a nematode worm)

Questions

1. Look at the diagram. Follow the life cycle from 1 to 9 to see how the worm enters the body and completes its life cycle. Write down the stages in the correct order.
2. Hookworms are 1 cm long and spend time in the blood system, the respiratory system and the digestive system. In which system does a hookworm (a) enter the body and (b) reproduce?
3. How does the worm travel from one person to another? What precautions could be taken to prevent its spread? (See also pages 152–3.)
4. What damage are the worms likely to do in (a) the blood system and (b) the digestive system?

Pinworm Pinworms are a kind of roundworm. They are called threadworms, as they look like pieces of white thread of about 1 cm long. They live in the rectum and cause itching around the anus. They are quite common in children. Infection is spread when a person scratches and picks up eggs from or near the anus. These can then be transferred back into the same person, or they can be passed on to another person by contact.

You can prevent this infection by washing regularly and by maintaining good hygiene. Special medicines can be given to cure an entire family.

Ascaris This is a pink or white roundworm, which is 20 to 30 cm long. The eggs are passed out in contaminated faeces. Young worms bore into the foot and grow and travel in the body in a similar way as hookworms do. A person may be infected with many worms at a time, causing a swollen belly and loss of energy. Special medicines can be given to cure the infection.

Worms causing elephantiasis Elephantiasis is caused by very small worms that live in the lymph vessels (page 128) and cause them to become blocked. The lymph cannot escape and so the affected part, usually a leg, becomes grossly swollen. Tiny larvae are released into the blood where they are sucked up by flies such as certain kinds of mosquitoes, and injected into new hosts.

Adaptations of parasitic worms

1. They have hooks or suckers for attachment to the gut wall.
2. Shape: round (nematodes) or flat (flatworms) and with a thin wall for absorbing food from the gut.
3. Covered by mucus and special substances which stop them from being digested.
4. Produce very many eggs, which pass out with faeces.
5. Cause ill health but rarely death.

Note The blood fluke *Schistosoma* (a flatworm) lives and mates in the blood vessels of the wall of the gut. Later eggs are released in the urine (page 152). Eggs of other blood flukes can pass out in the faeces.

Protection of humans

The eggs of parasitic worms pass out of the body of an infected person in the faeces or urine. If these wastes are disposed of in a safe way using long drops or toilets and treatment at the waterworks (page 140) infections can be controlled. But if there is poor sanitation – with people urinating into waterways, or leaving faeces exposed – water and food can become contaminated. This allows infection to be passed on to a new person (see also pages 152–3 and 154–5).

Parasites spread by mosquitoes

Malaria Malaria is caused by a protist called *Plasmodium*. It is injected into the blood along with the saliva of an infected *Anopheles* mosquito. The malarial parasite lives in the blood. Each organism enters a red blood cell and multiplies. The blood cell then breaks open to release new parasites which can infect more cells (below). The release of the parasites causes an attack of fever in the patient. This pattern is repeated every few days.

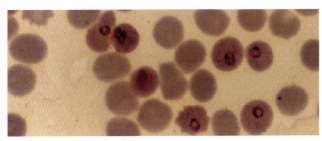

'Ring-stage' of a malarial parasite in the red blood cells

Special drugs can be given, though some drugs are now ineffective against some strains of malarial parasite. In the Caribbean we have the *Anopheles* mosquito, but few of these are infected. If more of the mosquitoes become infected, and the drugs for treatment become ineffective, we could have serious problems.

Yellow fever and dengue These are caused by viruses and spread by *Aedes* mosquitoes. Yellow fever can be controlled by vaccination. Dengue is similar to malaria, but the person often also has a rash. The person is very weak, and should rest and have plenty of fluids. Paracetamol can be given for fever and pain.

Elephantiasis This is caused by minute worms (page 150). The tiny worm larvae are spread by blood-sucking mosquitoes such as *Culex* mosquitoes.

Protection against mosquitoes

1. Avoid being bitten by using insect-repellent cream or spray, and being well covered after dusk. Mosquito netting may also be needed.
2. Use drugs to prevent or cure the attack. New drugs are always being developed against the malarial parasite, and it is hoped that in the future scientists will prepare a vaccination (page 161) against malaria.
3. Report any recurrent fever which begins one to two weeks after travelling in malarial areas, to the health authorities, in case it is malaria.
4. Spray kerosene onto stagnant water to kill mosquito larvae and pupae.
5. Spray insecticides to kill adult mosquitoes.
6. Reduce breeding places by draining swamps and lagoons. Also dispose of tins, bottles, and broken pots, etc. which could accumulate water for breeding.

Parasitic bacteria

Leptospirosis Leptospirosis is caused by a bacterium. We can catch the infection from rats, and farm animals such as goats, sheep, horses, cows, dogs and pigs. Germs are passed out in the urine and faeces of the infected animals. These animals are the vectors for the disease.

Symptoms In an infected human the disease at first can feel like flu, but becomes more severe. There is fever, headache, pains all over the body, chills and vomiting. The person has little energy and the eyes become inflamed. In advanced cases the eyes become yellow and blood is passed out in the urine. Leptospirosis can cause death.

Infection You may catch the disease in four ways by:
1. eating food contaminated with rat urine and faeces
2. handling infected farm animals
3. bathing in ponds and pools containing animal faeces
4. working in muddy fields without boots.

Eighty per cent of the people infected in the Caribbean are farmers, garbage collectors or people who work in abattoirs.

Prevention
1. Do not swim or bathe in still or stagnant water.
2. Boil water taken from ponds and rivers if you want to use the water for drinking, or to wash fruits and vegetables.
3. Keep your home free from rats and mongooses by using baits and traps.
4. Do not allow garbage to accumulate. Burn, bury or deliver it to garbage trucks regularly.
5. Wear gloves when handling farm animals.

Questions

5. Describe the life cycle of a named human endoparasite. How is it adapted to its way of life?
6. This is the front end of a tapeworm. How do the labelled structures help it in its way of life?
7. Why is it important to get rid of mosquitoes?
8. What problems *might* be associated with each of the following? Explain your answers.
   (a) Walking barefoot on damp soil.
   (b) Eating vegetables without first washing them.
   (c) Eating a mango taken from the ground.
   (d) Being bitten by a mosquito.
   (e) Handling a farm animal.
9. Prepare a chart giving advice on how to avoid becoming infected by ecto- and endoparasites.

What are the dangers of poor sanitation?

Sanitation means the disposal of sewage, which includes biological wastes (faeces and urine) and dirty water from homes and buildings. If the sanitation system is good, all of these wastes empty directly into pipes and they are taken to a waterworks (page 140) where the sewage can be treated so that the water can be safely re-used.

If the sanitation system is poor or people urinate and defecate into rivers and open drains, then the wastes may accumulate on the ground, in the streets or in the water. As a result any germs or worm eggs in the faeces from infected persons may get into our food and drinking water, either directly by dirty hands, or by the activities of flies and other animals. Worm infections described on page 150, and those described below can be spread in these ways. Prevention is described on page 153.

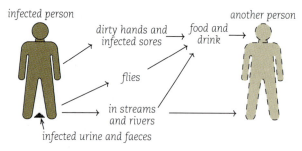

The dangers of poor hygiene

Bilharzia

Bilharzia is also called schistosomiasis because it is caused by a flatworm *Schistosoma*. This worm is called the blood fluke as it lives in blood vessels in the wall of our intestines. The infection causes diarrhoea, loss of blood, ill health and weakness.

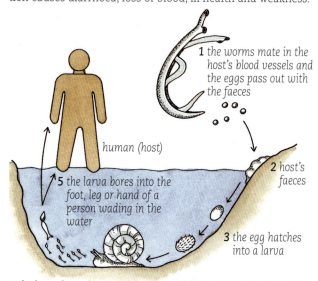

The fluke has a special water snail as its vector. In the Caribbean this is *Planorbis* which has a flattened coiled shell. Eggs leave an infected person in the urine. The eggs hatch and grow inside the snail. Larvae are released which bore into the feet of a person wading in the water.

The blood fluke has only been found in a few Caribbean countries, but many other countries have this particular water snail. If the worms were to spread to these countries there is a risk of bilharzia increasing.

Tapeworms

Another flatworm is the tapeworm. It is several metres long and has many small segments containing eggs.

Segments pass out with the faeces and are eaten by pigs or cattle. Cysts then form inside these animals. If meat from these animals is eaten without being well cooked, the cysts become tapeworms in the human's gut.

Dysentery

Dysentery is caused either by a bacterium, or more commonly by a protist (*Entamoeba*, page 13). These organisms live in the faeces of infected people.

They are spread by contaminated water and careless handling of food. They cause diarrhoea, dehydration, and blood in the faeces. Extra fluids (boiled water with a spoon of sugar and a pinch of salt) should be given to prevent serious dehydration – especially in young children.

Typhoid

Typhoid is caused by a bacterium, a kind of *Salmonella*. The infection causes high fever, delirium and intestinal upsets. The bacterium is spread from faeces by flies, contaminated sewage and careless food handling by infected people. Contaminated water, milk and shellfish also spread the disease.

The guidelines for hygiene should be followed. Vaccinations against typhoid can be given which last for three years. If a person does catch typhoid it can be treated with antibiotics.

Cholera

Cholera is caused by a bacterium. If you contract cholera, you will have severe diarrhoea, which will cause dehydration and collapse. If the body fluids are not replaced, the dehydration leads to death. Boiled water with a spoon of sugar and a pinch of salt should be given. Medical help should be sort – especially important for children.

The bacterium is spread through polluted water or raw fruits and vegetables. The guidelines for hygiene should be followed. A vaccination can be given which lasts for six months. If a person does catch cholera it can be treated with antibiotics.

Food poisoning

Food poisoning most often refers to food that is contaminated by bacteria which have produced poisonous chemicals called **toxins**. The major causes are staphylococci and salmonellae bacteria. Staphylococci often come from infected boils or cuts in the hands of people handling the food. Salmonellae are often present in the animals from which the food was derived, especially chickens. Both kinds of bacteria grow and reproduce in poorly cooked food and food that is left warm.

When food is left uncovered it attracts flies. These flies may previously have been feeding on faeces where they may have picked up germs which are then transferred to our food. Food should be covered and kept cool. Food which smells bad must not be eaten. Information on preserving food and keeping it safe is given on pages 158–9. Worm eggs can also be spread via food due to poor hygiene (page 150). Other causes of food contamination are given on pages 77 and 147.

The main symptom of food poisoning is diarrhoea, often with abdominal pain, vomiting and fever. It is important to drink a lot of liquid to replace what is lost. Recovery usually occurs without the need for antibiotics. If the symptoms continue for three days, you should see a doctor. For a child, take action after one day. Rehydration fluid such as boiled water with a spoonful of sugar and a pinch of salt, can be given. An upset tummy can also be a symptom of some diseases caused by bacteria and viruses.

Preventing the spread of infection

Follow these guidelines for hygiene.

Personal hygiene
1. Wash your hands after using the toilet.
2. Wash your hands before handling food.
3. Keep your nails clean.
4. Wash yourself at least once a day.
5. Keep your clothes and hair clean and tidy.
6. Do not walk barefoot on wet soil.
7. Use a handkerchief when you sneeze.
8. Do not spit.
9. Protect food from flies.
10. Wash fresh food before eating it.
11. Boil drinking water when the authorities advise this.

Community hygiene
1. Use proper toilet facilities for urinating and defecating.
2. Do not bathe in water used for excretion.
3. Dispose of sanitary towels properly.
4. Do not use human urine and faeces as manure.
5. Do not throw rubbish in the street.
6. Keep your rubbish in dustbins.
7. Make proper use of rubbish collections.

Can sewage be useful?

We have seen some of the problems associated with the incorrect disposal of sewage. Could these biological wastes and dirty water be put to any use?

When organic waste such as faeces is allowed to decay, without the presence of oxygen, certain gases such as methane are produced. This process of decay can be carried out in special containers and the gas produced is called **biogas**. This biogas can then be burned as a fuel to release energy. In several projects around the Caribbean, biogas has been produced and made use of as a source of energy in the home. These biogas containers could be used more widely. Biogas can be used directly for cooking in a way similar to the use of LPG (page 189) or it can be used to generate electricity which then has additional uses. The solid waste material which is left behind after the biogas has been made can be used as a fertiliser. The process of decay would have killed any germs or worm eggs and so the fertiliser is safe.

Another ingenious use of faeces is to collect the wastes produced by chickens and to use them to feed fish. This is especially useful on **fish farms** where freshwater fish are reared and later sold. In Asia chickens are caged in pens directly over the water or on the pond banks and the excreta and waste food falls into the pond where it is eaten by the fish (usually *Tilapia*). Chicken manure is very high in phosphates which are taken up by aquatic plants which are also eaten by the fish. Pigs, which the Chinese consider are 'costless fertiliser factories moving on hooves' can be used in a similar way. Trials using chicken and pig manure carried out in Jamaica have led to increased fish yields.

Farmyard manure, the excreta of farm animals can also be recycled. Manure should not be used directly on the soil, but can be added to compost heaps where it will decay and any organisms that could cause diseases will be killed.

In some countries, cow manure is dried and used as a fuel.

Questions

1. 'Flies feed on faeces and food.' What problems may this cause?
2. Some Caribbean countries have *Planorbis* snails, but not the blood fluke which causes bilharzia. Why is it very important that the worms are not introduced into these countries?
3. Look at the guidelines for personal and community hygiene. Give a reason for each of the suggestions, describing how each action would help to prevent the spread of disease.
4. You have heard that there is a cholera outbreak in an area of Africa which you are going to visit. What precautions would you take so that you would be less likely to catch the disease?

How can we control pests?

Pests affect people and their crops and farm animals. They cause economic losses. Some pests can carry parasites and pass these on to humans, but pests are different from parasites in these main ways.

Pests	Parasites
Pests live *near* other organisms	Parasites live *on* or *in* a particular host organism
Pests do not depend directly on the other organism for food	Parasites get food directly from the host; they depend on the host
Pests are a nuisance because they spoil or steal our food and can transmit parasites	Parasites themselves harm their host and can cause death
Pests move freely from place to place	Parasites stay with their host

Some common pests are shown below.

What measures help control pests?

Sanitary measures Sanitary means 'clean and healthy'. If surfaces and cupboards are kept clean, any germs or pests which might settle there will be removed. If food is put away or covered up there will be no food for pests to live on. Dark and dirty places are ideal breeding places for pests. Also follow the guidelines for personal and community hygiene on page 153 for reducing the spread of infections.

Look at the dirty kitchen on page 155. List all the ways in which it is dirty or unsanitary and describe how each fault could be corrected. On pages 156–7 we will learn more about keeping our surroundings clean and thereby reducing the danger of spreading organisms that cause disease.

Biological measures Biological measures use other organisms to control particular pests. This means that only the pest we want to get rid of is affected and we do not spread dangerous chemicals around.

Pests	Where they breed	What trouble they cause	How to control them
Cockroaches	Cockroaches breed in dark places, in drawers and cupboards and in cracks in the floor. The young nymphs are very small and can get into very small cracks. They prefer dirty places with food remains.	Cockroaches eat almost anything and so they carry bacteria around on their bodies. The bacteria can cause food to go bad. They also leave their droppings in the food and make it unpleasant to eat.	We can stop cockroaches from feeding by keeping all food in airtight containers or plastic bags. Drawers and cupboards should be cleaned out regularly. Pesticides can be sprayed to kill them, taking care to first cover any food.
Flies	Flies lay eggs on food left open to the air. They are especially attracted to decaying food. Leftover food that is not covered will soon be teeming with maggots (the larvae of the fly).	Flies feed on faeces and on human food. They carry bacteria on their bodies and in their saliva and faeces. They put these on the food, and so make it unsafe to eat. The bacteria can cause decay and disease. The maggots also eat our food.	Wrap up the food or cover it with another dish. Food kept in a food safe, made of fine netting, will also be protected. Catch flies on sticky flypaper, or spray pesticides to kill them, taking care to first cover any food.
Rats	Rats breed in dark cupboards and under floor boards. They also live on ships which are carrying food. They prefer crowded, dirty conditions.	They feed in refuse dumps and sewerage systems, and so may carry bacteria to our food. Rats can pass on leptospirosis and bacteria. They also carry rat fleas which can bite humans and may pass on plague bacteria. They can also damage wooden buildings.	Food should be stored in airtight containers. Houses and grain stores can be rat-proofed with barriers to stop the entry of rats. Poisoned food (bait) can be put down to kill them, and traps and cats can be used to catch them.
Mosquitoes	Mosquitoes breed in stagnant water in ponds and swamps. They also breed in old tins and pots which contain decaying plant material. Eggs hatch into aquatic larvae which grow into pupae.	Female mosquitoes suck blood from humans which causes irritating bites. They also spread diseases. *Anopheles* mosquitoes spread malaria parasites (page 151). *Culex* mosquitoes spread minute worms causing elephantiasis. *Aedes* mosquitoes spread viruses causing dengue and yellow fever.	We can stop them from feeding by using repellent cream and by keeping ourselves well covered in the evenings. We can stop them from breeding by draining swamps and removing containers that could become filled with water and then serve as a breeding place. We can kill larvae and pupae with kerosene, and adults with insecticides.

One example is to introduce a predator to eat the pest, such as having a cat to eat mice and rats. Another example is to introduce ladybirds to control aphids (greenfly) which suck plant juices and may spread diseases. Ladybirds eat aphids, so if we increase the number of ladybirds, we decrease the number of aphids.

We can also control pests by introducing an infection which will kill the pest. For example, the virus which causes myxomatosis was introduced to kill large numbers of rabbits in Australia. In this case it led to additional problems because so many rabbits were killed that it upset the balance of nature. Animals which used to eat rabbits started to eat farm animals instead. When using biological control it is important to try and predict the results of our actions.

We can also attack the breeding process of the pest. For example, if we produce a lot of sterile males and then let them go free to mate with the females, these females will not produce any young and so the number of pests will be reduced.

Chemical measures Pesticides are chemicals used to kill pests. The two most important ones are **insecticides** to kill insects, and **herbicides** to kill herbaceous weeds. These have to be used carefully, so as not to harm humans or plants and animals that aren't pests.

Handling pesticides

Pesticides come in different forms, for example pellets for killing rats and slugs, aerosol sprays for killing cockroaches and mosquitoes, or powders which have to be made up into solutions before spraying them onto plants.

Do
1. Choose the correct pesticide for the purpose.
2. Where necessary, make up the pesticide to the correct strength in water before use.
3. Cover all foodstuffs before using pesticides.
4. Wear gloves and other protective clothing when making up solutions and spraying crops.
5. Read and follow all instructions carefully.
6. Spray when the air is still, so the pesticide will not be blown away.
7. Wash spraying equipment thoroughly after use.
8. Lock away pesticides when not in use.
9. Wash your hands carefully after handling pesticides.

Don't
1. Don't drink, eat or smoke when spraying.
2. Don't breathe in the fumes of the spray.
3. Don't spray towards humans.
4. Don't use up spray left over from the last time.
5. Don't spray crops close to harvesting time.

Problems

1 Cost Pesticides are expensive, so it is very important to chose the right ones and to apply them correctly.

2 Poisonous Pesticides such as Deildrin and Paraquat, which have been banned in the USA and UK, are still used in some Caribbean countries. They can kill us by penetrating the skin, or being accidentally breathed in, or eaten in food, or drunk in water (page 147). Pesticides must be carefully labelled and stored in their original packaging.

3 Non-biodegradable Chlorinated insecticides such as DDT have also been banned in most countries because they do not break down easily (are non-biodegradable) after being sprayed. The DDT residues therefore accumulated in plants and animals, causing deformities in fish and birds. High levels of pesticides in our food may cause cancer.

4 Resistant strains Insects, weeds and fungi being treated with pesticides change so that they become resistant to the chemicals (no longer killed by them). There is a constant battle to produce new pesticides which are effective and safe.

Questions

1. Describe the characteristics of a pest. How does it differ from a parasite?
2. Give three examples of pests and three examples of parasites.
3. Summarise in your own words the conditions which encourage the breeding of household pests.
4. List six problems which pests can cause people.
5. We control pests by (a) stopping them from feeding (b) stopping them from breeding and (c) killing them. Choose one named pest and give details on how it is controlled in these ways.
6. How do we control pests by (a) sanitary (b) biological and (c) chemical means?
7. Give reasons for each of the **do** and **don't** rules for the handling of pesticides.
8. What are (a) the advantages and (b) the disadvantages of using pesticides?

How can we keep our surroundings clean?

How can we recycle rubbish, or dispose of it safely?

Activity | What rubbish do you produce?
The rubbish around our house or school is called **domestic waste** or **refuse**.
1. Explore around your house or school, and look at the refuse. *Do not touch it.* Look in the waste paper basket, the dustbin, the kitchen, and outside the building.
2. List the main things in each place that people throw away.
3. See if you have listed: food wastes, dirty water, tins, bottles, waste paper and cardboard, plastic bags, broken furniture, scrap metal, broken appliances and utensils, old clothes, shoes and toys.

4. For each kind of refuse, decide:
   (a) what happens to it after it has been thrown away
   (b) what problems would arise if it was not disposed of properly (see also pages 150 and 152).
5. Could any of the refuse be made useful?
6. When is your rubbish (refuse) collected?

Activity | What happens to our biological wastes?
1. (a) What kinds of toilets do you have at your home and school?
   (b) Are there enough toilets for everyone to use?
   (c) Are they in good working order and kept clean?
2. Is there running water to flush away the wastes?
   - *If yes*, where do these biological wastes go to? What happens to them?
   - *If no*, what problems may arise? How can the urine and faeces be removed and got rid of safely?
3. What can we do to help keep the toilets clean and in good working order?

Activity | What wastes are produced by industries?
Look at or visit a nearby industry or factory.
1. Does the factory produce dirty smoke or bad-smelling gases? What effect might these gases have on local plants and animals (including people)? How could the amount of gas be reduced?
2. How are solid waste materials and rubbish dealt with? Are they neatly stored to be disposed of, or are they just piled around outside the factory?
3. Are liquid wastes or chemicals produced? Are these liquid wastes pumped untreated into rivers and/or the sea? What problems might this cause? (see page 146).
4. Does the factory release heat into the air? What effect does this have?
5. Collect information and make a display chart.

Activity | What about other wastes?
1. *Public places* Look at any places in your surroundings where rubbish is thrown away.
   (a) List the waste materials thrown away.
   (b) What dangers might these waste materials cause?
   (c) Could any of the waste materials be recycled?
2. *Roadways* What wastes are produced by cars, trucks and buses? How do they cause pollution? (See pages 193 and 267.)
3. *Market* List all the types of rubbish, and all the ways in which the market is dirty. What dangers might these cause?
4. *Supermarket* Look around inside a supermarket. How well are things kept clean? Is there evidence of pests, such as cockroaches and rats? (See page 154.) Look around outside the supermarket. List all the different kinds of rubbish you can see. Which kinds of waste materials will attract flies and dogs? What problems might occur?
5. Prepare a report on your findings.

What can we do with our rubbish?

Storage The first problem is how to store our rubbish safely. We need to have a dustbin with a secure lid (see below). This is to stop animals such as dogs and goats from emptying the dustbin and spilling out the waste materials. The lid is also needed to keep flies away as they lay eggs which develop into maggots and then more flies. The lid will also help to keep down the smell of decaying food from reaching the house.

Right and wrong uses of a dustbin

Sorting Some of these problems can be reduced by sorting our rubbish into two parts. Leftover food, and remains like banana and orange peel are **biodegradable**. This means they are decayed and broken down by the action of bacteria (page 243). Other rubbish cannot be decayed and broken down by bacteria. It is **non-biodegradable**, such as things made of glass, metal or plastic. Non-biodegradable materials can be put in the dustbin to be taken away by the rubbish collectors. There is also the possibility that some of this could be recycled.

Recycling non-biodegradable materials We can separate the different items.

Glass bottles should be cleaned and can often be taken back to a collecting station where money might be paid for them. They can be re-used by the factories. Screw-top jars are also very useful around the house for storing food.

Metal tins and old appliances that contain metal can be sold for scrap. The metal is melted down and used again.

Plastic cannot easily be recycled. However, plastic bags, for example from bread, can be washed carefully and used again around the home. There may also be a collection system for plastic bottles for recycling.

Newspapers and cardboard do decay slowly but they can also be recycled. The paper is reduced to wood pulp, and this can be re-used. This reduces the number of trees that have to be cut down to produce paper.

Recycling biodegradable materials We can put biodegradable materials from kitchen scraps into a bucket with a tight lid. The bucket can be emptied often, and the decaying material put into a **compost heap** (below). We are making use of the decay process so that, instead of just producing smells and possible health hazards, we are producing compost to use to improve the texture of the soil and to provide natural fertiliser.

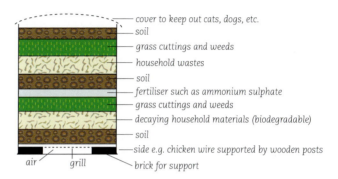

A compost heap is made up of alternate layers of soil, decaying household materials, grass cuttings and weeds. The soil provides the bacteria that decay the waste materials. As the decay process continues the compost heap becomes very hot. This kills the bacteria that produce bad smells.

The compost heap should not be in a closed container, it should have space all around it. It is often made of a wooden frame with wooden slats with spaces in between. The heap must also be turned over from time to time. The reason is that it must be well ventilated because the bacteria that cause decay need oxygen in order to continue their work. The heap should also be kept moist.

When the material has decayed it can be used instead of chemical fertilisers. This **compost** is good for the soil, because, like humus (page 233), it improves the texture of the soil and releases mineral salts to be used by the plants.

Questions

1. A lot of household rubbish could be burned. Give one advantage and one disadvantage of doing this.
2. Which items of household rubbish, that you might throw away, could you actually re-use in your home or at school?
3. Find out what is done in your community about recycling non-biodegradable materials. See if you can set up a project to recycle these materials.
4. Suggest six ways in which you and your classmates can help to keep your surroundings clean and safe.

How can we stop food from spoiling?

Micro-organisms such as bacteria and moulds spoil food and make it dangerous to eat. These are the same organisms that bring about decay. They decay our food!

The organisms secrete digestive juices onto the food and begin to dissolve it. Their waste products make the food smell and taste unpleasant. Some bacteria produce poisonous substances called **toxins** which pass onto the food. If this food is eaten it can cause **food poisoning** (see page 153).

Meat, milk products and shellfish are often affected. The contents of a bulging or 'blown' tin of food may cause food poisoning because the bulging shows that bacteria have been multiplying inside the tin and producing carbon dioxide gas to push out the ends. Such food must not be eaten.

Activity | How can we stop micro-organisms from growing?

We will choose bread mould as our micro-organism.
1. Toast a piece of bread to make it dry.
2. Cut it into six equal-sized pieces.
3. Dissolve salt in a test-tube full of water until no more will dissolve. Make another solution in the same way, but this time use sugar instead.
4. Set up six plastic bags as shown below.
5. Examine the bread each day for two weeks. Do not remove it from the bags.

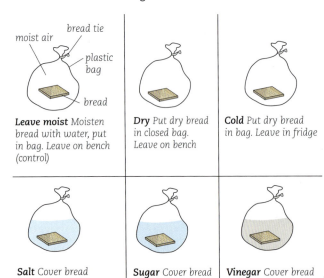

Leave moist Moisten bread with water, put in bag. Leave on bench (control)

Dry Put dry bread in closed bag. Leave on bench

Cold Put dry bread in bag. Leave in fridge

Salt Cover bread with concentrated salt solution. Leave on bench

Sugar Cover bread with concentrated sugar solution. Leave on bench

Vinegar Cover bread with vinegar. Leave on bench

6. Try and explain your results.
7. Which conditions (a) encourage or (b) discourage the growth of bread mould?
8. Throw the closed bags with bread into the rubbish bin.

What conditions are needed for micro-organisms to grow?

1. **Suitable temperature** Micro-organisms need a temperature which is not too low and not too high to grow. A temperature of 25 to 35 °C is ideal for the growth of most decay organisms.
2. **Oxygen** Most organisms that cause decay need oxygen for respiration. If oxygen is removed then they respire anaerobically (page 106) and ethanol or lactic acid is produced, which stops further decay.
3. **Water** Moisture is needed for the germination of the spores and the growth of the micro-organisms.
4. **Absence of dangerous chemicals** Micro-organisms are living organisms which carry out all the usual life processes. Certain selected chemicals can kill them, for example salt, sugar, vinegar or alcohol.

In food preservation we remove one or more of these conditions and so control or kill the micro-organisms and preserve our food.

How do we control the growth of micro-organisms?

1. **Unsuitable temperature** Low temperature Freezing does not kill micro-organisms, it just slows down their growth rate. The lower the temperature the slower the growth. The micro-organisms can resume their normal rate of growth at room temperature and can then cause decay.

    High temperature Most micro-organisms die if kept above 40 °C for several minutes. However, the spores of some bacteria can survive this treatment.
2. **Remove oxygen** When food is put into cans the air is removed before the lid is put on. Foods are often preserved in some other way, for example by adding sugar, before being canned.
3. **Remove water** Drying Heating the food dries it out. Crops such as tobacco leaves, and also fish can be dried in the sun (solar drying, page 196). Dried foods keep for a long time. They also take up less space and are therefore easier to transport and store. As soon as they have water added to them, they will begin to decay.
4. **Concentrated solutions** Food can be put into concentrated salt or sugar solution. The water inside any micro-organisms will be drawn out as it travels by osmosis (page 69) from an area of high concentration of water (in the micro-organism) to an area of lower concentration (in the solution).
5. **Add selected chemicals** Chemicals are added which kill the micro-organisms, but should not harm humans. For example, acetic acid (vinegar) is used in pickling, chlorine is used to purify water and sodium benzoate is used for preserving meat.

Food preservation methods

Heating
Principle Many micro-organisms killed, and too hot for others to grow well.

If food is cooked and heated in some way, the heat can kill off some of the micro-organisms.
The food can then be canned or packaged. Liquids can also be boiled to kill the micro-organisms.
As the food or liquid cools back to room temperature, the remaining micro-organisms increase their activity.

Cooling
Principle Too cold for micro-organisms to grow well: slows growth.

Food can be packed around with ice or put into a refrigerator. Food in the main part of the fridge keeps fresh for a few days, in the freezing compartment for a few weeks and in the deep freeze for a few months.
When frozen food has been defrosted, it will begin to decay more quickly.

Drying
Principle No moisture for growth. Proportion of salt and sugar increases, which helps preservation.

Drying removes water and is the most effective preservation method. Food can be left to dry in the sun. The food becomes *dehydrated*. Micro-organisms are killed, and spores which land on the food cannot grow. When water is added again the food starts to decay.

Pickling
Principle Vinegar (weak acid) kills micro-organisms.

Food can be pickled by soaking it in vinegar. Vinegar is weak ethanoic (acetic) acid. It is harmless to humans, but it kills the micro-organisms. The food is kept in the vinegar inside a closed bottle so other micro-organisms are unlikely to get inside.

Salting
Principle Salt draws water out of micro-organisms so they are killed.

Salt is rubbed into the food, or the food is put into a concentrated solution of salt (brine). This same principle is used when sugar is added, for example to preserve fruit.

Preservatives
Principle Selected chemicals kill micro-organisms.

Chemicals, such as sodium benzoate, are added to food. They are listed on the labels. Some preservatives may also be damaging (see page 77). Preservatives may be added to food which has been treated in other ways, for example, dried.

Questions

1. List three foods which could be preserved by each of the six methods shown above.
2. Look at the picture on the right. Describe how each of the foods has been preserved.
3. Discuss two methods of preservation which you could use yourself and explain the scientific principles underlying each one.
4. What could be the explanation for the following:
   (a) the ends of a can bulging out
   (b) frozen food in a fridge going bad
   (c) dried milk staying good, but liquid milk going bad (under the same conditions).

How has each item been preserved?

our environment | *health and hygiene* | *how can we stop food from spoiling?*

How can we be protected against disease?

A disease is anything which upsets the normal functioning of the body: it causes 'dis-ease'. The disease may be due to deficiencies in the diet (deficiency diseases). Or it may be due to faulty genes (inherited diseases). Or it may be the result of malfunctioning of certain body systems or organs (physiological diseases). Or it may be caused by another organism that has infected the body (infectious diseases caused by parasites or pathogens). The parasites may live inside or outside the body (pages 148–9), and may be visible to the naked eye (macro-organisms, pages 149–50 and 152) or very small (micro-organisms, page 151).

Examples of the various kinds of disease are given below, with references to information about them. In this topic we shall be mainly concerned with the spread of and protection against infectious diseases caused by micro-organisms living inside the body.

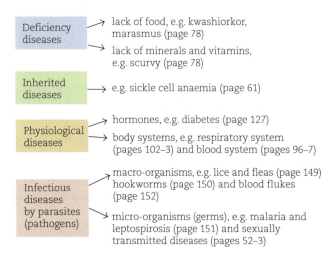

Different kinds of diseases

How infectious diseases are spread

Infectious diseases are spread by direct contact, through the air, by vectors and by contamination.

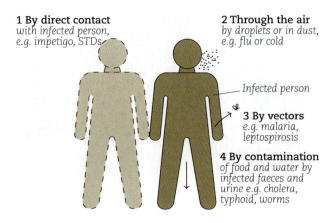

Protection against infectious diseases

1 **Avoid infection** We should avoid contact with infected people, although this is difficult because we may not know that they have, or are carrying, a disease. We should keep our distance from a person sneezing out germs. We should try not to be bitten by mosquitoes or rats and be very careful when handling farm animals and pets (page 151).

We should wash our hands after using the toilet, also wash fresh food before eating it (pages 152–3) and follow other guidelines for hygiene (page 153).

2 **Kill the vector** Common vectors which carry organisms that cause disease are mosquitoes, rats, farm animals (page 151) and certain water snails (page 152). It is important to kill or treat infected animals to break the life cycle of the parasite.

3 **Use drug**s A doctor can prescribe drugs that can prevent or cure a disease. Different drugs are effective against different diseases. Drugs kill the invading organisms without harming the patient. For example, drugs such as Antipar can be given to kill worms parasitic in the alimentary canal, and special shampoo can be used to kill lice.

In the fight against bacteria there is a special group of drugs called **antibiotics**. Two examples are penicillin and tetracycline. They have to be given in the correct dosage and for a particular length of time. If not, they may not be effective, or the bacteria may become resistant (no longer killed by the drug). Gonorrhoea and syphilis (page 52), typhoid, cholera and some kinds of dysentery (page 152) are caused by bacteria and can be cured by antibiotics. Antibiotics are sometimes given when a person has flu. They do not kill the virus which caused the flu, but stop the person suffering from a bacterial infection as well.

4 **Fight the infection** Micro-organisms (germs) such as viruses, bacteria and some protists can be attacked by the white cells (phagocytes and lymphocytes) in our blood (page 92).

The **phagocytes**, which are produced in the bone marrow, are able to surround and 'eat' any invading germs (below).

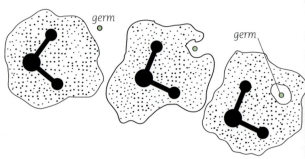

A phagocyte – a type of white blood cell which kills germs by 'eating' them

The other white cells, the **lymphocytes**, are formed in the lymph glands (page 128). Large numbers of lymphocytes are produced when germs get into the body.

1. Germs such as some bacteria and all viruses carry proteins called **antigens** on their surfaces. These proteins are special to that particular kind of bacterium or virus. The lymphocytes respond to the antigens by producing corresponding **antibodies** which can neutralise the effect of the antigen. The germs are killed.
2. Some bacteria release chemicals called toxins into the blood which cause the symptoms of the disease. The lymphocytes respond to the toxins by producing corresponding **antitoxins** which neutralise the effect of the toxins. So the person does not become ill.
3. The problem with AIDS is that HIV infects the lymphocytes and prevents them from protecting the person against that virus and other infections such as TB.

A lymphocyte – a type of white blood cell which kills germs by producing antibodies and antitoxins

Immunity

We can be protected against disease (become **immune**) if we receive or make antibodies and antitoxins.

Passive immunity
- mother's antibodies for young baby
- ready-made antibodies

Active immunity
- response to natural infection
- response to vaccination

Methods of achieving immunity

Passive immunity This is protection which is given to the person in the form of antibodies, but it does not last long.

Mother's antibodies Antibodies from the mother can pass through the placenta (page 63) into the developing baby and are also present in the breast milk. These antibodies protect small babies against the infectious diseases the mother has had. However, the immunity is not long-lasting and the baby must also develop its own active immunity.

Ready-made antibodies A doctor can inject you with ready-made antibodies against specific diseases such as cholera and a kind of hepatitis. This protection only lasts for about six months. For longer-term protection you need to develop your own active immunity.

Active immunity This is protection developed by the body in response to infection or vaccination. It is usually long-lasting.

Natural infection People who have had a certain viral or bacterial disease will have made antibodies or antitoxins to counteract the disease. Some of these antibodies or antitoxins will remain, and others can be quickly made to give protection against that disease on another occasion.

The degree of protection or immunity which we have depends on several factors. For some diseases, for example measles and mumps, once we have had them we develop life-long immunity to them. When we get flu we produce antibodies against that particular kind of flu and become immune to it. But there are so many kinds of flu that we are more likely to catch one we are not protected against!

Vaccinations The person is given vaccinations, which are inoculations of dead or weakened viruses, or bacterial toxins. These do not cause harm but they do make the body produce antibodies or antitoxins. 'Top-ups' or boosters are given at later dates. The table of vaccinations recommended in a Caribbean country such as Jamaica is given below.

Time	Vaccine	Method
3rd–7th month of pregnancy	Tetanus	Injection of tetanus toxin
Birth to 3 months	BCG (Tuberculosis)	Injection of weakened bacteria
3 months, 5 months, 7 months	Three doses of polio vaccine and DPT*	Polio: by mouth (weakened virus) DPT: injection of toxins (T) and weakened bacteria (D and P)
12 months	Measles	Injection of weakened virus
18 months, 3 to 6 years	Two polio boosters	As before
9 to 12 years (girls only)	German measles	Injection of weakened virus

*DPT = diphtheria, pertussis (whooping cough) and tetanus

Questions

1. What are the four kinds of disease? Give one example of each.
2. List the ways in which we can be protected against infectious diseases.
3. Distinguish between (a) antigens and toxins (b) antibodies and antibiotics and (c) passive and active immunity.

How can we avoid accidents?

An unsafe kitchen

Activity | How to avoid accidents

1. List as many dangers as you can see in this kitchen. Next to each one, write down what you would do to correct the problem and stop it from causing an accident.
2. Which of these accidents might also occur (a) in other parts of the house or (b) in the laboratory (see also pages 164–6)?
3. Which substances commonly used in the home and garden are dangerous? (Find out by looking at the labels on the containers and writing down the substances.) Wash your hands afterwards.
4. How are (a) young children and (b) older people at special risk of accidents?

Activity | Guidelines for safety

Give a reason for each of the following rules.

1. Label all household chemicals clearly.
2. Keep chemicals in their proper bottles.
3. Put pills, medicines and household chemicals where children cannot reach them.
4. Keep pills in 'child-proof' bottles that they cannot open.
5. Keep household chemicals where adults can reach them easily.
6. Keep poisonous chemicals such as insecticides and rat killer locked up.
7. Mend fraying insulation on electrical appliances.
8. Check that the thermostats on irons are working.
9. Use the right number of plugs in each electrical fitting (do not 'overload', page 180).
10. Light gas burners and cookers safely.
11. Mend any unevenness in floor surfaces.
12. Pick up things from the floor.
13. Have adequate lighting in all rooms.
14. If liquids are spilt on the floor, mop them up so the floor is kept dry.
15. Keep floors clean but do not polish them so much that they are slippery.
16. On a stove, do not have pot handles projecting over the edge.
17. Only handle hot utensils with protective gloves.

Accidents	How to avoid them
Bites and stings Many animals bite, for example ants, centipedes, mosquitoes, wasps, pets, rats and farm animals.	Handle animals carefully, wear gloves if necessary. Do not annoy pets or insects. Use repellent cream against mosquitoes.
Broken bones Falling over, or pulling something heavy onto yourself could break a bone.	Look where you are walking. Uneven or poorly lit floors, or utensils left on the floor, may cause falls.
Burns and scalds Burns are caused by touching hot objects. Scalds are caused by steam.	Take great care when touching hot liquids, utensils or the stove. Do not bump against pot handles on the stove.
Cuts Cuts break the surface of the skin and so open it to germs.	Handle all knives carefully. Cut on a flat surface. Point the blade away from you when you cut.
Electric shock Caused by an electric current passing through the body. Can cause unconsciousness or death.	Do not put fingers in electrical sockets or handle electrical appliances while hands are wet. Do not touch electrical wires. (See also page 180.)
Eye injuries Solid objects or chemicals may fall into the eye. Tears are a natural body response.	Avoid very dirty or dusty places. Point boiling liquids away from you. Handle all chemicals carefully.
Explosions May result from leakage of gas, or delay in lighting burners or cookers.	Check if the gas is leaking from cylinders. Check that the pilot light is on. Have your match ready to light the burner or cooker.
Fire Caused by overheating of an iron, electrical faults, burning liquids or gases setting other things alight.	Check iron thermostats, check for any overloading (page 180), take great care with matches and lamps.
Poisoning Children may eat pills, thinking they are sweets. They may also drink household chemicals. Adults may take overdoses of drugs.	Store pills, household chemicals, etc. safely and in clearly labelled containers. Throw away unwanted drugs. Take special care with sleeping pills.
Sprains Caused by pulling on joints unnaturally or by twisting an arm or leg when falling.	Look where you are walking. If you are falling, try not to let your weight fall onto your wrist or ankle.

Note See pages 170–1 for first aid for these and other problems.

Activity | Design and make a pot-holder
1. We use a pot-holder to handle hot utensils. What characteristics should a pot-holder have?
2. Find a variety of recycled materials and make your own pot-holder.
3. Compare your pot-holder with those made by others in the class. Decide which pot-holder would be most effective.
4. How expensive was it to make your pot-holder?

Questions
1. Choose a room in the house, such as the bathroom or kitchen. List the possible hazards and how to avoid them.
2. What special hazards exist in the garden and garden shed?
3. In small groups, design a poster to warn people against household hazards. Think of different ways to get your message across.

Working safely

Why is safety so important?

You need to be aware of risks in any situation, so that you can take the necessary safety precautions. For example, if you are walking along a road, it is safer to walk so that you are facing oncoming traffic, so that you can see what is approaching.

The same thing applies when you are working in a science laboratory. You need to be aware of and apply the safety rules, and it also helps if you know something about very basic first aid (see pages 170–1), so that you can assist a fellow student if they have an accident. (Your teacher will direct what action should be taken.) You need to make sure that the ways in which you handle chemicals and equipment are safe and will not cause injury to you or fellow students.

You also need to think very carefully about how you behave if you are out of school on a field trip or in a workshop. These issues are discussed on page 167.

Safety at home is also of crucial importance. Cooking is usually very safe, but a gas cooker uses flammable fuel and cooking oil can catch fire if not handled safely. Electrical appliances make use of mains electricity and this also needs to be handled with care. Wiring has to be safe, with effective insulation where it is needed (see pages 180–3).

Think safety at all times – it may help to save your life.

Activity | Safety rules

1. Write down 10 rules for the way you should work in a science laboratory.
2. Discuss these rules in small groups and then with your teacher. Are these rules included in the list of safety rules for the laboratory?
3. Do you think that you need to apply similar safety rules when you are at home?

Whenever you handle equipment or chemicals you need to find out how to handle them *before* you use them. Concentrated sulphuric acid is a very strong oxidising agent and also a very strong dehydrating agent. It reacts very vigorously with water, with the evolution of a great deal of heat. You add the acid *very slowly* to water – *never* add water to the acid. When food is being cooked you need to make sure that it is fresh and properly cleaned and prepared (see pages 153 and 158) and that the gas supply is safe. If you use a screwdriver, make sure that it is the right size and type before you use it.

These are examples of simple safety rules – which often amount to the application of common sense. If you follow them carefully, you will help to prevent injury to yourself and others.

What are some of the dangers?

We can divide the dangers in the laboratory and the home into two types:
(a) those involving chemicals
(b) those involving equipment or/and machines.

This is not a hard and fast distinction and there may be examples of dangers which include both chemicals and equipment. You should remember that the term 'chemicals' covers a very wide range of materials, including water and all the food items we use in cooking, including cooking oil. You do not expect to think of water as a potentially dangerous substance – you drink it every day and without it you would die. However, water and electricity are a dangerous combination, since water conducts electricity. You should never use electricity if you have wet hands. Look carefully at the water and electrical outlets in both your science laboratory and your kitchen and bathroom at home. Are they close together? There are safety regulations about the minimum distance between the two types of outlet, both in the laboratory and at home. The term 'equipment' can also cover a wide range, including simple items of glassware in a laboratory as well as sophisticated items of electrical equipment such as computers.

Activity | Safety with chemicals

1. Make a list of 15 to 20 chemicals which you have used in the laboratory. Try to make the list as varied as possible, by including relatively simple chemicals such as hydrochloric acid and sodium hydroxide as a start. Think too about chemicals such as the organic compounds ethanol and propanone. You might also want to include a chemical such as potassium manganate(VII).
2. Find out, if you do not already know, whether each of these chemicals is **corrosive**, **toxic** or **flammable**. If you are not sure of the meaning of these words, check with your teacher. A chemical may possess one or more of these properties – or none of them. (Well-labelled bottles should provide the answers.)
3. Make a table showing the properties of the chemicals listed.
4. Divide the chemicals into groups, such as 'extremely dangerous', 'dangerous', etc. These days the labels of chemicals carry warning symbols set in square boxes. Some of these labels are shown on page 165. Other hazard labels are found as triangles. Make sure that you handle chemicals carefully in accordance with the label on the bottle.

Part 1

There are four categories of safety signs:
- **Prohibition signs** indicate that certain behaviour is not allowed
- **Warning signs** draw hazards to your attention. On bottles the picture will be in a square. On vehicles it will be in a diamond
- **Mandatory signs** dictate a specific course of action
- **Safe condition signs** show what course of action must be taken to ensure safety

Here are a few examples of **warning signs**

 Risk of fire
 Risk of explosion
 Biohazard

 Toxic hazard
 Corrosive substance
 Risk of ionising radiation

 Oxidising
 Harmful/irritant
 Dangerous for the environment

Here are a few examples of **mandatory signs**

 Eye protection must be worn
 Respiratory protection must be worn
 Hand protection must be worn

How should we handle corrosive substances?
Substances such as concentrated sulphuric acid and sodium hydroxide are very corrosive. (You may have been told about the 'soapy' feel of sodium hydroxide – do not get either the solid or the concentrated solution on your skin.) Look back at the table you prepared and make a separate list of the corrosive substances. When we handle any corrosive substance we make sure that it is in a container made of a material with which it does not react. The container should be stored in a cool dry place, out of direct sunlight, with access to good drainage (why do you think this is important?). Cool places can be hard to find in the Caribbean, but it is important that storage rooms for science laboratories should be well designed. When a corrosive substance is being used, we have to make sure that it does not spill onto surfaces or ourselves. The safety rules of the laboratory may tell you about the use of laboratory coats and safety goggles.

Activity | Safety with acids and alkalis
1. Describe briefly how you would mix concentrated sulphuric acid and water. Assume that you wish to mix 10 cm^3 of the acid with 200 cm^3 of water.
2. List the possible ways in which you could try to prevent the corrosion of metal surfaces by substances such as acids and alkalis.

Activity | Safety with equipment in a science laboratory
1. Make a list of at least 10 pieces of equipment which you use regularly in the laboratory.
2. For each piece of equipment, make a note of the safety precautions which you take when using it. (Remember that even a simple test-tube has to be handled correctly.)
3. Discuss these safety precautions in small groups with other students, and with your teacher.

There are other places where you should be aware of safety precautions, such as your home and workshops and rooms for Industrial Arts and Home Economics. Think about the safety rules which might be appropriate for your home.

Questions
1. Find out as much as you can about the different types of accident which can occur in science laboratories. How could these be prevented?
2. Do you think that pedestrians should know about safety rules when walking on the road? What should the basic rules be?
3. Why is it necessary to learn to drive before obtaining a licence to drive a car? Should the driving test be made more difficult?

Working safely

How can we make the laboratory a safer place?

Look at these pictures. The top one shows students working in an unsafe way, and the bottom one shows students working safely.

An unsafe science class

A safe science class

Question

1 Spot six differences between the two pictures. Write down a safety rule which could be applied to each difference.

How can we make our homes safer places?

Activity | Safety with common appliances and materials

1 Make a list of all the appliances you use at home, even if you use them only occasionally. Add to the list the appliances used by adults only. Do not forget to include items such as washing machines or lawnmowers.
2 For each appliance, indicate the energy source – e.g. mains electricity, battery, gas.
3 It is possible that the manufacturer of an appliance may have supplied some safety instructions for its use. Try to find these instructions and make a note of them. If these are not available, discuss with your parents, other students and your teacher the safety precautions you would take when using each appliance. Make a note of the safety precautions you agree on. (These may be very simple. For example, if you have an electric kettle, you need to make sure there is water in it before switching it on.) You may already have been taught about these safety precautions – and the reasons for them – by your parents.
4 Make a list of all the materials (chemicals, etc.) you use at home. These might include water, soap, bleach and lubricants (if, for example, you have a bicycle or a sewing machine).
5 For each material, identify any particular precautions you have to take when using it. Make a note of these precautions and the property of the material which makes the precaution necessary.
6 Make a note of the way in which each material is stored.

You and your parents can make sure that your home is a safe place to live, while at the same time enjoying all the advantages of modern appliances and materials.

Questions

2 Which appliances used at home have a sharp blade? What precautions should you take when using one of these?
3 Some pieces of equipment used in the garden are made of a metal such as iron. What precautions can you take to make sure these are properly cared for?
4 Many people have skin which is sensitive to chemicals such as detergents. How can such people protect their skin while washing crockery and cutlery, for example?
5 Assume that you are using an electrical appliance such as a vacuum cleaner or kettle, and that it does not work when the electricity supply is connected to it. What steps would you take to try to find out why it did not work? Do you think you should simply call an electrician to solve the problem?

Part 2

How can we be safe on trips outside the school?

You may already have been on trips (sometimes called field visits) outside the school. If so, you will know that there may be many reasons for such trips. For example, in Social Studies, it may be helpful to visit a particular historical site or a government department or ministry. In Science, studying the ecology of an area may involve going outside the school to a river or a wooded area. For these trips to be of value there has to be careful and thorough preparation of the subject matter, including worksheets for completion by participants. Another major consideration is safety, since accidents can occur all too easily on these trips.

Activity | Safety issues for field trips

1. Work in small groups. Assume that your class is to undertake a field trip on a Science topic. Use group discussion to identify a suitable topic (from the CXC Integrated Science syllabus). Make sure that your teacher agrees with the choice of topic.
2. Once there has been agreement on the topic, the group should then identify the precise purpose for the trip.
3. The group should then consider and prepare an agreed list of all the safety precautions that will need to be considered before the trip can be approved. For example:
   (a) Should the parents of each student provide formal written approval for their child's participation in the field trip? This may not seem like a safety precaution, but there might be medical reasons why a particular student should not go on a particular trip. You may also want to consider who will be legally responsible if there is an accident on the field trip as a result of which a student sustains an injury.
   (b) How many teachers should accompany 25 students?
   (c) What will be the means of transport to the site of the field trip? Who will be responsible for arranging this?
   (d) Is any special clothing required?
   (e) Will it be necessary for each student to bring their own food and water?
   (f) Should the teacher(s) carry a first aid kit?
4. The groups should make sure that they have covered as many safety considerations as possible for the journeys to and from the site, as well as while working at the site.
5. The groups can now discuss their lists with the teacher and with each other.

Questions

6. If there is to be a field trip to a site close to your school, would you recommend that students walk to the site? What safety precautions should the group take in walking to and from the site?
7. If a group of students is to travel by bus to and from a site for a field trip, what safety rules should apply while they are on the bus?

How can we work safely in a workshop?

Look at these pictures. The top one shows students working unsafely in a workshop, and the bottom one shows students working safely.

An unsafe workshop

A safe workshop

Question

8. Spot six differences between the two pictures. Write down a safety rule which could be applied in each case.

our environment | safety | working safely part 2

How can we control fires?

How can we start a fire?

On page 35 we looked at the flame of a Bunsen burner. The amount of air supplied affects the type of flame formed. You can control the air supply to the burner, but you cannot get a flame simply by turning on the gas supply: in other words, it is not enough to have a supply of fuel (gas in this case) mixed with air. The fuel/air mixture will burn only if you start the flame by using a source of ignition such as a burning match.

The flame of a Bunsen burner and the flame of a gas cooker may be thought of as **controlled fires**. In each case, you control the amount of fuel supplied in any given period of time. For the Bunsen burner, you also control the amount of air supplied (see page 191). It is not always possible to control either the amount of fuel or the amount of air supplied to a fire, and this can be very dangerous. Bush fires, which start when vegetation is very dry and can be set off by lightning, for example, may burn out without causing any problems. But if they begin to burn out of control they can be dangerous to all forms of life. You may have heard about the very dangerous bush fires which happen each year in the USA and Australia.

Activity | Fuels and fires

All fires need a fuel and a supporting atmosphere for the fuel to burn in. This might be air or oxygen, for example.
1. Make a list of as many different types of fires as you can. Identify the fuel in each case.
2. Can these fires be controlled simply by limiting the amount of fuel supplied to them?
3. Fires can be started in different ways. You may light a piece of paper with a burning match. A gas cooker may have a pilot light. Write down, for each fuel you can think of, how a fire using this fuel can be started.
4. How do bush fires start?

Your teacher will demonstrate how some fires are started, using different materials. Since fires can be dangerous, make a note of the safety precautions taken for this demonstration. Materials which could be used include crude oil, paper, methylated spirits, wood charcoal and kerosene. Some of these materials catch fire very easily, while others have to be heated strongly before they will catch fire.

Ignition temperature The temperature at which a substance catches fire, or **ignites**, is called its **ignition temperature**. Any substance with a low ignition temperature (such as many organic liquids) should be kept away from flames in the laboratory, since they ignite very easily.

This is very different from a metal such as magnesium, which has to be heated strongly before it will begin to burn. You will probably have seen your teacher heat magnesium ribbon, holding it with a pair of tongs. Once the metal begins to burn, a very intense white light is given out. The ignition temperature of solid fuels is higher than that of liquid fuels. Some substances, such as ether (dioxyethane), begin to burn at temperatures lower than their ignition temperature.

Flash point The flash point of a liquid is the lowest temperature at which the vapour of the liquid can form an explosive mixture with air. Ether has an ignition temperature of 180 °C but its flash point is –45 °C. The very poisonous substance carbon disulphide has an ignition temperature of 100 °C and a flash point of –30 °C.

Why are fires dangerous?

1. Flames may be so intense that there is danger to all forms of life.
2. Fires may also cause so much damage to buildings that these become dangerous to humans. Fires in forests and on farms may damage crops and trees. Any fire can cause pollution of the air.
3. Some substances burn with the formation of poisonous gases, and these endanger life.
4. The fuel for a fire may be used up very quickly and the oxygen in the air will be removed. People and animals in or near the fire can suffocate through lack of oxygen.

If fires start, it is important to stop them from spreading. There are two ways of doing this.
1. Cut off the supply of fuel. This can sometimes be done quite easily – by turning off a gas supply, for example.
2. Cut off the supply of air or oxygen. It is possible to do this using **fire extinguishers**, or by using very large quantities of water. (You may have seen your local fire service putting out fires by this last method.) The water also cools the area, thus helping to make sure that the temperature is kept below the ignition temperatures of potential fuels.

Questions

1. Are there such things as useful fires? List any you can think of.
2. The flash points of benzene and propanone are – 11 °C and – 18 °C respectively. What would be a good method for storing these liquids? Why?
3. How common are bush fires in your country? Try to find out if any records are kept about such fires, for example, where they occur, what caused them, and how they were brought under control.

How can we put fires out?

As with our health, **prevention** is often better than cure. In other words, it is better to try to prevent fires from starting than to try to stop them once they have started. To do this, you need to think about and implement simple safety measures.

Gas fires If gas is the fuel, the gas supply should be safe. You need to check the gas cylinder for any leaks, and ensure that the lines carrying the gas supply have no leaks. If you detect a leak in a gas line it may be necessary, for example, to replace a length of tubing.

Flammable liquids Never heat flammable liquids directly with a flame in the laboratory. Use a water bath instead. If you need to heat the liquid to a temperature above the boiling point of water, use a sand bath. Take care when using fuels such as kerosene, which is used for lighting – it is easy to knock over a kerosene lamp.

Electrical fires Electrical fires can be avoided if safety procedures are followed. You should check any electrical circuit you use in the laboratory. At home, regular checks of wiring and fuses (or circuit breakers) can be useful.

Once a fire has started, there are three things to consider urgently and possibly act on:
1. The source of **fuel** for the fire
2. The supply of air or **oxygen** to the fire
3. The **heat** needed to get the fuel to its ignition temperature.

In your school science laboratory, there may be a number of items to help you deal with a fire:
1. A fire blanket
2. A fire bucket (with sand in it)
3. A fire extinguisher.

Questions

4. What are the properties of the fire blanket and sand that make them useful?
5. What type of fire extinguisher is there in the laboratory you work in?

Fire extinguishers

1. **Water extinguishers** These act by making use of the cooling action of water. There are two types:
   (a) **Soda acid**, in which sulphuric acid reacts with a dilute solution of sodium hydrogen carbonate. The carbon dioxide formed expands and forces the solution out of the nozzle.
   (b) **Carbon dioxide**, in which a cylinder of the gas inside the extinguisher is broken and the gas's expansion forces water out of the nozzle.
2. **Foam extinguishers** These act by excluding air from the fire. The fire stops as soon as the oxygen is used up. The extinguishers contain solutions of chemicals which react together to form a foam, which is stable.
3. **Carbon dioxide extinguishers** As you will find out on page 191, carbon dioxide does not support combustion. Fire extinguishers that supply carbon dioxide (which is stored at high pressure) therefore help to exclude oxygen from a fire. The gas is released from the extinguisher by a trigger mechanism.
4. **Powder extinguishers** Powder extinguishers usually contain a very unreactive substance. For example, some contain sodium hydrogen carbonate, which will react to form the very stable sodium carbonate.
5. **Liquid extinguishers** These extinguishers contain liquids which form vapours with a high density and therefore exclude oxygen. One liquid used is bromochlorodifluoromethane (BCF).

A modern fire extinguisher that can be used for all types of fire. Where are the fire extinguishers in your school?

Questions

6. What would you do if a pan containing hot cooking oil caught fire at home?
7. Do you know the safety rules for your science laboratory? Where are they kept?
8. Would you use a water fire extinguisher on an electrical fire? Why or why not?

How do we give first aid?

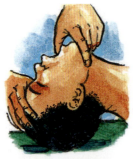

(a) Lay the patient face upwards. Tip the head well back and pull the lower jaw forwards and upwards. This will force the tongue forwards and open the air passages.

(c) Pinch the patient's nose shut. Take a deep breath, then open your mouth and seal your lips against the person's mouth. Breathe out firmly but gently into the person's mouth and so into his or her lungs.

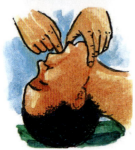

(b) Sweep around deep inside the patient's mouth with your finger to make sure that nothing is blocking the windpipe.

(d) Lift your mouth off, and look at the person's chest. If you have been successful you will see that it has risen and now is falling as air comes out of the lungs.

Repeat steps (c) and (d) until the person starts breathing for him- or herself.

Mouth-to-mouth respiration (artificial respiration)

Mouth-to-mouth resuscitation, also called **artificial respiration,** must be done within 3 minutes of a person stopping breathing. There are several reasons why a person might stop breathing, these include: heart attack (page 97), drowning (page 256), electric shock, poisoning, suffocation and choking. Get a doctor's help as soon as possible.

Bites and stings Mosquito and other insect bites Relieve the pain with calamine lotion or surgical spirit. Do not scratch. Treat any infected bites with gentian violet and cover if necessary. If reactions to poisonous bites (for example, some spiders) are severe, take the patient to a doctor.

  Stings Wasp: treat with vinegar. Bees: treat with sodium hydrogen carbonate solution. Jellyfish: scrape off stings with sand, wash with salt water.

Broken bones Move the person as little as possible. Treat any severe bleeding (see cuts). A splint can be bandaged to the side of the limb to stop it from bending. A sling can be made from a large piece of cloth. Keep the patient warm. Do not give anything to eat or drink. Get a doctor's help.

Burns and scalds Small Cool by putting in cold water or rubbing with ice. Clean the area but do not put on Vaseline. If a blister forms, leave it uncovered and do not break it. Burns can be covered with a clean dry cloth or dressing.

  Large If a person's clothing is on fire, first smother out the flames with a blanket. Cool and clean the area with cold water. Remove clothing unless it is stuck to the wound. Lightly bandage. Raise the burned part and give the person small sips of cold water to replace body fluids. Treat for shock. Get a doctor's help. Do *not*: burst blisters. Do *not* put on anything fluffy like cotton wool, or apply ointments or Vaseline.

Aims of first aid
1. To help prevent injuries from getting worse.
2. To summon a doctor or ambulance if necessary.
3. To make the patient as comfortable as possible.
4. To provide reassurance.

Warning Students wishing to give first aid are advised to follow a special course in first aid.

Choking If something (such as food or a toy) blocks the air passages the person will have trouble breathing. If the person can cough, they may cough out the object. If not, feel around in the throat to see if you can remove the object. Sit the person in a chair, leaning forward. Hit them sharply between the shoulder blades. Give artificial respiration if necessary. Get a doctor's help.

Cuts Small wounds stop bleeding as a clot forms. Wash the area and remove dirt. Clean with an antiseptic, dry and cover with a clean dressing. If the wound is deep (for example from a nail) or if it becomes inflamed, see a doctor who may give antibiotics and a tetanus injection.

  Large There may be severe bleeding. Raise the injured part of the body and pick out any pieces of glass, etc. Press on a major artery between the heart and the wound to stop blood flow. Press hard on the wound with a clean pad, and bandage tightly to stop blood flow and to prevent entry of germs. Add more bandages if blood seeps through. Get a doctor's help.

Electric shock Before touching the patient, switch off the current, or separate him or her from the source of electricity. He or she may be 'live' and you could also get an electric shock if you touch the person. Use a wooden broom (or other non-conducting object) to disconnect the source of electricity. Give artificial respiration if necessary and lay the person on his or her side in a comfortable position. Get a doctor's help.

Eye injuries *Chemicals* Wash the eye in running water. Cover with a pad and see a doctor.

Small objects Use the corner of a clean cloth or paper tissue to remove the object. If necessary, pull the eyelids down and then up to dislodge the object.

Fainting Patient feels dizzy and becomes unconscious. Caused by insufficient blood flow to the head. Fainting can be brought on by an emotional scare, as when seeing blood at an accident. Sit the patient down with head between the knees, or lie them down and lift their feet. Blood gets back to the head. People in a faint should regain consciousness within a few minutes.

Heatstroke Follows exposure to high temperatures. The body's heat regulating mechanisms (page 119) are upset so the body temperature rises from the normal 37 to 40 °C or higher. The person looks flushed, with hot, dry skin and rapid pulse and may become unconscious.

Remove the person's clothing and sponge them with cold water or wrap in a cold wet sheet. Fan them. Lay them on their side in a comfortable position. Get a doctor's help.

Heat exhaustion Less serious than heatstroke. Person has a damp skin and may feel sick and dizzy with headache and muscle cramps. Lay patient down, with feet raised. Loosen clothing and give salty water to drink (one teaspoon of salt in a litre of water).

Poisoning Common household chemicals taken accidentally by children or others can be harmful, such as bleach, aspirin, lavatory cleaner, washing-up liquid, weed killer, insecticide, rat pellets, paint thinner, and kerosene.

Follow any instructions on the container from which the poison came. Otherwise do not try and make the person vomit. Wash around the mouth but do not give water to drink. Give artificial respiration if necessary. Lay them on their side in a comfortable position. Get a doctor's help.

Shock Shock can follow a severe burn or blood loss. Person is pale, faint and sweating, with a cold skin. Lay them on their back, head low and legs raised. Loosen tight clothing and wrap in a coat or blanket. Do not give anything to eat or drink. Get a doctor's help.

Sprains Joint is pulled unnaturally. Common in the ankle or wrist due to falls in the home or on the playing field. The joint will swell and be painful and bruised. Sponge with cold water. Bind the joint in its proper position with light bandaging. Rest it for as long as necessary.

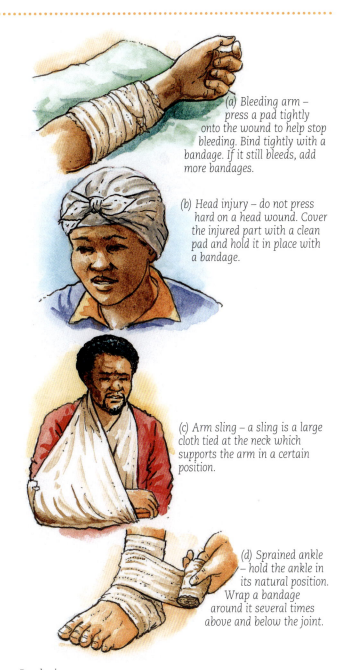

(a) Bleeding arm – press a pad tightly onto the wound to help stop bleeding. Bind tightly with a bandage. If it still bleeds, add more bandages.

(b) Head injury – do not press hard on a head wound. Cover the injured part with a clean pad and hold it in place with a bandage.

(c) Arm sling – a sling is a large cloth tied at the neck which supports the arm in a certain position.

(d) Sprained ankle – hold the ankle in its natural position. Wrap a bandage around it several times above and below the joint.

Bandaging

Questions

1. What principles of breathing (page 98) are used in mouth-to-mouth resuscitation?
2. Give reasons for (evaluate) the first aid methods for the treatment of: (a) burns (b) cuts (c) electric shock (d) sprains.
3. List three circumstances in which bandages could be used. Why are they needed in each case?

Which substances conduct electricity?

Activity | How do you test for conduction of electricity?

1. Set up a circuit as shown in the picture below.

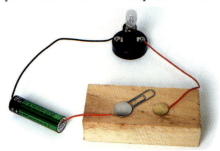

2. The apparatus in the picture may be represented using standard symbols like this. Which symbols represent which parts of the circuit?

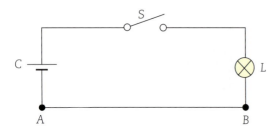

Circuit diagram – AB is a removable length of wire, C is a 1.5 V cell, L is a light bulb, S is a switch

3. Assuming that all parts of the circuit are working, and that there is good electrical contact, what happens when the switch is in the ON position? If the bulb lights up, what does this tell you about the wire used in the circuit?
4. Remove the length of wire AB. You now have an open circuit. Attempt to close the circuit by inserting other objects and materials between the points A and B. Try string, a ruler, a nail, a piece of tin foil, erasers and pens.
5. Make a table like the one below and record your results.

Test object	Material of object being used	Does the bulb light up?		Conclusions	
		Yes	No	Good conductor	Bad conductor
Nail	Iron				
Tin foil	Aluminium				
Pen					
Eraser					

6. The diagram below shows how liquids can be tested for the conduction of electricity. Try a few liquids such as oil, methylated spirits, sea water (salt water) and tap water to see if they can conduct electricity. Does it matter if the wires dipped into the test liquid are moved further away or nearer to each other?
   Caution: Do not let the wires touch each other.
7. Repeat the test on salt water using a 4.5 V or 6 V battery. Add all your results to the table.

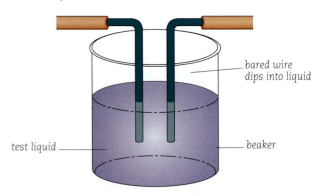

Use this apparatus in your circuit to test if liquids conduct electricity. Do not let the wires touch each other

Your table shows which materials, for example iron, permitted electricity to flow in the circuit and made the bulb light up. These are **good conductors of electricity**.

Your table also shows the materials that did not make the light bulb glow. These are called **non-conductors** or at best poor conductors of electricity. The poor conductors include rubber, plastic and wood.

These materials were shown to be poor conductors of heat on page 114, and are useful as such. As poor conductors of electricity, they are also very useful. Plastic and rubber are used for covering wires to isolate them electrically.

Do liquids conduct electricity?

Mercury is a liquid metal, and it is a very good conductor of electricity. But what about tap water?

Some substances, in particular water and some solutions in water, appear to be very poor conductors of electricity when tested with a low voltage. But at a high voltage, tap water (which usually contains a tiny amount of dissolved salts) becomes such a good conductor of electricity that it can be very dangerous. Therefore, electrical switches and points (sockets) are not usually found in bathrooms. Switches and sockets are also kept well away from the splash area of taps. Look around your own home and see where the light switches are placed. If there is one in the bathroom, what precautions are there to prevent electrical contact with your hand?

Electric shock

If an electric current flows through the human body it can cause muscles to seize up, and the heart may stop, which could be fatal. This happens if good electrical contact is made between parts of the human body and a power source. Wet hands used on switches and on electrical appliances can provide the fatal contact. People can also get electric shocks in a thunderstorm, especially if there is good electrical contact with the ground due to bare feet or leather-soled shoes (which absorb water). Sheltering under trees in a thunderstorm is dangerous because the trees are tall, and attract and conduct high voltage electricity.

First aid to be given in the event of electric shock is described on page 170.

What makes electricity 'flow'?

Electricity was known to the Greeks in about 600 BC. In fact, the word electricity is derived from the Greek **elektron**, which means *amber*. Thales, a Greek philosopher, noticed that when amber (a waxy substance derived from plants) was rubbed with a piece of woollen cloth, the amber would attract small bits of dust, feathers or straws. The amber became electrified.

Current electricity The electricity that we find most useful is the type that travels along conductors. This is called **current electricity**. It can be produced from chemical cells or by mechanical means involving the use of magnetic forces, for example a bicycle dynamo (see page 186).

Static electricity Many substances can be electrified by rubbing them. Electric charge produced by this means is called **static electricity**. **Static** means *stationary* or *at rest*. Lightning is a result of static electricity being discharged between two charged clouds or between a cloud and the Earth. (The charges are thought to be produced by friction between water droplets and warm air.)

If you are in a motor vehicle in a thunderstorm, you are fairly safe because the rubber tyres prevent electric current flow to the ground. However, for the same reason the motor vehicle, with its many moving parts, can build up high electrical charges of its own. While these charges may not be a danger to you, sparks might be produced that could ignite flammable vapours such as gasoline. As a precaution, gasoline trucks have an earth chain suspended from the chassis to carry any sparks of electricity to the ground as shown in the photograph.

Why are metals such good conductors?

Electricity is defined as being a *flow of charged particles called electrons*. Electrons are the negatively charged parts of atoms

An earth chain hanging from the back of a tanker

and molecules (page 19) that orbit around the positively charged nucleus. The flow of electricity depends on how freely the electrons can move. In metals, electrons are very free while in non-metals such as plastic, they are not. An exception is graphite (pencil lead). Graphite, a form of **carbon**, is a non-metal but it is a very good conductor.

What are semi-conductors?

In the early 1900s, certain substances were discovered which are on the borderline between conductors and non-conductors. These substances are called **semi-conductors**. They have a small number of free electrons in them at normal temperatures. This number may be increased or decreased by the addition of impurities. Semi-conductors are used in solar cells, transistors and computer chips. (See pages 29 and 197.)

Questions

1. Describe how you would distinguish between a good conductor and a poor conductor of electricity.
2. Explain briefly why some substances may be better conductors than others.
3. Discuss the safety precautions that should be taken when using electricity in the home.
4. Explain what an earth chain is and why it is carried by some motor vehicles.

What are the characteristics of electricity?

Activity | How does a flashlight work?
1. Take apart a flashlight (torch) to see how it works. Observe that it has a light bulb, on/off switch and two or more cells (incorrectly called batteries). See how they are positioned for the torch to work.
2. Remove one cell. Can the bulb light up with just one cell?
3. Replace the cell but place it the opposite way round. Can the bulb light up with the cells arranged in this way?

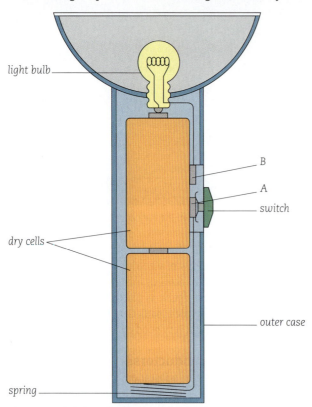

When the switch is pushed up, A connects with B, completing the circuit so that the bulb lights up

The electrochemical cell
Electricity is a flow of electrons (page 173) and it can be obtained from different sources. The **electrochemical cell** is one source. If the cells in the torch are arranged incorrectly, the torch does not work. This suggests that the electrons must flow in a specific direction through the cells. Look at one cell. It has two different connecting points, a **negative** and a **positive** pole. When the cells are arranged so that the negative pole of one cell is in contact with the positive pole of the other cell, the torch lights up. When two or more cells are connected in this way, they are said to be **in series**.

If you look at different sizes and types of cells, you will see that most are marked 1.5 V. This stands for 1.5 volts and is the usual voltage for the most common cells (see page 186).

What is a volt?
The volt is a measurement of how hard electricity is pushed around a circuit. It is like the pressure resulting from the height of a liquid in a tall tank. The size of the tank does not affect the pressure; only the height of water above the outlet does. The volt is also called the **electromotive force** (e.m.f.), and can be measured by a voltmeter.

Activity | How do we measure voltage?
1. Connect up a circuit like this.

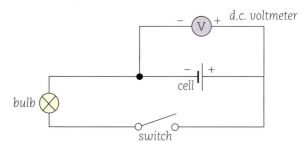

Use a torch bulb, a 1.5 V cell, and a voltmeter. The type of electricity produced by the cell is called direct current (d.c.) electricity because it flows in one direction only. Therefore, we use a d.c. voltmeter. The range of the voltmeter should be 0–3 or 0–5 volts. The switch is optional.
2. Record the reading on the voltmeter when the bulb is disconnected and when it is connected.
3. Repeat the experiment, but this time use two cells connected in series, part (a) below. Is the bulb brighter? What is the reading on the voltmeter? What is the reading on the voltmeter with the bulb disconnected?
4. Now, reconnect the cells so that the positive pole of one is connected to the positive pole of the other. This arrangement can be seen in part (b) below. This arrangement of cells is called a **parallel** arrangement.
5. Again, measure and record the voltage readings when the bulb is lit and unlit. Record anything else you notice.

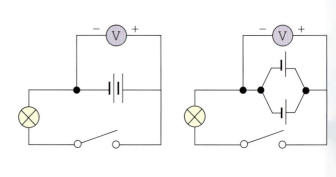

(a) Cells in series   (b) Cells in parallel

> **Question**
>
> 1 Copy the two sentences below and complete them in your notebook.
> When two cells are connected in series the voltage is ...
> When two cells are connected in parallel the voltage is ...

How do cells differ?
Many advertisers of cells do not tell you about volts; they are more concerned with how long the cells last, in other words how much electrical energy cells can give out altogether. To measure how much electrical energy can be obtained from a cell, we need to measure the electrons being produced by the cell. The instrument used is called an **ammeter** and the flow of electrons, which is called the **electric current**, is measured in **amperes**, amps for short. To find out the total quantity of electricity produced by a cell you must also know the length of time the current is flowing.

Quantity of electricity (Q) = electric current (I) × time (t)
or $Q = It$

Activity | How do we use an ammeter?
1 Set up the four circuits (a) to (d) shown below (in turn). Use 1.5 V zinc-carbon cells, a 0–3 or 0–5 voltmeter, and a 0–5 ammeter. A d.c. ammeter must be used. Like the d.c. voltmeter, it has negative and positive poles that must be positioned correctly, as in the diagrams.

(a) 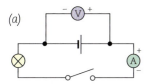   (b)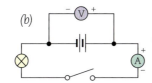

(c) 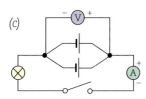   (d)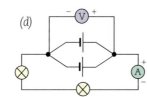

In order to make a good comparison between circuits, use new cells of the same make and size. Note that circuits (c) and (d) have two cells connected in parallel. Notice also the way the ammeter is connected up in an electrical circuit. The voltmeter is connected in parallel while the ammeter is connected in series.

2 For each circuit, record your observations of voltage, current and the brightness of the bulb in a suitable table.

	Voltage (volts)	Current (amps)	Brightness
(a) 1 cell, 1 bulb			
(b) 2 cells (series), 1 bulb			
(c) 2 cells (parallel), 1 bulb			
(d) 2 cells (parallel), 2 bulbs (series)			

3 Discuss your results in class. Can you account for the voltmeter and ammeter readings in each circuit?

Activity | How powerful is a lamp?
1 Set up circuit (b) from the previous Activity. Exchange the lamp for three or four others in turn. Make a note of the brightness as well as the current and voltage readings.
2 Torch lamps vary in quality since they are not manufactured to specific standards. Some appear to be more powerful and brighter than others. Did you notice the brightness change with the instrument readings, particularly the ammeter readings?
3 If you have a sensitive ammeter, try to calculate the exact power of the lamps as they perform in the circuit. This is done by multiplying the voltage by the current.

Power (watts) = current (amps) × voltage (volts)
$W = IV$

Generally, the higher the wattage of a bulb, the brighter it is. Look at the wattage of some light bulbs in your home. You will find that the brighter bulbs have high watt ratings (75 to 100 watts), while dimmer bulbs have lower ratings (15 to 40 watts). The voltage determines whether or not a light bulb can be used safely. Torch and other low voltage bulbs cannot be used safely with the mains as they only take about 3 to 6 V, but the mains supply voltage is either 110 to 120 V or 220 to 240 V, depending on the country you are in.

> **Questions**
>
> 2 How would you prove or disprove the claim that a particular cell is more powerful than another?
> 3 How would you get a 9 V supply from some 1.5 V cells?
> 4 Draw a circuit which is powered by three cells in parallel.
> 5 Draw a circuit showing an ammeter and a voltmeter correctly connected in a circuit.
> 6 How can lamps of different power ratings be distinguished?

Is resistance important?

Why does wattage vary?

You have probably concluded that something about the manufacturing process makes light bulbs different. But what specifically is the reason for the difference?

In a previous Activity (page 175), you observed that light bulbs of different brightnesses have different currents and wattages. The brighter lamps have a higher wattage and the dimmer lamps have a lower wattage. Look at your results. Draw a table like this and complete the final column.

Lamp number	Brightness rating	Voltage, V (V)	Current, I (A)	V x I (W)	Resistance V/I

When you calculate the ratio of voltage/current (V/I) the answer is called the **resistance**, R, and it is measured in ohms, with the symbol Ω. Resistance is a measure of the opposition to the passage of electricity through a conductor. All materials have a resistance at normal temperatures, even the best conductors. Extremely poor conductors, for example plastic and rubber, have resistances which are measured in millions of ohms (mega ohms). The passage of electricity through the filament of a lamp makes it glow (see page 175). Lamps which glow brightly have a lower filament resistance than lamps which are dim. You need to learn the formula:

$$\text{Resistance} = \frac{\text{Voltage}}{\text{Current}} \qquad R = \frac{V}{I}$$

If you know the values of any two parts in the formula, the third can be calculated. Remember that voltage is measured in volts (V) and current is measured in amps (A).

Example A light bulb operates at 120 V and a current of 0.5 A. What is the resistance of the filament? *Answer:*

$$R = \frac{V}{I} \qquad V = 120\,V, I = 0.5\,A \qquad R = \frac{120\,V}{0.5\,A} = 240\,\Omega$$

Example (a) What is the power in watts of the bulb in the previous example? (b) What is the resistance of a bulb with twice this power output? *Answer:*
(a) Power = Voltage × Current (Watts = volts × amps)
    = 120 × 0.5 = 60 W
(b) A lamp with twice the power output is 60 × 2 = 120 W.
    To find the resistance we must first find the current.

$$\text{Current} = \frac{\text{Power}}{\text{Voltage}} = \frac{120}{120} = 1\,A$$

$$\text{Resistance} = \frac{\text{Voltage}}{\text{Current}} = \frac{120}{1} = 120\,\Omega$$

Activity | Current, voltage and resistance

1. Make a coil by winding 30 cm of wire around a pencil. Place the coil in a circuit like the one below.

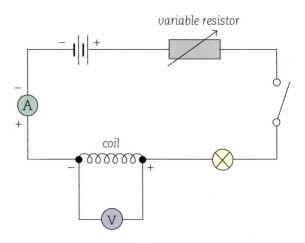

2. Measure the current passing through the circuit and the voltage across the coil at the same time. Slowly alter the current by adjusting the variable resistor. Take several readings of current and voltage. Record your results using a table like this.

Voltage (V)	Current (I)

3. Plot a graph of potential difference (voltage) against current. Use the x-axis for current and the y-axis for voltage.
4. Your results will produce a straight line (if there is a slight curve, this may be because the resistor has got warm and distorted the resistance reading slightly).

Ohm's law The straight line from the last Activity illustrates Ohm's law, which states that the *potential difference (voltage) in a circuit is proportional to the current*.

How are current and voltage distributed in a circuit?

Example Consider this circuit diagram.

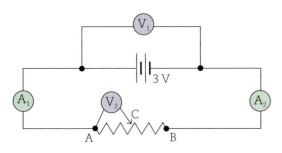

AB is a length of resistance wire 100 cm long, and C is a moveable point along AB. Note the two voltmeters V_1, V_2 and two ammeters A_1, A_2 in the circuit.
(a) What is the voltage reading on V_1?
(b) What is the voltage reading on V_2 when C is at the 50 cm mark?
(c) What is the reading on V_2 when C is at the 100 cm mark, i.e. when AC = AB?

Answer
(a) V_1 = 3 V
(b) If AB is a uniform length of wire, and if C is at the 50 cm mark, then V_2 reads 1.5 V.
(c) If AC = 100 cm, V_2 reads 3 V. So the voltage on V_2 is proportional to the distance of AC along AB. The 100 cm length of resistance wire may be thought of as one hundred resistors of 1 cm each. Resistors joined in series have a total resistance of the sum of the individual resistors.

Example In the next circuit (below), the resistors are R_1 and R_2. R_2 has a resistance of twice R_1. If the voltage in the circuit (V_1) is 3 V what is the voltage across R_1 and R_2?
Answer The voltages across R_1 and R_2 are in proportion to this resistance. The voltage across $R_1 + R_2$ = 3 V and since the ratio of resistances of R_2 to R_1 is 2 : 1 then the voltage ratio is also 2 : 1. This will be shown by the voltmeters. V_2 should read 1 volt and V_3 should read 2 volts.

Example R_1 and R_2 are the same resistors as in the previous example. Consider the current in this circuit.

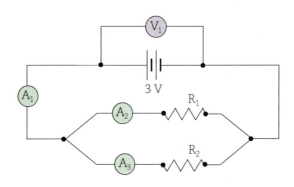

(a) If the supply voltage is 3 V, the voltage across R_1 is 6 Ω and across R_2 is 12 Ω, find the currents measured by A_2, A_3 and A_1 respectively.
(b) What is the total resistance in the circuit?

Answer
(a) Can you calculate the current on A_2 using this formula?

$$I = \frac{V}{R}$$

In this case, V = 3 V and R_1 = 6 Ω. What does A_2 read? For A_3, V = 3 V, R_2 = 12 Ω. What is the current recorded on A_3?
(b) The total current in the circuit is A_1, which is $A_2 + A_3$, so you should now be able to find this.

Can you use the formula to find the total resistance from the total current, A_1, and the voltage, 3 V? You will find that the total resistance is less than the sum of the individual resistances. This is because the resistors are connected in parallel, not in series.

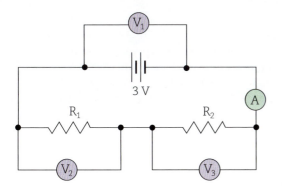

Questions

1. Draw a circuit showing three lamps connected in series. If one lamp is removed, how are the other lamps affected?
2. Draw a circuit showing three lamps connected in parallel. If one lamp is removed, how are the other lamps affected?
3. Which circuit, series or parallel, is best suited for household lighting? Give your reasons.
4. When two resistors are connected in parallel they give a total resistance that is less than either of the two resistors used. Explain why.

How much does electricity cost?

Activity | Can you understand an electricity bill?
1. Examine an electricity bill, preferably from your own home. What is the amount to be paid? Do you know how the figure is arrived at?
2. Is the total cost related to the amount of electricity used? Are there columns on the bill that give meter readings? Do the meter readings produce the figure in the kWh (units) column? What was the time interval between meter readings?
3. Compare figures in the kWh column on other bills for the same time interval. Give the amount on the bill related to the kWh figure.

If you compare figures with your classmates, you will see that the total charge relates to the figure in the kWh column. However, there is an extra fixed amount called a standing charge, which everyone pays.

A typical electricity bill from Barbados. How does it compare with yours?

What is a kilowatt hour?
The letters **kWh** stand for **kilowatt-hour**. The watt is a unit of power (see page 175). A kilowatt is one thousand watts. A kilowatt-hour represents a rate of working (power) of one kilowatt for one hour. This is a convenient quantity of electricity commercially, and is known as a **unit**. Apart from the standing charge, all electricity charges are usually based on the number of units, or kWh of electricity consumed.

Due to the increasing cost of fuel from which electricity is produced at power stations (see page 21), the flat rate charges per unit of electricity and other additional charges are often adjusted. For up to date information on electricity charges in your country you should ask the electricity company in your country or government information service. For example, in Trinidad and Tobago there is a stepwise scale in use for electricity bills.

Example What is the cost of 50 units of electricity at a flat rate of 25 cents per unit?

Answer $50 \times \frac{25}{100} = \$12.50$

How is electricity consumption measured?
As electricity passes through the wires in electricity meters, it produces a magnetic effect. This magnetic effect spins a disc. In some meters, this spinning disc is connected to dials on which the passage of electricity is recorded. In the more modern meters, digital meters, the disc gives a direct read-out.

A dial electricity meter

The meter itself uses up a very small amount of electricity. The speed at which the disc spins depends on the amount of current flowing in the wire and hence the quantity of electricity used. Look at an electricity meter at home or at school. The faster the disc spins, the more electricity is being used.

Can you read an electricity meter?

The direct reading or digital meter Look at the numbers displayed on the meter. Electricity consumption is found by reading and recording the figures over a certain time period such as a day or a week. The first reading is taken at the start of the period and the second reading is taken at the end. Subtract the first reading from the second to find the number of units used. The last figure on the right may be a part of a unit.

The dial or analog meter Can you read the numbers indicated by the dials in this picture?

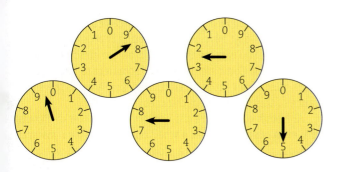

The answer is 9 872.5 kWh. Notice how the numbering is arranged. The dial on the far left is numbered in a clockwise direction, the next one anti-clockwise, the next one clockwise, and so on. Can you explain why this is so? Think about how a row of gears works.

Moving from left to right, each pointer spins 10 times more slowly than its neighbour on the immediate right. In reading the number given on the dial it is important to note in which direction the pointer moves. The number indicated by the pointer is the number to the left of the pointer if the movement is clockwise, and to the right of the pointer if the movement is anti-clockwise.

Activity | Did you understand?

Draw a row of five dials like those above. Think of a number with five digits and pencil in the pointers appropriately to indicate the number you have just thought of. Ask your teacher to check your answer.

Which appliances use the most energy?

Examine the table showing electrical equipment used in the home and some of their specifications. Which appliances have the highest wattage? Which have the lowest wattage? Which get the longest continuous use? The greatest energy consumption over a period of two months may be caused by a machine of moderate wattage, which is being used continuously. What might that be?

Appliance	Typical wattage	Assumed usage per week	Est. kWh per month	Est. cost per month
Air conditioner	1 300 W	28 h	156	
	750 W		90	
Blender	290 W	45 min	1	
Broiler	1 375 W	40 min	28	
Coffee maker	850 W	6 h	21	
Fan	85 W	28 h	10	
Floor polisher	325 W	$2\frac{1}{2}$ h	3	
Freezer				
0.25 m^3	297 W	all day,	200	
0.40 m^3	385 W	every day	270	
0.65 m^3	495 W		360	
Hair dryer	1 250 W	1 h	3	
Hot plate	1 500 W	1 h	6	
Clothes iron	750 W		$7\frac{1}{2}$	
	or	$2\frac{1}{2}$ h	$7\frac{1}{2}$	
	250 W		12	
Light	60 W	28 h	7	
Amplifier/ record player	35 W	10 h	1.4	
Refrigerator				
0.25 m^3	100 W	all day,	80	
0.60 m^3 frost free	425 W	every day	220	
Electric stove				
Small ring	1 250 W		70	
Large ring	2 500 W	14 h	140	
Oven	3 800 W		210	
Television (19–23")	80–190 W	28 h	10–25	
Toaster	1 140 W	$1\frac{3}{4}$ h	7	
Washing machine	280–375 W	$1\frac{1}{2}$ h	3	
Water heater	1 600 W	21 h	240	
	4 500 W			
Vacuum cleaner	540 W	1 h	2	

Questions

1. What is the unit used for measuring power?
2. What is the unit used for measuring energy?
3. How does the unit of energy relate to the unit of power?
4. What is the power of a machine which draws a current of 5 amps at 120 volts? How much energy does it consume if it operates continuously for 10 hours?
5. Assuming a cost of 30 cents per unit for energy, calculate the estimated cost per month of the household appliances in the table above.

How do we use electricity safely?

Activity | How is a fuse useful?

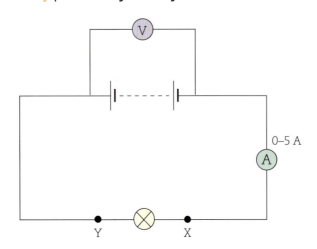

1. Set up this circuit. The voltage supply can be increased during the experiment by adding fresh 1.5 V cells. The voltmeter is optional since the voltage is known from counting the cells. A robust ammeter must be used, range 0 to 5 A. The bulb is a low voltage 1.5 to 2.4 V torch bulb.
2. Starting with the lowest voltage, 1.5 V, allow the bulb to burn for 1 minute and record the voltage and current. Increase the voltage by adding cells and again record the current.
3. Repeat the procedure until the bulb 'blows' and record the results in a table.
4. What were the current and voltage specifications of the bulb? Were these values exceeded? Imagine that the bulb was an essential piece of electrical or electronic equipment. What would have happened to it?
5. Disconnect the bulb at points XY and instead insert a strand of steel wool (may have to be held by crocodile clips). What happens to the steel wool? Why does the steel wool burn before other parts of the circuit?

What is the principle of the fuse?

A low voltage torch bulb is designed to operate efficiently at a low power. If the low power rating is exceeded, the filament overheats and breaks. This breaks the circuit, hence switching off current flow. Similarly, the steel wool will permit a certain amount of current to pass before becoming hot enough to glow and **burn out** as it reacts with the oxygen in the air. Pieces of very thin wire can be placed in the circuit with similar results to the steel wool. Thus the steel wool or thin wire are 'weak points' in the circuit, and are the first to give out when the current exceeds a certain value. So, if it is connected in series, it could be used to protect some delicate or expensive equipment.

This 'weak point' is called a **fuse**. Professionally made fuses are accurate in that they 'blow' at a fairly precise current level. They are safe to use because they are either made of low melting point metal which quickly melts, giving out only a little heat, or they are enclosed in a glass tube isolated from anything which could burn.

Fuses like this are easy to replace if they burn out. However, before replacing a fuse you must make sure that the fault which made the fuse burn out, has been traced and fixed. Never use a piece of ordinary wire or tin foil (aluminium foil) as a fuse because these may be unable to protect the circuit against overloading.

How can house wiring be overloaded?

Can the wiring which carries current to household appliances become overloaded? If so, what would happen to the wires?

Many house fires have been caused by overloading of house wiring. Modern houses have concealed wiring but in old buildings that have been re-wired, it is possible to trace the wires. If you can see the wires in an old building, notice the thickness. Compare those which connect to lamps, and those which connect to an electric cooker. What is the significance of this difference in thickness?

Although copper wire is a very good conductor of electricity, even copper wire has a small but measurable resistance. This can sometimes cause a problem when a high wattage appliance is unwisely connected to a light socket or to a too thin extension cord. The resistance to current flow can use up a high proportion of the electrical energy, resulting in the overheating of the wire.

An electrician working on house wiring

Part 1

Type of appliance	Wattage (W)	Operating voltage (V)	Required current (I)	Resistance of appliance (Ω)	Resistance of flex (W)	Power loss* in flex (W)	Current capability of flex
Light bulb	60	120	$I = \frac{W}{V}$ $= \frac{60}{120}$ $= 0.5$	$R = \frac{V}{I}$ $= \frac{120}{0.5}$ $= 240$	4	$\frac{4}{244} \times 100$ $= 1.6\%$ $\approx 1\,W$	2
Electric iron	720	120	$I = \frac{720}{120}$ $= 6$	$R = \frac{120}{6}$ $= 20$	4	$4/24 \times 100$ $= 17\%$ $\approx 100\,W$	2

*The power loss in the flex is worked out as $\frac{\text{resistance of flex}}{\text{total resistance}} \times 100$

Activity | Electrical appliances

1. Examine the information supplied in this table. Two circuits are described, each using a length of thin flex with a resistance of 4 Ω.
2. In which circuit is the current limit exceeded? What are the power losses by the flex in each circuit? What form do you think this energy would take? It is very unlikely that the flex would safely get rid of over a hundred watts of power. For this reason, the manufacturer places a limit on the amount of current that the flex should carry. For the electric iron's circuit a much thicker wire is required than for the light bulb's circuit.

House wiring To guard against situations where the house wiring could be damaged, modern house wiring systems are adequately protected by fuses and/or trip switches. Fuses limit the current to those values for which the particular circuits were designed. Trip switches are spring-loaded devices which respond to the heating effect and the magnetic effect of electricity. Look at the house wiring circuit diagram below. The lighting circuit is protected by a 5 A fuse (note the symbol for a fuse). Should an appliance requiring more than 5 A be connected to this circuit, then the fuse will 'blow' and the wires will be protected from overheating and damage. On the other hand, the main ring circuit has much thicker and stronger wires to meet the current demands of electric cookers, electric irons, or washing machines.

In many modern household circuits, fuses have been replaced by trip switches, which flip off when the current exceeds a certain value. These switches are easily reset.

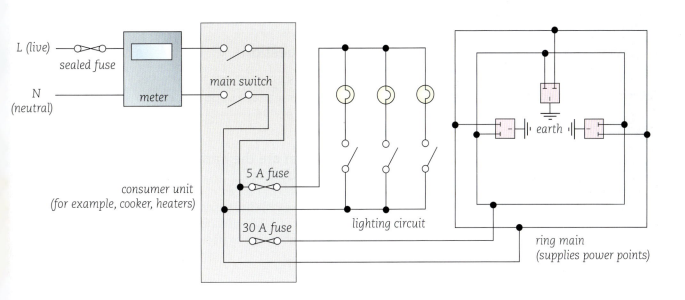

our environment | electricity | how do we use electricity safely? part 1

How do we use electricity safely?

How is the user protected from electricity?
The dangers of electric shock have already been mentioned (page 173). A voltage as low as 60 volts has caused death. The dangers of electricity must not be underestimated.

Most household electricity supplies in the Caribbean operate at 120 V: twice the possible fatal voltage. In addition, supplies of 220 V and 240 V are common.

Many appliances, for example electric irons, cookers, or kettles, have metal parts or metal covers that could easily come into contact with the current-carrying or 'live' wire. The appliance would continue to function normally but would deliver a dangerous shock when the metal parts were touched.

Wiring plugs This problem of 'live' metal appliances is solved by using three-pin plugs, which include a pin for a third wire which is connected to the metal frame of the appliance and then to earth. If the live wire accidentally touches the metal cover a large current will flow to the earth terminal. This blows the fuse and breaks the circuit. In the diagram on page 181, the **ring main** in a house is shown with the earth connection. Appliances that use a lot of electricity should always be fitted with a three-pin type of plug fitted with an earth pin and possibly a fuse. Whether the square- or round-pin type of plug is used, there is a colour code for connecting wires into the plug, which must be strictly followed.

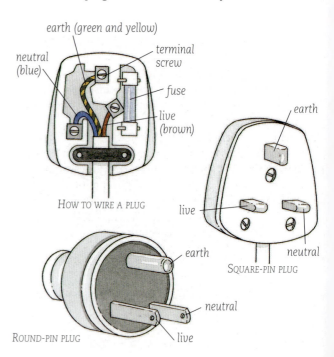

Two common household appliances. What safety precautions should you take when using them?

Question

1. Find out the correct way to wire up (a) a square-pin plug and (b) a round-pin plug. What colour wire should go to each pin?

What are the hazards of household appliances?
Look back at the list of household appliances on page 179. Make a table like the one at the top of the next page and enter each type of equipment in the correct column. Some appliances can be placed in more than one column.

Part 2

Name of appliance	Electrical requirements		Functions performed by appliance			
	Heavy user > 250 W	High voltage > 200 V	Light	Heat	Turning power	Audio/Visual
Washing machine	✓			✓	✓	
Table lamp			✓			
Television						✓

Heavy users of electricity These need to have **thick wires** connecting to the electricity supply (discussed earlier). They should also be **earthed** and protected by **fuses**. Fuses can prevent extensive damage to equipment when a mains fault develops.

Transformers Some household appliances in use in the Caribbean have been made for the European market. These tend to have a higher operating and therefore more dangerous voltage (240 V rather than the Caribbean 120 V). The voltage has to be increased by using a transformer for proper operation. There must be a match between appliance and transformer so that the correct amount of power is supplied to the appliance. Too little power or too much power could cause damage to the appliance or the transformer.

High voltage appliances The dangers of high voltage have already been mentioned and it is very important to be extremely careful when handling high voltage appliances.

Light bulbs Electric light bulbs are moderate to low users of electricity, but occasionally they present problems because they may overheat. Bulb wattages of 100 W or more produce a lot of heat. So, you must be careful when choosing lampshades that may melt and catch fire if used with too hot a bulb. Recommendations for the use of some plastic lampshades are given when they are sold, for example, to be used with a 75 W bulb (maximum), and these guidelines should be strictly followed.

Electric hotplates The rings of an electric cooker can become orange-yellow hot, i.e. more than 1 000 °C. Paper, oil, or cooking fat will burst into flames when in contact with such a hot surface. When switched off a hotplate can retain heat for a long time and, unlike the gas fire, can continue to cause burns when touched.

Motors Many appliances have electric motors. Electricity is needed to supply the necessary turning power for these motors. Once they are switched on, the motors should spin. If a motor gets stuck, the wiring may become overloaded and burn out. Prompt action may, however, save an expensive machine from severe damage.

If the motor in a food mixer, vacuum cleaner, or washing machine becomes stuck, switch it off immediately and try to find the problem before switching it on again.

Sound systems High levels of sound can cause deafness and other problems (see page 267). Many people have been persuaded to buy 50-watt hi-fi sound systems where 10 watts would be satisfactory. Even the personal cassette player is being abused. Because of the closeness of the earphones, very high sound levels can be produced which may affect your hearing.

Electric lighting The effect of high light intensity is discussed on page 133. In recent years special effects lighting has become popular. This often involves the prolonged use of ultraviolet light, which can cause damage to the delicate cells in the eyes (retina). This is especially so when the general lighting level is low and the pupils are wide open (see page 280). Flashing lights, which correspond to electrical rhythms in the brain, can cause mental disturbances, particularly in people who are prone to epilepsy.

Fluorescent lamps contain a mixture of different substances, which enable white light to be produced (see page 133). Extremely poisonous beryllium compounds are usually present in the mixtures. You should be very careful if you handle broken fluorescent tubes.

Television sets Television sets contain transformers which produce very high voltages (for example, 5 kV), for the cathode-ray tube to operate (see page 285). Tampering with the internal mechanism of the TV set can be extremely dangerous. Sitting too close to a TV set for too long can cause eye strain and may expose the viewer to radiation which may damage the delicate cells of the eye.

Questions

2. What is the principle of the fuse? How is it used?
3. Discuss the need for an earth wire as a safety precaution.
4. How does the size of conducting wire relate to the electrical power it can carry safely?
5. Discuss the correct use of an electrical appliance you have learnt to operate, mentioning any hazards associated with its use.

Magnets and electromagnets

According to legend, a man accidentally stumbled upon bits of dense rock, which strangely enough attracted each other. This type of rock was called **lodestone**. For a long time lodestone would have been treated as an object of curiosity with no important application. Lodestone is a type of iron ore called **magnetite**.

When humans learned how to separate iron from its ore, they found that lodestone was a particularly rich source of iron and that the iron held the property of attraction, which was eventually called magnetism.

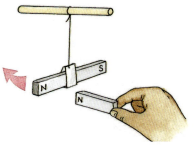

(a) *Repulsion between like poles (two north poles or two south poles)*

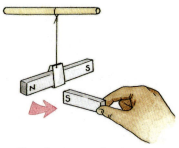

(b) *Attraction between unlike poles (one south pole and one north pole)*

What important uses could you find for magnets?

You may be surprised to learn that the use of magnets led to a great advancement in travel and exploration. Someone discovered that a long thin piece of suspended lodestone moving freely eventually came to rest pointing in a north–south direction.

Activity | How would you investigate the properties of a bar magnet?

1. When suspended and moving freely, why does a bar magnet always come to rest pointing north–south? Does the same end always point north? How does the bar magnet behave when another magnet is brought near it?
2. What kinds of objects are attracted to magnets?
3. Are all metallic objects attracted to magnets?
4. Are there different types of magnets?

You will discover that an ordinary iron nail though it is readily attracted to a magnet, does not itself become a magnet. This is because of the type of iron the nail is made from. Nails are made from 'soft' iron, which is a relatively pure form of iron. However, steel needles or masonry nails, used for penetrating concrete blocks, will retain some magnetism after they have been in contact with a magnet. Soft iron is used to make temporary magnets of the type usually found in electromagnets (see below). Cast iron, which is a very hard, impure and brittle form of iron, and various alloys of iron, called **steel**, are used to make permanent magnets. These are magnets which retain their magnetism.

There is a special place for iron in the study of magnetism. The Earth has a core of iron: that is why the Earth itself behaves like a big magnet with magnetic poles in the north and in the south. These magnetic north and south poles are near the geographic north and south poles respectively. When a suspended magnet or compass needle comes to rest, they line up along the **magnetic lines of force**, which run from the magnetic north pole to the magnetic south pole.

Activity | Can magnetic lines of force be shown with a bar magnet?

Perform the following experiment following the sequence shown in the diagram.

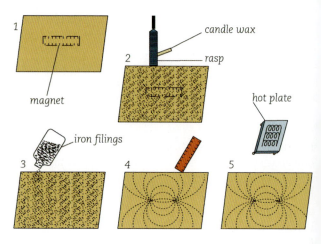

1. Place a bar magnet under a stiff piece of card.
2. Grate a thin layer of candle wax onto the card.
3. Sprinkle iron filings lightly but evenly over the wax.
4. Tap the card gently with the edge of a ruler and observe the pattern formed by the filings.
5. Pass a Bunsen burner or electric hot plate rapidly over the grated candle wax to melt it and so fix the pattern. Repeat steps 1 to 5 with two or more magnets in close proximity to the first.
6. Examine and comment on your results.

Are magnetism and electricity connected?

Yes, they are! This discovery in 1821 by Hans Christian Oersted caused a great technological leap forward. Just think, what would we do without electricity? In 1831 Michael Faraday found that electricity was generated when a coil of wire was moved in a magnetic field. One of the simpler applications of this principle can be found in the bicycle dynamo (page 186–7).

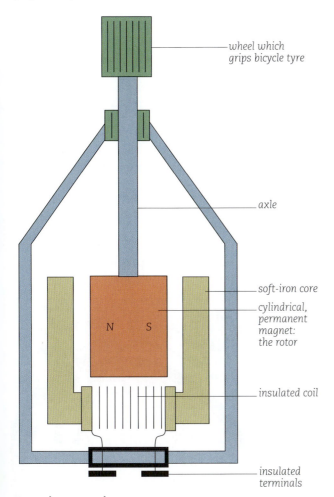

How a dynamo works

In the same way that magnetism is used to produce electricity, electricity can be used to produce magnetism.

What is an electromagnet?

The kind of magnet that is made when there is a flow of electricity through a circuit is called an **electromagnet**. This is a temporary magnet, which lasts only as long as there is a flow of electrical current. Electromagnets are found in transformers, electric motors and in most electrical devices which transform electrical energy into mechanical energy.

The substation transformer, electric clock and pole transformer all have electromagnets

Questions

1. How would you compare the magnetic lines of force between (a) two like poles and (b) two unlike poles?
2. Are other elements besides iron magnetic?
3. What coins would you expect to be magnetic?

our environment | electricity | *magnets and electromagnets*

How do we get and use electricity?

How do batteries produce electricity?
The most common type of battery is the dry cell battery. This is found in portable appliances, for example radios, tape recorders and torches. A **cell** is a chemical method of producing electricity, so it is referred to as an **electrochemical cell**. In a cell, chemical reactions take place to produce electricity.

Activity | Looking at a battery
1. Take apart an old (discarded) 9-volt battery. Use a screwdriver to prise the metal cover apart, and a pair of pliers to fold it back. **Be careful**.
2. Count the individual cells that make up 9 volts.
3. If possible, take apart other old batteries. Confirm that all batteries are made up of cells. Wash your hands afterwards.

What are cells made up of?
Probably the most common type of dry cell is the zinc-carbon cell. It consists of a carbon rod, a zinc case and a paste made with ammonium chloride, starch and some powdered carbon and manganese(IV) oxide called the **electrolyte**.

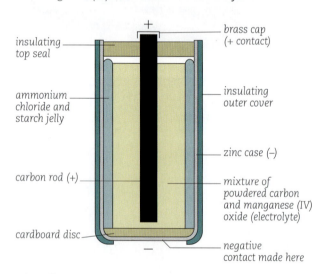

A dry cell

How does it work? Electrons build up on the negative pole of the cell; this is the zinc case in the zinc-carbon cell. Atoms which are short of electrons build up on the positive (carbon) pole. If a wire (for example, in a circuit) is joined between the poles of the cell, electron flow occurs which is capable of doing work. It is important to connect the cells the right way round in a circuit. The zinc-carbon cell produces current at 1.5 V.

There are other types of electrochemical cells, such as the high-energy **alkaline** cell and the **lead-acid** cell, which is rechargeable.

There are also some rechargeable dry cells. Lead-acid cells are used in car batteries, which produce the electrical energy needed to start the engine.

Current flow is in one direction in cells. Therefore electricity produced from an electrochemical cell is called **direct current** or d.c. electricity. But does all electricity flow in only one direction?

What kind of electricity comes to our homes?
You might think that the electricity which comes to your home from an electricity generating station (a power station) is the same as that from batteries, only at a higher voltage. Indeed there are many similarities. However, closer observation will show you that there are differences. For example, d.c. meters cannot be connected to mains electricity because there are no negative and positive poles. Transformers (discussed below) are very useful with mains electricity but not with d.c. electricity.

Mains electricity is called **a.c.** or **alternating current** electricity because it alternates or shifts its direction of flow. In so doing, the electric current flow is interrupted and the effect is a series of pulses. In mains electricity these pulses are produced 50 or 60 times per second.

How can a.c. electricity be produced?
If it is possible, examine a bicycle dynamo.

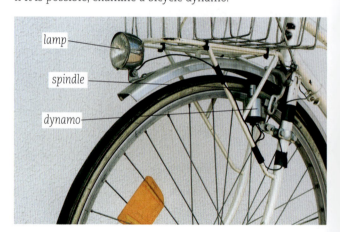

In a bicycle dynamo, a spindle which can turn rests on the bicycle wheel. The wheel turns the spindle which is attached to a bar magnet at the other end of the shaft. The magnet spins close to the coil of insulated wire. The turning magnet produces current flow in the coil (see also page 185).

As it turns, the poles of the magnet are reversed and this reversal in a magnetic field causes a reversal in the direction of current flow. If a bicycle dynamo is connected to a loudspeaker and is turned, the speaker cone can be seen to vibrate in and out. If the vibration speed is increased, the speaker produces a low hum. This indicates the stop-and-go, pulsating nature of the electric current being produced.

A bicycle dynamo produces a.c. electricity. It has the same characteristics as the electricity which is supplied to homes, schools, etc. by electricity generating stations.

The a.c. electricity produced is at a constant voltage, usually 120 V (Trinidad) or 110 V (Barbados and Jamaica) and 220 V in many other Caribbean countries. The rate of reversal or alternation in current flow is kept to a constant frequency of 50 Hertz (Hz) (cycles per second) or 60 Hz in Trinidad and Jamaica. A maximum current of 30 A is normally available to a household.

Electricity generation

The principles of generating and supplying electricity in a power station are the same as for the bicycle dynamo. Giant dynamos (generators) are needed, along with a source of turning power, transformers and transmission lines (a distribution system). In Caribbean countries, most of the electricity supplied nationally is from one or more power stations, and is produced by dynamos.

Generators (dynamos) Mechanical energy is needed to rotate either magnets or coils of wire to generate electric current (in the same way that the bicycle wheel rotated the dynamo).

This mechanical energy may be produced in different ways, and the most efficient method used nowadays is the **steam turbine**. Heavy oil (which is usually cheaper than other fuels) is burnt in a furnace or steam generator (see page 21).

Water circulates through a boiler, absorbs heat and becomes superheated steam under great pressure. The steam is led through pipes to the turbines where it turns the blades. Since the dynamo is attached to the turbine, electricity is generated as the dynamo is turned in the same way as with the bicycle dynamo.

Generating electricity In a steam turbine the energy transformations that take place are:

chemical → heat → mechanical → electrical.

Energy is transformed more efficiently than in an internal combustion engine, where the changes are:

chemical → mechanical → electrical.

When electricity is produced at a power station, dynamos produce a very high voltage: 4 to 11 kilovolts (kV). Before the power leaves the station, it must be **stepped-up** to as much as 60 kV, especially if it is to be sent over long distances. Transformers play a key role in these voltage transformations.

The transformer

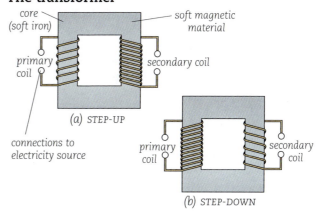

How step-up and step-down transformers are made

A transformer consists of two separate coils of wire called the **primary coil** and the **secondary coil**. These are wound close together around a soft iron core. When a.c. electricity flows around one coil, it produces a rapidly changing magnetic field. Because the other coil is near, it is affected by the rapid change in magnetic field, and a.c. current flow begins in the other coil. If there are a different number of turns and different sizes of wire in one coil compared to the other, voltages can be *increased* (stepped-up) or *decreased* (stepped-down).

In a power station, the step-up transformers are used to increase the voltage. At the same time the current is reduced. The total power (voltage × current) remains basically the same. If power were transmitted as high current and low voltage, large diameter wires would be necessary – otherwise, a lot of energy would be wasted in heating the wires (page 181). But if the voltage is high and the current is low, much thinner wires are satisfactory, because power losses are small. The thinner wires are also more economical.

Before electricity can be used in the home, high voltages from the transmission lines must be brought down to the much lower domestic levels. This is done by a series of step-down transformers. These may be located at an electricity sub-station if the area to be supplied is large, or else they may be fastened to 'electric' poles if the area is small.

Questions

1. What are the differences between the two types of current electricity?
2. What are the sources of (a) d.c. electricity and (b) a.c. electricity?
3. Explain how power losses in electrical power transmission systems (a) occur and (b) are kept to a minimum.

How can we make crude oil useful?

How can we get what we need from crude oil?
Crude oil is a mixture of many compounds. Most of these compounds are hydrocarbons (see page 24).

Activity | A sample of crude oil
Your teacher will give you a small sample.
1 What colour is it?
2 Does it pour easily?
3 Describe the smell of the oil.
4 Pour a small quantity into a bottle top or crucible lid. Try to light it. What is the colour of the flame?

Crude oil is not very useful as it is. We need to separate the different substances in the crude oil so that they can be used for different purposes.

Crude oil is extracted from the ground or from under the sea, and is then sent to an **oil refinery**, such as the one in the south of Trinidad. The compounds in crude oil are separated out by a process that is called **fractional distillation**. The process takes place in a **fractionating column**, also known as a **pipe still**.

Questions
1 What is the boiling point range of (a) the naphtha fraction and (b) the heavy gas oil fraction?
2 What are some of the uses of (a) the petroleum gases (b) kerosene and (c) gasoline?
3 Why is the process called *fractional distillation*?

Activity | Fractional distillation
1 Look at the diagram showing the apparatus used to demonstrate fractional distillation. The most important part is the piece labelled 'fractionating column packed with glass beads'.
2 Make a drawing of the apparatus.
3 The fractionating column is used to separate parts of the crude oil (called fractions) in the same way that a pipe still is used in a refinery. Your teacher will demonstrate how the apparatus is used. What do you see when the crude oil is heated?
4 Your teacher will collect a number of different fractions from the apparatus.
5 Examine each fraction separately. For each:
 (a) record the colour and smell
 (b) test how easily it pours
 (c) find out how easily it burns
 (d) record the colour of the flame.
6 Record your results in a table. Include, for each fraction, the boiling point range of the fraction.

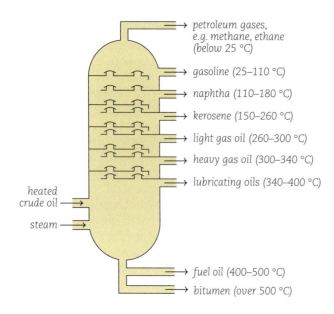

A pipe still

The crude oil is heated and then passes into the pipe still. Compounds with a low boiling point, such as gases, rise to the top. Just below the top we find liquids with a low boiling point, such as gasoline. Further down, compounds with higher boiling points, such as heavy gas oil, are found. The substances with boiling points higher than about 400 °C are found at the bottom of the column.

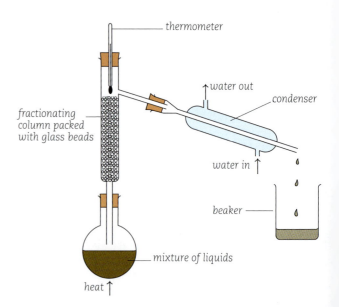

Fractional distillation apparatus

> ### Questions
>
> 4 If there is an oil refinery in your country, try to obtain information about it. What are its main products?
> 5 How do the properties of the fractions change as the boiling point ranges increase?
> 6 What is the purpose of the condenser?
> 7 What are the differences between the process you have seen in the laboratory and the process used in a refinery?

How do we use gasoline?

The gasoline fraction obtained from crude oil has a boiling point range of between about 25 °C and 110 °C. The most important use of gasoline is as a fuel (see page 192), but it is difficult to use gasoline in the form in which it is obtained from the pipe still. The fuel obtained burns too vigorously with air, so that there is a **knocking** sound in the engine.

In the past, one of the solutions was to add a compound called **tetraethyl lead** to the fuel as an 'anti-knock' agent. The use of this compound has been phased out because of the dangers to the environment of lead compounds. Most of the gasoline sold in the Caribbean is now lead-free.

A more acceptable solution to the problem of 'knocking' is to change the chemical composition of the compounds in gasoline. Gasoline can be 'cracked' to form more useful compounds. The original material is heated and the vapour obtained is passed over a **catalyst**. The vapours then pass through a fractionating column and over another catalyst. This increases the **octane number** of the gasoline.

Octane number The efficiency with which the gasoline burns in a car engine is measured by the octane number. Gasoline with a low octane number does not burn very efficiently. Examples of the sorts of molecules which help to make low and high octane numbers are shown below.

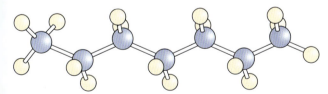

Heptane – makes low octane numbers

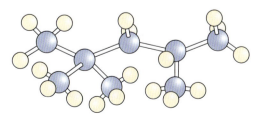

Iso-octane – makes high octane numbers

> ### Questions
>
> 8 Heptane is rated 0 on the octane number scale. Iso-octane is rated 100. Can you work out the octane number of gasoline containing equal amounts of heptane and iso-octane?
> 9 What can cause 'knocking' in an engine?
> 10 Why are some refinery products 'cracked' using catalysts?

Why do we use liquefied petroleum gas (LPG)?

Liquid fuels such as gasoline and kerosene are very common in the Caribbean and elsewhere. Electricity is often expensive (except in Trinidad, which has its own resources of oil and natural gas) and many people use gas for cooking and heating water. You may have seen the gas cylinders (20 lb or 100 lb) that many people use. Parts of Barbados are supplied by a mains supply of natural gas, so there is no need for separate cylinders in these areas.

Look back to the previous page and the diagram of the pipe still. The fraction with the lowest boiling point range is that containing the petroleum gases, including methane, ethane, propane and butane. (Check the chemical formulae of these gases.) Propane and butane may be liquefied under pressure. Cylinders containing LPG will usually contain a mixture of the two gases. The substance in the cylinders is a liquid, kept under pressure. When the pressure is released, the liquid becomes a gas, and this is used as a source of energy when it is burned. People using gas cylinders have to take care when they are using them, making sure that there are no leaks. When they go away for a long period, they have to turn off the gas supply at the cylinder.

Neither propane nor butane has a very distinctive smell. So it is necessary to add a trace of a chemical with a very strong and distinctive smell to the LPG. In this way you can always smell even relatively small quantities of the gas, which is very important for detecting a leak. If you were to light a match in a place where a lot of gas had gathered, there could be a loud explosion!

LPG is a relatively cheap and convenient form of energy. The price of LPG depends on the price of crude oil.

> ### Questions
>
> 11 Many gas cookers and water heaters have 'pilot' lights. Do you think these waste energy? How could you check this?
> 12 Is it likely to be more economical to buy 100 lb gas cylinders than 20 lb cylinders? How can you check your answer?
> 13 If gas cylinders are used for supplying energy in your school laboratory, what safety precautions are taken to protect the gas supply?

How can we get energy from fuels?

Where do some common fuels come from? When you use a Bunsen burner in the laboratory or travel in a bus or car, you are making use of a fuel as an energy source. The fuel may be liquefied petroleum gas (LPG) for the Bunsen burner and gasoline or diesel for the car.

Activity | Different fuels
List all the different fuels you know, indicating whether they are solid, liquid or gas. Pages 188–9 may help you.

Solid fuels Some of the fuels you have listed are solids. These are found in the ground both on land and under the sea. One example of a common solid fuel is **coal**. This is a black solid, found in many parts of the world in large quantities, for example, in the USA, Russia and China. The substance often called coal in the Caribbean is in fact **charcoal**. This is derived from another common solid fuel, **wood**. When wood is burned, heat energy is given out but waste materials are formed also. Wood can be changed to charcoal by heating it in a very limited supply of oxygen for a long period of time. The photograph shows charcoal being made. Another solid fuel, found in Jamaica, for example, is **peat**. It is found in quite large quantities near Negril and Black River.

Activity | Heating a solid fuel
1. Heat a small sample of wood in a test-tube, using the apparatus shown in the diagram below. You are heating the wood in a limited supply of air.

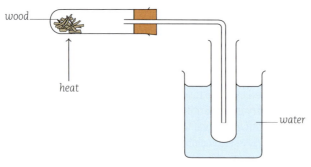

Apparatus for heating a solid fuel

2. Examine the contents of the test-tube and the other substances collected once there is no further change.

The use of wood and charcoal as fuels can lead to environmental problems unless great care is taken. If trees are cut down for wood (or charcoal) and not replaced, the soil will be affected because there may be erosion problems. Creatures will lose their habitats and there will be a range of impacts on the entire ecosystem. Too many trees cut down over a large area (e.g. a tropical rain forest, such as in the Amazon basin), can affect rainfall over the entire area and beyond.

Burning wood to make charcoal ('coal') to use as fuel

Liquid fuels These are commonly used in the Caribbean. As we have seen, the most widely used are those obtained from crude oil, such as gasoline and kerosene. Crude oil is a **fossil fuel** because it was formed by the action of heat and pressure on the remains of small plants and animals in the earth. This process took a very long time – perhaps up to 400 million years. Crude oil is found underground. There are important underground oil fields (which contain natural gas as well) off the east coast of Trinidad.

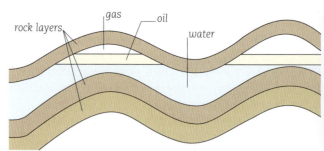

An anti-clinal trap – this is how oil forms under the ground

Using fossil fuels
Liquid fuels are used mainly for transport (for example, buses, cars, aeroplanes) and for generating electricity. Gases are used for domestic purposes such as cooking and heating water, as well as in various industries.

When fuels are burned, energy is released. We can think of the energy obtained from a burning fuel as being the result of energy that has been stored in compounds formed over millions of years. Fuels are used up when they are burned. This means that we cannot get back the original substances – new substances have been formed and energy has been given out. These new substances are mainly gases such as carbon dioxide and water vapour. The gases escape into the atmosphere and can change the composition of the atmosphere over time. There are two important issues to think about.

1. **What will happen when the fossil fuels run out?**
   There is only a certain amount of crude oil and coal in the ground. We do not know how long the supplies of these fuels will last – perhaps no longer than the end of this century. Perhaps not even that long. We need to consider how to use these fuels as efficiently as possible. At the same time, we should look at alternative sources of energy.
2. **How does the burning of fuels affect the atmosphere?**
   Most of the products of burned fuels go into the atmosphere as gases. In many cities, such as Los Angeles or Mexico City, there may be very large quantities of these gases in a relatively small volume of air. This can lead to serious pollution of the air. The worst effect is smog, which is a mixture of smoke and polluting gases, and naturally occurring fog. Smog can hang over an area for a long time.

The quality of the air we breathe can have a serious effect on our health. There are also other problems associated with the amount of carbon dioxide in the atmosphere. Many scientists now believe that there is a general warming effect, called **global warming**, as a result of excess carbon dioxide in the atmosphere. This is a very important issue for everyone (see pages 193 and 318).

Questions

1. Are there any fuel resources in your country? What are they? How much fuel has to be imported each year?
2. List any alternative sources of energy you know about, such as solar power, hydroelectricity, etc. How are they used in the Caribbean?

Activity | How does gas burn?
1. Light the Bunsen burner. Adjust the air control so that there is little or no air being burned with the gas.
2. What is the shape and colour of the flame?
3. Adjust the air control so that the air supply is at maximum.
4. What effect does this have on the shape and colour of the flame?
5. Draw a diagram of each of the flames you have seen.
6. With the air supply at maximum, take a piece of wire gauze and hold it carefully at the tip of the flame. What do you observe?
7. Move the gauze carefully down until it is just touching the top of the intense blue part of the flame. Then move the gauze slowly towards the top of the burner.
8. What does this tell you about the heat energy in the different parts of the flame?

When the air supply is at maximum there are two parts of the flame. The hottest part is that in which burning (combustion) is complete. The air supply allows the fuel (gas) to burn completely.

What happens when we burn a fuel?
When the fractions obtained from the fractional distillation of crude oil are burned, heat energy is obtained. The fuel which you use for heating in the laboratory is the gas supply to the Bunsen burners.

Do we need oxygen to burn things?
Fuels are usually burned in air. Air contains about 20% oxygen by volume – much of the remaining 80% or so is nitrogen. Oxygen is the most important part of the air for burning. Your teacher will prepare oxygen in the laboratory by adding hydrogen peroxide to manganese(IV) oxide powder.

Activity | Oxygen and burning

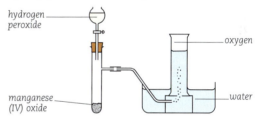

Apparatus for preparing oxygen

1. Collect enough oxygen in a gas jar to test a number of burning materials.
2. Investigate how materials such as splints and small candles burn in oxygen. You can also burn a small piece of magnesium ribbon – but be very careful.
3. Is oxygen necessary for burning?
4. Record your results in a table.
5. Try to burn the same materials in other gases, such as carbon dioxide (prepared by reacting dilute hydrochloric acid with calcium carbonate).
6. Is carbon dioxide as effective in supporting burning as oxygen is?

Questions

3. Heat energy is produced in a car engine when the fuel is burned. Which other forms of energy are produced? Is the heat energy converted to other forms of energy?
4. Welding is a technique used for joining two pieces of metal or alloy together. Which gases are used for welding?
5. A candle flame may look similar to one type of flame of a Bunsen burner? Why?

What are the effects of using cars?

The **internal combustion engine** was invented towards the end of the 19th century and has made a huge impact on the ways in which we get around. The basic principle of the engine is that a fuel such as gasoline or diesel is burned inside a cylinder, and the heat energy released is converted to mechanical energy, which drives the vehicle. The manufacture of cars has become a major part of the global economy, since they are the preferred means of transport for most people who can afford them around the world. Many countries have public transport systems involving buses and/or trains. Is there a public transport system in your country?

Activity | Forms of transport

1. The class will be divided into groups by the teacher. Each group will focus on one form of transport and record how many students use this form of transport to get to and from school on a regular (daily) basis. Record these results in the form of a table. Discuss these results with other groups and find out which is the most common form of transport used by students.
2. Make a list of some of the common forms of transport used in your country. (You might include people using their feet to walk, and those using bicycles.) If the form of transport requires an engine, find out what type of engine is used. If you live on the coast or near a river, make sure you include boats. Some of the forms of transport do not make use of an engine. What is the source of energy in these cases?
3. Try to find out how important one of these forms of transport is in your country. For example, how many cars are there in the country? You can compare the number of cars to the size of the total population (this can be used as a measure of the wealth of a country). If there is a public transport system, try to find out how many people use it every day.
4. People also travel to and from your country. Try to find out how many travel by sea and by air each month.

Vehicles need fuel. The cost of imported fuel is a major component of the total annual import bill for most Caribbean countries. Countries such as Trinidad and Venezuela are fortunate to have reserves of oil and natural gas.

What are the products of the internal combustion engine?

When an internal combustion engine is in use, fuel is burned. The fuel is a hydrocarbon. There are four stages to the combustion process. In the first stage, fuel is mixed with air in the cylinder. This fuel/air mixture is then compressed during the second stage. In the third stage, the mixture is ignited. As a result, hot gases are produced and these force the piston down the cylinder. In the final stage, the hot gases leave the cylinder through the exhaust valve, while the piston is pushed back up the cylinder by a crank.

When hydrocarbons are burned in air or oxygen, you would expect that oxides of carbon and hydrogen would be formed. The oxide of carbon that is formed when there is complete combustion is **carbon dioxide**. If some of the fuel is not completely burned, **carbon monoxide** is formed. You will have seen vehicles emitting very dark smoke – this contains unburnt **carbon**. The air consists mainly of two gases, oxygen and nitrogen. In the combustion process, some of the nitrogen is oxidised to form one or more oxides of nitrogen. The hydrogen present in the fuel is oxidised to form water (steam).

Can the products of the internal combustion engine be harmful?

New substances are formed when fuels are burned. So far, we have looked at the products of burning fuels in an internal combustion engine. The same principle applies to any fuel which is burned in order to obtain energy.

Most of the fuels we burn are what are called fossil fuels – see pages 188–91. The energy stored in the chemical bonds of the fossil fuels can be released when the fuels are burned, usually in air. This releases a great deal of heat energy and this can then be used to drive machines.

Carbon dioxide is formed when fuels containing carbon are burned completely. You may have heard of the term **global warming**. There appears to be some strong evidence that the average temperature is slowly increasing in many parts of the world. (Not everyone agrees about this, since the climate of the earth changes over time – the last Ice Age may have finished as recently as 10 000 years ago!) Much of this temperature increase is thought to be caused by the increasing amounts of carbon dioxide being released into the atmosphere by the burning of fuels – including wood from the clearing of tropical rain forests, such as that in the Amazon basin. (This is often called the **greenhouse** effect, because the gases trap the sun's heat, turning the Earth into a giant 'greenhouse'. See also page 318.)

In all industrialised countries the internal combustion engine is a major source of this carbon dioxide. At the time of writing, many countries had accepted the Kyoto Protocols, according to which they agreed to try to reduce carbon dioxide emissions. However, the USA had decided not to participate in this agreement, which presents a major problem because the USA is the source of a large amount of carbon dioxide.

Carbon dioxide causes problems on a global scale, but the other oxide of carbon, carbon monoxide, can be the source of more local problems. Carbon monoxide is a very poisonous gas. It reacts with haemoglobin in the blood to form carboxy-haemoglobin. This prevents oxygen being carried around in the blood. If you breathe carbon monoxide rather than oxygen, your body will be starved of oxygen and as a result you will die. Wherever there are very large numbers of vehicles in a relatively limited space – for example, in a crowded city street – you will find high concentrations of carbon monoxide. Ongoing exposure to such an atmosphere is dangerous to health, as is exposure to too much particulate matter (particles such as smoke particles), particularly for those with chest conditions such as asthma.

Nitrogen makes up about four-fifths of the volume of the air we breathe. Oxides of nitrogen are formed when the air/fuel mixture is burned in an internal combustion engine. These oxides can rise in the atmosphere and combine with water in the atmosphere to form acids. You may have heard of **acid rain**, which has caused major problems for both vegetation and water sources such as lakes in parts of North America and Europe.

Nitrogen and sulphur oxides formed by the combustion of fossil fuels are major sources of acid rain. Acid rain is not a major problem in the Caribbean, but the impact of global warming could be serious, particularly if one result is a significant rise in sea level over time. (Why do you think global warming could affect sea level?) There is more discussion on these topics on pages 318–19.

A busy city street – Bridgetown, Barbados

Questions

1. Can you find out what the major differences between a spark-ignition and a diesel internal combustion engine are? Do they have different impacts on the environment?
2. What is the difference between a gasoline car engine and a motorcycle engine?
3. Suggest ways in which you could try to measure the impact of exhaust gases from cars and trucks on the environment.

Can we solve some of the problems?

'Think global, act local' is a phrase which has been used a great deal in the past few years. It is often applied to problems and issues affecting the environment. It is important that we should think about the big issues such as global warming – but at the same time we need to act in such a way that we can have an impact on the local environment.

For example, we know that pollution of the environment can occur because of the products of the internal combustion engine. While as individuals it will be difficult for us to have an impact on the world as a whole, we can encourage our friends and families to think carefully about some of the issues. This might involve finding information about alternative approaches to transport, including some mentioned above. If your family is responsible for a car, it is important that it runs as efficiently as possible, which means that it should be regularly maintained. At the same time, we can take carefully considered decisions about when to use a car as opposed to public transport. Cars are very convenient, but there is always a price to pay for convenience.

Governments can take decisions which have a long-term impact, such as the banning of the use of 'anti-knock' agents which contain poisonous lead. Most if not all Caribbean countries sell lead-free gasoline now – this was not the case a few years ago. Governments can also use legislation to encourage the use of alternative energy sources. For example, cars powered by a mixture of gasoline and ethanol (derived from sugar) have been used for many years in Brazil.

Activity | Discussion

1. Should buses and trains be given preference over cars in towns, since they carry many more people than a car?
2. Suggest other ways in which government policy could reduce the negative effects of cars on the environment.
3. There is increasing concern about the impact of lack of exercise on health. Would you try to encourage people to walk or use a bicycle for short journeys?
4. Are there problems of air or noise pollution as a result of cars in any of the towns that you know? How would you try to solve such problems?

How do we use solar energy?

Activity | How can we trap radiant energy?

1. Get two identical clean tin cans with a capacity of 250 to 500 cm³. Tins originally used for condensed milk or fruit juices would be best as these will not have had their tops removed.
2. Widen the holes in the top, just enough to let in water. Remove any paper labels and then blacken one tin by holding it over a candle flame or a sooty Bunsen flame.
3. Put equal quantities of cold water in each tin and place them in the sun. Measure the temperature of the water at the start of the experiment and then at intervals of every 2 to 5 minutes.
4. Plot your results on a graph with time on the x–axis and temperature on the y–axis. It is best to plot your graph while the experiment is in progress so that you can decide when to stop.
5. Can you explain why the curves are different? Why did you choose to stop the experiment when you did?
You saw on pages 114–15 that the three ways by which heat is transferred are **radiation, conduction** and **convection**. In the above Activity, each of these ways is important. First, the radiant energy of the sun is absorbed by the walls of the tins.

If you performed this Activity carefully, there would be no difference between the experimental conditions for the two tin cans except that one is blackened. A blackened surface is better at absorbing heat (i.e. **radiation**). Heat travels by **conduction** through the metal to the water inside. **Convection** is the main method of heat transfer in the water itself, because water is a poor conductor of heat.

How does the sun get its energy?

A famous 19th century scientist, Lord Kelvin, thought that the sun was 20 million years old and would soon burn itself out. Scientists at the time thought the sun must be made of some sort of combustible material like coal.

Today, scientists believe that the sun is much older and has a longer future. It derives its energy not from combustible material but from matter, which is changed into energy by a process known as **nuclear fusion**. Hydrogen atoms are the sun's fuel, and release enormous amounts of energy (see page 21). There is enough hydrogen present in the sun to keep it going at its present rate for another thirty thousand million years.

Because the Earth is so far from the sun (150 million km), only 1/2 000 000 000 of the sun's energy reaches the Earth's surface. This still gives more than 5 kilowatt hours per square metre per day in the Caribbean (see below). Most of the energy reaching the Earth's surface is in the visible, infrared and ultraviolet parts of the electromagnetic spectrum (see page 320).

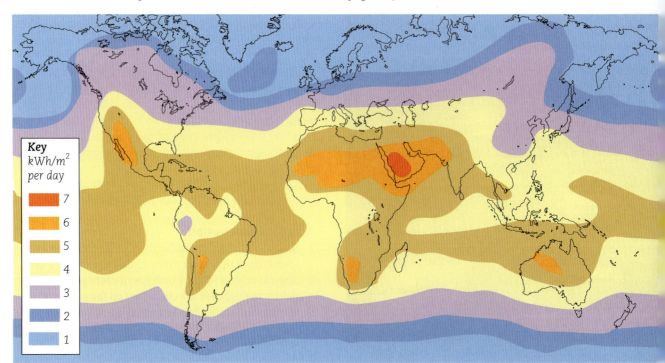

Global solar radiation – annual average in kWh/m² per day

For thousands of millions of years this energy has been poured upon the Earth and life has evolved to use this energy. Plants trap and store the sun's energy by photosynthesis (page 66), thus changing it to a more utilisable form. Energy stored up long ago continues to serve us today in the form of coal derived from land plants and petroleum (crude oil) derived from marine plants and animal life (page 188).

How is the sun's energy useful today?
The heating effect of the sun evaporates seawater, giving clouds and rain (see water cycle, page 139). Rainwater causes rivers to run; the water can be collected into dams for generating hydroelectric power. Visible light energy can be used to power solar batteries (see page 197).

One of the major demands for energy worldwide is energy for domestic uses. The greatest demand by far is for heating – mainly for cooking and hot water. Power stations use oil to generate electricity. Before 1973, a low oil price ensured that the electricity reaching our homes came at a low price. However, since 1973, dramatic increases in oil prices have put an end to low-cost power and much attention has been turned to the *direct* use of solar energy. Given the high costs of electricity from oil, it is cheaper to use solar energy in the Caribbean. Solar energy is also more trouble-free and dependable than other types of heating.

How does a solar water heating system work?
Basically, the system consists of two parts: a **collector** and a **storage tank**.

The storage tank The storage tank is usually a steel tank lined with glass to prevent rusting and it is strong enough to withstand the pressure of water. The tank is insulated on the outside with, for example, glass wool or polyurethane. The tank is connected at two points to a collector.

The collector This is a flat surface, which collects heat from the sun and passes the heat on to the pipes, which are bonded to the plate. Water passes through these pipes and absorbs the heat.

The system in the picture is called the **thermosiphon system** because the water circulates by convection currents. Water in the collector becomes less dense when it absorbs heat; it therefore rises into the storage tank. At the same time, cooler water flows from the bottom of the tank to the collector again. This system does not require any external source of power to circulate the water.

However, the tank needs to be placed higher than the collector for the system to work. If the tank is placed at a lower level, an electrically driven pump must be installed to provide the circulation.

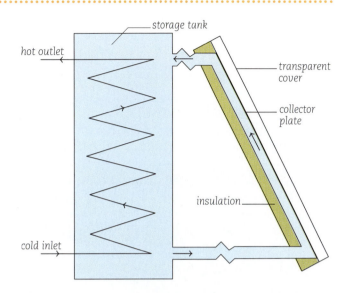

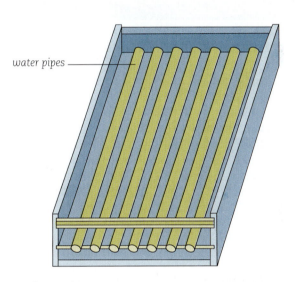

A solar panel

Questions

1. How does the sun get its energy?
2. Explain how solar energy from prehistoric times can be used now.
3. Explain how the solar water heater unit is able to conserve heat.
4. How is convection applied in the solar water heater unit?
5. Examine the chart on the opposite page. Can you explain why the Caribbean Islands lie in an area of high solar radiation?

How can we make better use of solar energy?

How could you decide on a water-heating system?

First you must decide how much hot water you need for your home. For a small family this could be 200 to 250 dm^3 (litres) per day. Then you must also consider the cost.

What does it cost to heat water electrically? In the Caribbean, 'cold' water is not usually less than 25 °C. Hot water is about 60 °C. The water must therefore be heated through 35 °C. It takes 4.2 joules of energy to raise one gram of water by 1 °C. (A joule is a quantity of energy equal to one watt per second, a kWh = 3.6×10^6 joules.) It takes 4 200 joules (4.2 kJ) to heat 1 litre (= 1 kg) of water by 1 °C. Therefore the energy required to heat 250 litres of water is 250 × 35 × 4.2 kilojoules.

When we know how long the heater can be used for, we can work out the number of kilowatt hours or **units** needed and we multiply it by the cost per unit, to find the total cost. Towards the end of 2000, the cost per unit of electricity (in Barbados) was around 17 cents (US) so that the cost of heating water for one year was around $630.00 (US).

What does a solar heating system cost? The sun's energy is free. However, the apparatus used to collect this energy may be too expensive for the average Caribbean family to buy.

The minimum cost of installing a solar water-heating unit is over $1 000 (US). So, a solar heating system has to be considered as an investment, which once installed will in the long run save money.

In the Caribbean, electricity produced from imported oil is expensive and because of the abundance of sunlight, locally produced solar water heaters are a good idea.

How else can we make use of solar energy?

Solar still The sun's energy in the form of heat can be used to produce pure water from tap or seawater in a **solar still**. Solar stills are used to produce distilled water in schools. A simple still consists of a box containing water under a clear glass top so that the sunlight can get in. The bottom of the box is blackened to absorb heat. As the water is heated, vapour condenses on the slightly cooler glass top and drops run down the sloping surface into a container.

Solar drier The sun's energy has always been used for drying foodstuffs such as meat, fish and fruits (page 158). The **solar drier** is an improvement on traditional systems. It is a box containing stones or bricks, with a shelf where materials for drying are placed. The stones absorb the sun's heat during the day and release their heat to circulating air, which dries the foodstuff.

Solar furnace

A solar furnace in France. There are 20 000 small mirrors, which direct the sun's energy into the furnace in the centre. In a few moments, a temperature of around 3 000 °C can be produced. This temperature would take hours to produce by usual methods

Activity | Collecting solar energy

Use a concave (shaving) mirror to focus the sun's rays. Compare this with the heat produced by using a magnifying glass of similar size. Can you design an efficient solar energy collector using aluminium foil?

You will notice that the sun's rays are brought to a focus by a concave mirror just as easily as with a convex lens. A large mirror is much lighter and easier to make than the corresponding lens, so a collector such as a curved mirror is the best way of obtaining high temperatures from sunlight.

Can the sun produce electricity directly?

The energy in sunlight can be turned directly into electrical energy by the **photovoltaic or solar cell**. Although its efficiency is low, it is robust and reliable.

It is usually made from two very thin layers of silicon. The two layers of silicon are treated to give them special properties, which enable them to transfer electrons and thus behave as a cell to produce electricity (see page 173). When sunlight reaches the junction between these two layers, it makes the electrons flow. These electrons are collected and flow through as an electric current.

At first solar cells were used mainly in space, to supply energy to satellites. They are expensive to produce and are not considered economical except in remote areas.

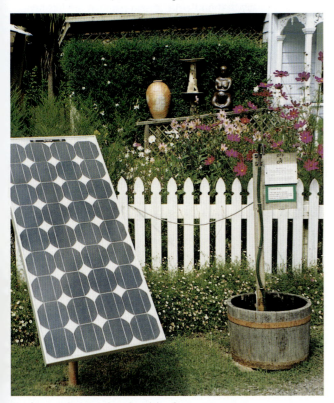

A solar panel generating electricity for a water pump

Is wind energy linked to solar energy?

Winds are caused by the rotation of the Earth, as well as by the uneven heating of the atmosphere by the sun. Sails have been used for hundreds of years to exploit wind energy; so too have windmills. The very high cost of oil in relatively recent times has encouraged new research into this ancient energy source. Very many significant strides have been made and wind energy is making a significant contribution to the energy needs of many countries.

The turning motion of the massive blades on these wind turbines is transformed into electricity in generators just behind the blades

Questions

1. Why do you think that stones are used in the solar drier?
2. How can a reflecting surface be used to concentrate the sun's energy?
3. What are the advantages of solar cells over ordinary cells (batteries)?
4. How is wind energy related to solar energy?

How reactive are some of the metals?

Activity | What happens when metals react with acids?

1. Pour dilute sulphuric acid into a test-tube to a depth of about 4 cm.
2. Cut off a 3 cm length of magnesium ribbon. If there is no magnesium ribbon, use a small quantity of magnesium powder.
3. Add the magnesium to the acid. Observe any changes which occur.
4. Now repeat the Activity but test the gas given off by putting a burning splint into the mouth of the test-tube. This is a test for hydrogen. This may give you an idea of the dangers of handling hydrogen gas. Any gas which explodes when it is mixed with oxygen and a flame is applied is likely to be very dangerous.

Hydrogen is given off in this reaction and we can write a simple word equation for the reaction:

magnesium (s) + sulphuric acid (aq) →
magnesium sulphate (aq) + hydrogen (g)

It is useful for us to know whether metals will react with acids, since many substances we use in our everyday life contain acids. For example, if we use a metal container to move a liquid around, we need to know whether the metal will react with acidic liquids.

5. Repeat steps 1 to 4 of the Activity above using dilute hydrochloric acid. Record your results in a table. List the metal, the state in which the metal was used (e.g powder, ribbon, turnings), whether or not a gas was given off and the colour of the solution at the end of the experiment.
6. Repeat steps 1 to 4 above, again using dilute sulphuric acid, but this time use separate small amounts of the following metals: aluminium, copper, tin, iron and zinc. Add the results to your table. Your teacher may demonstrate the reaction of silver with dilute sulphuric acid.
7. In each case, if there is no reaction with the acid when the metal is cold, warm the test-tube gently and test for hydrogen if any gas is given off.
8. Write a word equation, including state symbols, for any reaction you observe.
9. Repeat steps 6 to 8, using the same metals, but using dilute hydrochloric acid. Record your observations in a table. Write word equations for the reactions you observe.

You have now investigated how some metals react with two *dilute* acids. *Concentrated* acids may also react with the metals. Since the concentrated acids have to be handled with care, your teacher will demonstrate their reactions with certain metals. Observe whether any gases are evolved, and then watch your teacher test them. Check the colour of the gases carefully. In order to identify the gases, you may need to use a splint and also red or blue litmus paper. (Should the litmus paper be dry or moist?)

Hydrogen is not usually one of the products of reaction when concentrated acids react with metals. Concentrated sulphuric and nitric acids behave more like **oxidising agents** than acids – this makes them different from concentrated hydrochloric acid, for example.

Activity | What happens when metals react with dilute nitric acid?

1. Repeat the experiments which you carried out with dilute sulphuric and hydrochloric acids, but use dilute nitric acid instead.
2. If a gas is given off, try to identify it. Remember to check the colour of any gas.
3. Record your observations in a table.
4. Write word equations for the reactions.

Hydrogen is not usually given off when nitric acid reacts with a metal. Depending on the concentration of the acid, one of the oxides of nitrogen may be given off. One way of explaining this is to say that dilute nitric acid reacts more like an **oxidising agent** than an acid. (See the comments on the properties of concentrated acids above.) The other dilute acids act as sources of hydrogen when they react with metals.

There are also differences in the way that the concentrated acids react, as we have already seen. Most metals react with the concentrated acids, even though the gases given off are not what you might expect.

Your teacher may demonstrate the reactions of the metals with concentrated nitric acid. Some of these reactions have to be carried out in a **fume cupboard**, because of the dangerous gases given off, such as **nitrogen dioxide**. Record your observations of any reactions you observe and write word equations for the reactions.

Questions

1. Some metals react quite vigorously with acids, and others hardly react at all. Prepare a list of the metals you have tested in order of the vigour of their reactions with the dilute acids.
2. Do you know of any metals which are more reactive than those you have tested? Which are the most reactive metals you know?

Part 1

Questions

3 If you had a copper saucepan, would you expect it to react with it with (a) tea (b) vinegar and (c) juice from a citrus fruit such as orange or grapefruit?
4 Can you think of any reasons why aluminium saucepans are more common than copper saucepans? Which is the more reactive metal?

Activity | What happens when metals react with water?

You have already investigated the reactions of some metals with some acids. As a result of your investigations you can write a list of metals with the most reactive first and the least reactive last.

Would you expect the metals which react with the acids to react with water also? The dilute acids contain water, so we might expect the acids to be more reactive than water. You can test your predictions by investigating the way in which metals react with water.

1 Pour distilled water into a test-tube to a depth of 4 to 5 cm. Add a short length of magnesium ribbon which has been cleaned.
2 Observe whether a reaction takes place. If a gas is given off, test for hydrogen. If no gas is given off, warm the test-tube and note whether a reaction takes place.
3 Write a word equation for any reaction you observe.
4 Repeat the experiment, using separate small quantities of the following metals: aluminium, copper, iron, tin and zinc. Your teacher may demonstrate the reaction of silver and water.
5 Record your results in the form of a table, including the name of the metal, its state, and whether a gas was given off.

You should find that most of the metals do not react with cold water, and that there is little reaction with warm water. This is what you might have expected, since acids are likely to be more reactive than water.

Although we use water for many different things, steam is commonly used in industry. You may have seen steam boilers in the sugar industry, for example. Water is strongly heated and converted to steam, which is then used as a convenient form of energy. The metal tubes in a steam boiler must not react with either water or steam – can you think of a suitable metal or alloy for making these tubes?

The reaction of iron or zinc with steam Your teacher will demonstrate the reactions of iron or zinc with steam, using the apparatus shown in the diagram at the top of the next column.

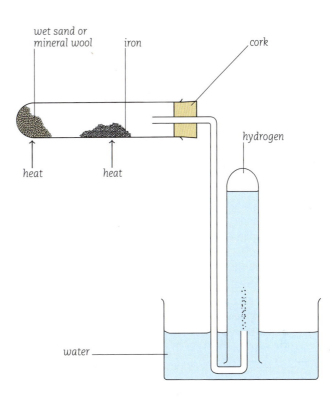

Apparatus for the iron/steam reaction

Water contains hydrogen and oxygen, so when it reacts with a metal you might expect the compound formed to contain the metal and hydrogen or the metal and oxygen. When steam reacts with a metal, an **oxide** (i.e. a compound containing the metal and oxygen combined) is formed and hydrogen is given off. When iron reacts with steam, magnetic iron oxide is formed. The word equation for the reaction is:

iron (s) + water (g) → magnetic iron oxide (s) + hydrogen (g)

Questions

5 Which of the metals you have tested reacts (a) most vigorously and (b) least vigorously with water?
6 Why is it important to record the physical state of a metal and the amount of metal used in these investigations?
7 Sodium and potassium are metals which react very vigorously with water, giving off hydrogen. What are the names of the metal compounds formed in these reactions?
8 Calcium is a metal which reacts with water. In this reaction, a metal compound is formed which is used for the detection of carbon dioxide. What is the name of this compound? How does it react with carbon dioxide?

How reactive are some of the metals?

Activity | How can we prepare oxygen?

You have already prepared oxygen gas – see page 191. Revise this quickly as follows.
1. Put a few cm³ of hydrogen peroxide solution (20 volume) into a test-tube.
2. Add a small quantity of powdered manganese(IV) oxide. Test the gas evolved with moist red and blue litmus paper and with a glowing splint.
3. Record your observations.

Oxygen gas is colourless and has no smell. It has no effect on litmus, but it will relight a glowing splint. This is a test for oxygen and a good way of telling it apart from hydrogen. Hydrogen explodes quietly, with a 'pop', when a burning splint is placed in it. The explosion is not so quiet if there is a lot of hydrogen!

You have investigated how some of the metals react with water (or steam) and acids. You wrote down which metals are more reactive and which are less reactive. You will now investigate what happens when some of the metals react with oxygen. Since the very reactive metals such as sodium and potassium react violently with oxygen, you will not investigate these reactions.

Activity | What happens when magnesium reacts with oxygen?

1. Prepare a number of boiling tubes full of oxygen using the apparatus shown below. Seal the tubes with bungs until they are needed.

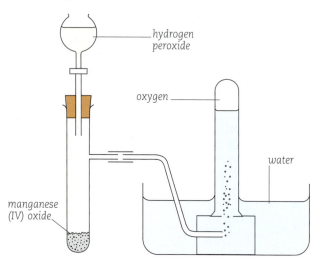

2. Take a 3 cm length of magnesium ribbon and clean it carefully. The easiest way to do this is to dip it into dilute acid, remove it quickly and then wash it in distilled water, allowing it to dry afterwards.
3. Fold the cleaned ribbon until it will fit onto a small deflagrating spoon. Heat the ribbon in a Bunsen flame and, as soon as it begins to burn, put it into one of the boiling tubes full of oxygen.
4. Do *not* look directly at the magnesium as it burns.
5. Allow the solid which is formed to cool down to room temperature. Transfer the cooled solid to a small beaker (50 cm³) and add a small quantity of distilled water. Mix the solid and liquid thoroughly.
6. Add a few drops of universal indicator solution to the mixture of solid and liquid.
7. Record all your observations.

Magnesium reacts vigorously with oxygen, burning with a very intense light. Air contains about 20% oxygen. The magnesium reacts with this oxygen, and also with some of the nitrogen in the air. The white solid formed is called **magnesium oxide**. The word equation for the reaction is:

magnesium (s) + oxygen (g) → magnesium oxide (s)

When this solid is added to water and then tested with universal indicator solution, the indicator turns blue, which means that the mixture in water is alkaline (see page 220).

Magnesium oxide is almost insoluble in water. The oxides of the more reactive metals are more soluble in water, forming **hydroxides**. Some of the less reactive metals form oxides which are insoluble. All of these oxides are bases – see page 220.

Activity | What happens when other metals react with oxygen?

You should still have a number of boiling tubes full of oxygen. You need to use safety goggles in this Activity.
1. Investigate the reaction of the following metals with oxygen: aluminium, copper, iron, tin and zinc.
2. Take a small quantity of the metal, preferably in the form of powder or turnings, and heat it strongly in a small deflagrating spoon until it glows. As soon as it glows, put the deflagrating spoon into the oxygen in the boiling tube.
3. Repeat steps 5 to 7 of the previous Activity. If the oxide does not seem to dissolve, record that it is insoluble in water.
4. Write word equations for the reactions.
5. Write a list of the metals you have tested in order of the vigour with which they react with the oxygen, starting with the most reactive and ending with the least reactive. You should find that this is very similar to the list you came up with when investigating the reactions of metals and acids.

Part 2

Questions

1. Write the word equation for the preparation of oxygen.
2. Why is it more effective to use metal powder or turnings in the last Activity?

Is there a pattern in the reactions of metals?

You have now investigated the reactions of some metals with acids, water and oxygen. It is difficult to compare how reactive these metals are with the different substances, since water is relatively unreactive. But it is probably true to say that the most reactive metals are always more reactive, while the least reactive metals hardly react at all, except with concentrated acids.

It is possible to arrange the metals in order of their reactivity with different substances. This is called the **activity series**, although it is probably more accurate to call it the **reactivity series** of metals. We can write such a series like this:

sodium
magnesium
aluminium
zinc
iron
lead
copper
silver

increasing reactivity

As you might expect, the most reactive metals form compounds very easily – for example, sodium has to be stored under oil because it would react with the oxygen in the air. The least reactive metals form compounds much less readily. When metals are found in the Earth's crust, they are often found as compounds. The least reactive metals may be found as the metal alone, as is the case with gold or silver. On pages 202–3 you will look at some of the reasons for using metals. Very unreactive metals or mixtures of metals are needed for everyday objects such as coins.

What happens when metals compete?

Some metal oxides can be reduced to the metal by reaction with charcoal (carbon). For example, lead(II) oxide can be reduced to lead by reaction with charcoal. Reduction reactions like this are also possible using an oxide of one metal and a more reactive metal.

Iron is above copper in the reactivity series. If we heat a mixture of powdered iron with black copper(II) oxide, the mixture begins to glow even when the Bunsen flame is taken away. If the mixture is then allowed to cool, copper and an oxide of iron are formed. We can write a word equation for this:

iron (s) + copper oxide (s) → iron oxide (s) + copper (s)

The more reactive iron has taken the oxygen away from the copper.

This type of process can be useful, for example, in the **Thermit process**. In this process, aluminium is heated with the oxide of a metal which is less reactive.

The Thermit process: the heat of combustion of powdered aluminium and iron oxide produces a temperature of about 2 600 °C, used for welding

Copper sulphate solution is blue. If you add an iron nail to some copper sulphate solution and leave it to stand for some time you will observe some changes. The iron nail turns brown on the surface because a layer of copper is deposited on it. The blue copper sulphate solution may begin to turn green.

iron (s) + copper sulphate (aq) → iron sulphate (aq) + copper (s)

Questions

3. Will iron 'win' if it is competing with lead rather than copper?
4. What would you expect to see if zinc is added to copper sulphate solution?

Why do we use metals and alloys in the home?

Activity | Looking at metals and alloys
1. Study the objects shown in the photograph carefully.
2. Make a list of all the objects and guess what they are made of.
3. Add to the list other objects found in the home and made of similar materials.

You may have noticed that many of the objects shown in the photograph above are made of metal, even if parts of these objects are made of other materials, such as plastics – see pages 204–5.

Metals are very common elements. You have already investigated the properties of some of the metals – see pages 198–201. For example, **copper** is a very good conductor of both heat and electricity. You may have noticed that wires used in electrical circuits are often made of copper, although they have a plastic or cloth coating.

Although most of the metals you have investigated are not very reactive, there are others which are highly reactive. These easily form compounds such as oxides. When we use a saucepan we do not want the pot to react with the water it contains. When we use a car we do not want the metal of the engine to melt!

What are alloys?
We have to find materials which have the right properties for a given task. When we use a saucepan, we need a metal that is a good conductor of heat. It must also be unreactive so that it does not react with water or acidic or alkaline solutions. If we are unable to find a metal with the right properties, we use an **alloy**. An alloy is a mixture of two or more elements, at least one of which is a metal. An alloy is generally more useful than the individual elements.

For many years, people have investigated what happens when metals are mixed in the liquid state and the mixture is then cooled. This is one way of making an alloy such as **brass** (an alloy of copper and zinc) or **bronze** (an alloy of copper and tin). One of the most important alloys known is **steel**, which was originally an alloy of iron and carbon (which is a non-metal). There are now many different steels in use, and they may contain other metals, such as nickel or chromium.

Metals and alloys are used to make particular objects because they have the right properties for the job. For example, both copper and aluminium are good conductors of electricity and they are both used for electrical wiring. Because the pure metals are soft and weak, electrical cables used for high voltage transmission have a steel core, to provide strength.

Brass is useful because it is harder than copper but also resistant to corrosion. **Duralumin** has a low density, like aluminium, but it is much stronger and more resistant to corrosion. Generally, alloys are harder than the pure metals and have lower melting points. **Solder** is another important alloy. It is made from tin and lead and can be melted easily by a soldering iron, so that it can be used to join wires in electrical circuits. It has a low melting point.

Questions

1. What is an alloy?
2. Why are metals such as sodium and potassium not used in the home?

What are some of the uses of metals and alloys?

Most cooking utensils are made of metals and alloys. For example, a lot of cutlery is made from stainless steel. Many chairs have a metal frame for strength and a plastic seat so that the chair is comfortable and lightweight. Light bulbs contain a metal filament, and electric kettles have a metal heating element. Cars contain a great deal of metal in their engines and in their bodywork (although there is increased use of other materials nowadays). Steel bands use 'steel' pans made from old oil drums, which are made of alloys. The pans carrying the top notes are often chrome-plated.

Activity | Different types of metals and alloys

1. Examine very carefully all the objects you use at home apart from those used for cooking and cleaning. Make a list of all those that are made of or contain metal.
2. Now look at the objects used for cooking and cleaning. Make a list of those made of or containing metal.
3. Repeat this exercise for any garden or farm tools.

You will probably find that you have a long list by now! You will not be able to tell for sure which metals and alloys have been used to make these objects. But you can make a guess, based on characteristics such as colour, strength and weight.

4. Think about the properties needed for a particular object. Pick a few of the objects you have listed and suggest the properties they need. For example, a metal chair frame needs to be strong but not too heavy. It also needs to be resistant to corrosion.
5. Select a few objects and suggest which metals or alloys they contain. For example, a lightweight, silvery, cooking utensil may contain aluminium. Ordinary electrical wiring is probably copper. This Activity should help to give you some idea of the importance of metals and alloys in our lives.

Questions

3. What are the properties needed for a metal or alloy used in a car radiator?
4. Name three important alloys and their uses.
5. Why should tools used in the garden or on the farm be resistant to corrosion?

Which metals and alloys are used for cooking and canning?

Cooking Most common cooking utensils are made from metals or alloys. There are exceptions to this, though. For example, many frying pans now have a non-stick coating applied to them. Special glass (such as Pyrex) is used for saucepans and dishes which can be heated in an oven. (What does this tell you about the properties of this glass?) Ordinary glass cannot be used in this way because sudden temperature changes cause it to break. You should never pour boiling water into an ordinary glass container.

Two metals very commonly used for cooking utensils are copper and aluminium, as we have seen. Copper utensils tend to be very expensive. In the Caribbean, we often use pots made of cast iron or aluminium. The metals or alloys used in cooking utensils need three properties in particular: they must be **good heat conductors**, they must be **unreactive** with water and dilute acids or alkalis, and they must **not be toxic** (poisonous). Copper is fairly unreactive and aluminium forms a layer of oxide which is very unreactive.

Canning Canning is one of the ways in which food can be preserved. The can in which the food is kept has to be **resistant to corrosion**. For example, fruit juices may be canned, but these juices contain acids. This means that the cans cannot be kept for too long. Common canning materials are aluminium or steel coated with tin. The tin protects the other metal/alloy from corrosion.

Questions

6. What properties are needed for metals or alloys used for canning food?
7. Why do you think that plastics have not yet been used for making cooking utensils such as pots and pans?
8. Do you think that cans used for preserving food or juices should be recycled? Should there be a cost for doing this?

Why do we use plastics in the home?

Plastics are now very common in our lives, although this has happened only in the last 30 years or so. Most plastics are **polymers** – see page 25. Polymers are very large molecules which are formed by a process called polymerisation. Many small molecules are joined together to form one very large molecule. Polythene is a polymer formed when many molecules of the compound ethene are joined together.

Activity | Which plastics are commonly used?

1. Examine the photograph. Which of the objects is likely to be made partly or completely of plastic?
2. Make a list of other objects you use at home which are made of plastic.
3. Choose a few of the objects and suggest some of the properties of the plastics used.

You will see that a wide variety of plastics exists. Some can be made into very thin ('cling') films, while others are quite strong. Some can be moulded easily, others are hard, and so on. Each type of plastic has different properties.

Plastics are examples of synthetic polymers. They contain the elements carbon and hydrogen, and may also contain other elements, such as chlorine or nitrogen or oxygen. Crude oil – see pages 188–9 – is a source of the raw materials for plastics.

Some of the common plastics are:

1. **Polythene** This can take at least two forms. The low-density form is used to make detergent bottles, plastic bags and insulation for electrical cables. High-density polythene is used for making objects such as washing-up bowls and dustbins.
2. **Polyvinylchloride (PVC) or polychloroethene** This is extensively used for making wrapping film and pipes and gutters for houses.
3. **Polystyrene or polyphenylethene** This is used to make food containers and disposable cups. It can be formed into expanded polystyrene, which contains a large quantity of air (see page 116). Polystyrene is very light and a good insulator and is used for packing delicate articles.
4. **Polyacrylonitrile** This can be spun into long fibres and used as a substitute for wool. The resulting fabrics are sold under names such as Orlon and Acrilan.

The starting point for each of these four plastics is ethene.

Two other useful plastics are:

5. **Nylon** This contains nitrogen and oxygen, as well as carbon and hydrogen. It is used as a fibre in clothes and ropes. Another form is used for making domestic fittings such as hinges for cupboards.
6. **Terylene** This contains oxygen as well as carbon and hydrogen. Woven terylene is used in making sails for boats – it is strong and does not rot easily. It is also used, in the form of Crimplene, in making clothes.

Questions

1. What is a polymer?
2. Name three natural polymers – see page 25, for example.
3. Name the pieces of apparatus you use in the laboratory that are made of plastic. Would you use a plastic beaker for boiling water?

What are the properties of some of the plastics?

You have seen that there are many different plastic objects used in the home and elsewhere. As with metals and alloys, plastics are used in the manufacture of specific items if they have the required properties. Investigate as many plastics as you can, so that you can see how different they are.

Activity | Testing plastics

1. Collect samples of as many different plastics as you can. For example, you could start with an empty washing up liquid bottle and a piece of plastic film used for wrapping food.
2. Examine the obvious properties of each sample. These should include the colour and approximate density of the substance. Note that the colour of the plastic does not make any difference to the properties of the plastic. Test the plastic to see if it is hard or soft.
3. Record your observations in the form of a table. If you have separate small pieces of plastic, heat a *very small* piece *very carefully* in a fume cupboard. Some plastics give off poisonous fumes when heated. Record the effect of heat on any samples you test.
4. Test the electrical conductivity of each sample using a simple electrical circuit like the one on page 172. Record the results of your experiments on conductivity in your table. Plastics have very different properties from those of metals and alloys.
5. If you have samples in the form of a film or thin sheet, test these samples by stretching them. Pull a sample between your hands as a first test. You can also use a simple apparatus like this to test elasticity.

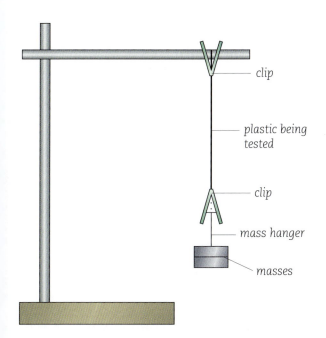

Add more masses to the mass hanger to stretch the plastic. Keep records of your results

6. Investigate the reaction of solvents such as water, ethanol, propanone, ammonia solution and dilute acid with small pieces of the plastics.

Questions

4. What are some of the advantages and disadvantages of using synthetic fibres in clothing?
5. How do supermarkets package food in your country? What happens to the packaging when the food has been eaten?

Why are plastics so useful?

You have seen that there are many different types of plastics, each with different properties. Objects made of plastic are now used where once these would have been made of wood or metal. Why is this? There are several possible reasons, including:

- The plastic object is cheaper than the same object made of wood or metal.
- The plastic has properties better suited to the task.
- Plastics have been developed with unique properties – for example, the transparent film used for wrapping food is unlike almost anything else.

Problems with plastics The major problem with plastics is waste and litter. Many of you will have visited a beach which is in a very poor state because people have left plastic food packaging, bottles and ice-cream containers behind. The packaging used for food and drinks is too often just thrown away once the contents have been consumed. Since most plastics do not break down into simpler materials when they are buried in the soil, this old packaging can lie around for years. Fortunately, more **biodegradable** plastics are beginning to be available now. These break down over time.

One of the advantages of using metal objects is that used items can be collected and used as 'scrap'. Many objects made of iron or aluminium can be used in this way, so that the metal is recycled. When you are thinking about the materials that were used to make the items you use at home, you will see that a balance often has to be found between factors such as cost, convenience, the properties of the material and possible waste problems.

Questions

6. Do you know of any schemes for collecting and re-using metal cans or plastic objects in your country? Find out what you can about these schemes. Who pays for them?
7. Should we aim to develop and use more biodegradable plastics?
8. Find out if plastic items are cheaper than similar metal items. Why is this?

What do we use for washing and cleaning?

What are soaps?
Soaps used today are often salts formed between sodium and complex organic acids. For example, common soap is made of a compound called sodium octadecanoate (sodium stearate). Soaps are made by the reactions of an alkali such as sodium hydroxide with an organic acid.

How were soaps made traditionally? In ancient times, the alkali was obtained from wood ash, which contains potassium hydroxide. The soluble hydroxide was washed out of the solid ash. The usual source of the acid was lard (tallow) obtained from animals such as goats. The alkali and the acid were boiled together to form soap.

Making soap today The basic principles behind the making of soap have not changed much. However, these days we expect soap not only to wash well, but also to have other properties, such as a scent and a smooth texture. Compare bath soap with a bar of washing (laundry) soap – what are the differences?

In the modern manufacture of soap, sodium hydroxide is boiled with a mixture of oils and fats. Coconut oil is an ingredient often used. The soap formed by the reaction of the alkali with the acids in the oils and fats is then separated from the solution by adding salt (sodium chloride) to the boiling solution. We say that the soap has been **salted out**.

When the solution is allowed to cool, there are two separate layers. In the first layer, there is the soap, sodium octadecanoate, while the second layer contains the solution, sodium chloride and glycerol: this mixture is called **lye**.

The word equation for the manufacture of soap is:

glyceryl octadecanoate(aq) + sodium hydroxide(aq) →
sodium octadecanoate(aq) + glycerol(aq)

Activity | Making soap
1. Place suitable quantities (your teacher will tell you how much) of concentrated sodium hydroxide solution and an oil in a 250 cm^3 beaker. Take care in handling the alkali.
2. Boil the mixture for about 20 to 30 minutes.
3. Add solid sodium chloride to the solution and allow it to cool.
4. You should find that you have two separate layers. Remove the soapy layer, wash it with water a few times and then dry it. You need to wash it to remove any sodium hydroxide which may remain.
   Do not use the soap you have made for washing because it is very impure.
5. Take a small piece of the soap you have made and shake it well with water in a test-tube. Compare how it lathers with a piece of bath soap of similar size.

Questions
1. Why are perfumes added to bath soap?
2. Why is salt added to the boiling mixture of the alkali and oil/fat?
3. What happens to soap when it is shaken up with hard water? (See page 218.)

How is soap manufactured?
Coconut oil and sodium hydroxide solution are boiled together. Sodium chloride is added when the oil begins to **emulsify**. When the oil has been converted to soap, tallow, grease and other oils are added, as well as more sodium hydroxide. Boiling continues until the process is complete. (The process is called **saponification**.)

The soap and lye are separated and the soap washed because it may contain both sodium hydroxide and glycerol. The soap is then boiled and water added. More salt is added to make sure that separation is complete. The glycerol is recovered and may be used again. The soap may be bleached at this stage, or colouring matter and perfumes added.

The soap then goes through a process called **fitting**, in which the texture of the soap is made even throughout. Water is added to the boiling soap and the soap is then pumped away to be dried.

What are detergents?
Soaps are examples of **detergents**. These are useful as cleaning agents because of the properties of the molecules in them. Detergent solutions have three important properties which make them particularly useful for cleaning:
1. They reduce the surface tension (see page 255) between water and any insoluble materials, such as oil or grease.
2. They foam easily, so they spread well over surfaces.
3. They stabilise insoluble particles which are removed from your skin or from clothes.

Detergents consist of molecules which may be thought of as consisting of two parts:
- a water-attracting or **hydrophilic** part
- a water-repelling or **hydrophobic** part.

Soaps
Soaps have two major disadvantages when they are being used. The first is that they form insoluble compounds with some substances present in water, producing a 'scum': the grey flakes formed when cleaning. This is common in a number of Caribbean countries, where water is obtained from sources which involve limestone or coral.

The salts present react with the soap and an insoluble salt is formed. This means that a lot of soap has to be used to get a good lather. Secondly, soaps are not stable in acid solutions which may occur in industry.

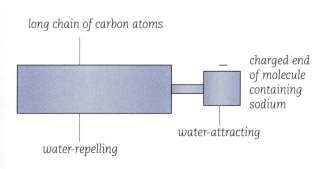

A soap molecule

Are there any alternatives to soaps?

Soapless detergents The demand for soap increased at the same time as more edible oils and fats were needed for food. There were smaller quantities of the oils and fats available for soaps. As a result, synthetic (soapless) detergents were developed. Many of these detergents are not affected by substances present in hard water – see page 218. They are often better for cleaning. Soapless detergents are often just called detergents.

The most important group of detergents are the **anionic detergents**, which have molecules similar to soap molecules. However, there is an important part of the molecule called a sulphonate (or sometimes a sulphate) group.

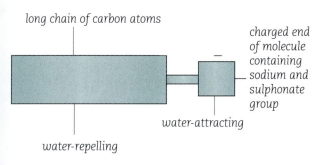

A detergent molecule

The action of a detergent may be shown in simple diagrams such as those below.

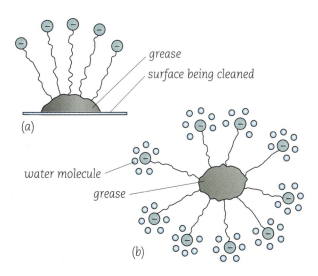

How a detergent works: (a) The uncharged end of the detergent molecule is attracted to the grease spot on the dirty surface. (b) The charged end of the molecule is attracted to water molecules, so the grease spot is pulled off and dissolves in the water

One of the problems associated with the early use of detergents was that of excess production of foam. This foam often did not break down and some rivers were polluted with large quantities of foam which killed plant and animal life. Detergents produced now do not contain the same material and foaming is no longer such a problem.

It is interesting to note that a packet of detergent may contain relatively little actual detergent. In many cases detergent may contribute as little as 20% of the mass of the contents of the packet. The rest of the mass is made up of others substances, such as water softeners, bulking agents, bleaches, enzymes and brighteners.

Detergents have been developed for all sorts of special uses, including washing clothes and dishes.

Questions

4 What is the main disadvantage of soap?
5 Are detergents affected by use in hard water?
6 Some people react badly to the effect of detergents on their skin. How could this reaction be prevented?
7 What are the main differences between soaps and detergents?
8 How would you test two different brands of washing-up liquid to find out which was more effective?

How can we keep ourselves clean?

Every day our skin becomes dirty. We pick up dust and dirt from our surroundings. We sweat and the perspiration dries on our skin. The top layer of our skin dies and the dead cells accumulate. Micro-organisms live on our skin (page 289) and decay the substances there, producing unpleasant odours.

What do we use to keep ourselves clean?

1. *Soap and water* As we have seen (page 207) soap acts to loosen dirt and grease. This means that when we wash ourselves with soap it takes away the surface layer of dirt from our skins. It cleans away the sebum (page 49) which might otherwise encourage pimples or acne. It also cleans away the sweat which might otherwise lead to an unpleasant stale smell.

    We need to use a gentle bath soap for our skin. Detergents or harsher soaps, such as laundry soap, may have too much alkali in them, which can damage sensitive skins. Some soaps have antiseptics added which help to kill germs.

2. *Deodorants* Deodorants are designed to take away odour (smell). They kill the bacteria which cause the smell. They are usually powders, sprays or roll-ons. They contain a substance which absorbs odours. They often have a perfume of their own which replaces the original odour.

3. *Antiperspirants* These are designed to block the sweat glands and so stop us from perspiring. This may stop the smell of decaying sweat accumulating but is not necessarily a good idea. We need to sweat when it is too hot and use this method to cool ourselves.

4. *Vaginal douches* These are specially prepared cleaning solutions that a woman can use to wash out her vagina. However, the solutions may kill the natural bacteria which can be useful and may wash any germs higher up into her reproductive system and cause infection.

5. *Antiseptics* These can either be fluids, creams or substances added to soaps. An antiseptic is used to kill bacteria and thus prevent infection. For example, antiseptic mouth washes can be used as gargles to kill off the germs that may cause a sore throat. Antiseptic creams are put onto a wound to keep it clean while the skin grows to heal it.

6. *Antibiotics* Antibiotics are used if the person becomes infected by bacteria. For example, there are antibiotic creams which can be used on sores which have become infected. The cream kills the germs and then promotes healing of the sore.

    Other antibiotics such as penicillin or tetracycline can be given as pills or injections. Antibiotics have to be prescribed by a doctor and used carefully according to instructions (page 160).

7. *Disinfectants* Disinfectants are usually too harsh to use on people. They are used to keep *things* clean such as toilet bowls, sinks and baths. The disinfectant is usually in the form of a liquid or cream which is able to kill germs which might otherwise infect humans.

Note Take special care when cleaning your eyes (use clean water without soap) and ears (use soapy water but do not push anything inside the ears).

Manufacturers make claims about their cleaning materials. Read the label on any cleaning material. How could you scientifically test the claims made?

Activity | Cleaning materials

1. Collect examples of the different kinds of cleaning materials described above. Examine the labels of each one to find out what is in it. Try to identify what kind of cleaning it does, for example removal of grease, or absorption of odours.
2. Make your own deodorant. Mix together three parts of talcum powder with one part of sodium hydrogen carbonate. Add a few drops of perfume and mix together. Put the mixture into a container and sprinkle it onto your body after washing. The sodium hydrogen carbonate will absorb body odours, while the perfume will give you a pleasant smell and the talcum powder will make your skin feel smooth.

Why do we need to clean our teeth?

When we eat, some of the food particles accumulate between our teeth. This is especially a problem with foods which contain starch and sugar. This builds up a layer called **plaque**. The bacteria that are in our mouths break down the food particles to produce acids. These acids attack the enamel of the tooth and may cause tooth decay. What we eat is also important – milk and milk products such as cheese contain calcium and phosphorus to help build strong teeth. Fluoridation can also help to harden the teeth so they are not easily decayed. There is more information on page 83.

After you have eaten something sweet, your teeth are attacked by acid for up to an hour. Therefore limit the amount of sweet things you eat, and clean your teeth after meals.

How do we clean our teeth?

1. *Using a toothbrush and toothpaste* Why do we use a toothbrush? It has stiff bristles which can get between the teeth. These dislodge some of the plaque. We should brush our teeth on both sides, away from the gums.

    Why do we use toothpaste? Toothpaste has a consistency to slide over the teeth, and with the help of a toothbrush gets into crevices between our teeth. It also contains chalk or other powder to clean the surfaces of our teeth.

    Toothpaste usually also contains an antiseptic which helps to kill the bacteria that are inside our mouths. It also has a flavour, such as mint, which helps to freshen the breath. This helps make toothpaste attractive to use.

    Many toothpastes contain fluorides which help to strengthen young teeth. However, only the very outer surface of the teeth is affected, compared with the greater effects of fluoridation of the water supply (page 83).

2. *Using a mouthwash* Rinsing out the mouth with sodium hydrogen carbonate solution after eating will neutralise (page 221) any acids produced by bacterial decay.

3. *Using a chewstick* This is a bit like using a toothbrush without the toothpaste. The fibres of the chewstick can dislodge some of the food particles between the teeth. Some people also use chewing gum.

4. *Using crisp food* Crisp foods such as carrot, celery or crisp fruits can help to dislodge food particles. But a toothbrush is more effective.

5. *Using dental floss* This is a strong narrow thread of nylon material. It is pushed down between the teeth, right to the gums, and it is able to dislodge particles of food that have been left behind by the toothbrush.

6. *Using cheese* Cheese helps to neutralise the acids that are formed during the decay of other foods. So if you are eating cheese, have it at the end of the meal.

Questions

1. In the Caribbean we tend to concentrate on brushing our teeth when we awake in the morning, but we do not place as much emphasis on brushing our teeth after meals. Is this good practice? Explain.
2. How effective are the different cleaning materials shown below? Design an Activity to find out.

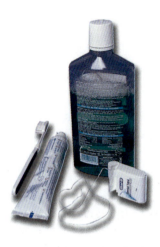

Materials for cleaning the teeth and mouth cavity. How could you test their effectiveness?

Activity | What is the pH of toothpaste?

The pH describes the acidity or alkalinity of a substance. When an alkali is added to an acid, **neutralisation** occurs. If there are lots of bacteria in our mouths acting on the food to produce acids, then what would you expect the pH of toothpaste to be? Let's find out.

1. Each student should bring a small sample of toothpaste to school. You could wrap the toothpaste in a small piece of clean white paper. Label your sample.
2. Mix a small amount of each sample with a little distilled water.
3. Use pH paper to find the pH of each of the samples. Draw up a table to show your results.

Questions

3. Are these the results which you expected?
4. What effect will the toothpaste have in your mouth?
5. Why did we mix the samples with *distilled* water?
6. Which of your toothpastes had added fluoride? Did this make any difference to the pH reading?

How can we keep household appliances clean?

Why do household appliances need cleaning?

You probably know that iron **rusts** when it is exposed to air and water for some time. This process is considered in more detail on pages 212–13. Rusting is one example of **corrosion**, and corrosion is a chemical reaction in which a metal reacts slowly to form new compounds.

Many household objects are made of metals or alloys – see pages 202–3. Some are made in such a way that the metal is protected from the air. Some metals **tarnish** when they are exposed to the air or to other gases. For example, silver tarnishes when it is exposed to air containing the gas **hydrogen sulphide**. Metal taps in kitchens and bathrooms may become tarnished and need regular cleaning. We can remove metal tarnish by using a metal polish that reacts chemically with the stain.

Activity | Household metals

1. Look back to pages 202–3. Copy the list of household objects made of metals or alloys.
2. Which of these objects must be cleaned immediately after they are used?
3. Which have to be cleaned after being exposed to the atmosphere for a long time?
4. Are there any objects that do not need to be cleaned?

Cooking utensils These need cleaning immediately after they are used. One obvious reason for this is that cooking utensils are heated and exposed to water and other chemicals in the food. The metals or alloys used for making the utensils are not very reactive, so there should be little corrosion or tarnishing of the surfaces. There will, however, be traces of food on inner and sometimes outer surfaces, and these must be removed for health reasons before a utensil can be used again. We keep the surfaces of cooking utensils clean not to make the kitchen shine, but to keep ourselves healthy.

One of the reasons for the continuing use of **aluminium** is that it is protected by a surface layer of oxide. This is very unreactive and difficult to break, so the metal surface is well protected. This reaction of the metal is very different from the corrosion (rusting) of iron, since the rust does not protect the metal underneath. If you have an aluminium saucepan at home, clean it very carefully with an abrasive such as steel wool and then dry it. What do you see?

Questions

1. Vegetable oil is commonly used in cooking. Is this likely to preserve the surface of metal cooking utensils?
2. What is meant by (a) rusting and (b) corrosion?

How can we clean used household appliances?

You may have washed items such as plates, knives and forks. The surfaces can usually be cleaned with a dishwashing detergent and hot or warm water. You may have to rub quite hard with a cloth or sponge to remove all the food remains, particularly if these have been left for some time.

Metal objects may need more than just detergent and water. You may have seen your parents use steel wool or an abrasive pad such as a Brillo pad. When you use steel wool you not only remove the food remains, but also scrape off some of the layer of metal or alloy. Steel wool is hard and the surface of the utensil is softer. You can see the scratches formed by the action of the wool on the metal surface.

Activity | Cleaning methods

1. Check the list of objects that need cleaning after use or after being exposed to the atmosphere.
2. How many different ways of cleaning metal items do you know? Make a list of these methods.
3. Which methods involve cutting into the metal's surface?

Cleaning materials. Detergent (washing-up fluid) is being used. Near the sink is a plastic scouring pad which would not scratch a pot's surface as badly as steel wool would

You may be able to try out some of the methods of cleaning you have listed. If you start with a dirty or tarnished object, you can try cleaning it using hot water and a cloth only, to see how effective this method is. You could then try other methods, starting with a detergent only. In other words, you are starting with the simplest method and then working up to more complex methods if the simpler methods don't work.

You may be able to compare different methods of cleaning by examining which method is most effective for particular objects. Also think about costs and possible health hazards – do you read the labels on bottles of cleaning agents?

Questions

3 Would you clean all metal objects with a metal polish? Consider the costs and possible health hazards.
4 Do shops or supermarkets in your country sell cleaning agents made from locally available materials?
5 How would you try to remove traces of burnt food from the surface of a cooking pot?

Cleaning by scouring

How do scouring pads work? Steel wool is a common scouring material. Using a hand lens, examine the surface of an object which has been cleaned with steel wool. You are likely to see a mass of scratches on the surface. You will see similar scratches when other cleaning pads are used, such as those made from hard nylon. Some cleaning pads are filled with detergent, so that you clean as well as scour the surface.

What are scouring powders made from? There are a number of substances sold in shops and supermarkets which can be used for cleaning the surfaces of household items. The surfaces to be cleaned will vary a great deal – they may include metal containers, sinks and baths. If you live in an area where there is hard water, baths and sinks will need constant attention. Scouring substances are usually powders, but they are often suspended in a liquid to form a paste. The powders are made from very hard substances, to which a bleach may be added.

Activity | Scouring powders

1 Find several samples of cleaning powders or pastes.
2 List these samples, and check the labels to find out what the cleaning substances contain.

Question

6 Why is bleach added to some scouring powders or pastes?

How do scouring powders work? Soaps and detergents work best on the surfaces of objects that need to be cleaned. Dirt and grime which is deeply engrained is more difficult to remove. This is true of dirt on clothes or on household objects. One problem which is often difficult to deal with if left for too long is that of dirt on sinks and baths. This may be due to one or more of a number of causes. For example, hard water causes scum, and this can build up on a surface if it is not cleaned off. Another problem is that of greasy objects being washed in a sink, but the sink is not cleaned regularly. Remember that you should not use abrasive substances such as steel wool on sinks or baths – you will ruin their surfaces. Scouring powders work in two main ways:

- Removing surface dirt and grime by the rubbing action of hard particles against the surface.
- Removing other substances by chemical action, if a bleach has been added to the powder.

Questions

7 What would you use to clean a black plastic sink? Why?
8 Why should you wash an object with water after it has been cleaned with scouring powder?

A clean and shining bathroom. Surfaces which accumulate scum and grime need to be washed with a scouring powder

What causes rusting?

Are air and water important for rusting?

You probably know what happens when a machete or cutlass is left out in the rain overnight. In the morning, you are likely to see a brown substance on the surface. The same thing happens if you leave an iron cooking pot with water standing in it for some time. The brown substance which forms on the surface of iron or steel objects is **rust**. Rust contains an oxide of iron and **rusting** is an example of **corrosion**. Unlike the oxide layer on aluminium, rust does not protect the surface of the iron or steel. We need to inspect iron and steel objects for rust regularly because they can be weakened by it, which makes them potentially dangerous.

Since rust is formed at the surface of iron, we can investigate the conditions necessary for its formation fairly easily. If we can find out how rust is formed, we may be able to find ways of preventing rusting. There are so many objects made of iron and steel that protecting them from rusting may save money and lives.

Activity | How does iron rust?

Investigate the ways in which iron rusts by examining what happens to an iron nail when it is subjected to different conditions.

1. Take four iron nails, each of which can fit into a test-tube or boiling tube. Clean each of the nails carefully using emery paper. Make sure that the surface of each nail looks clean.
2. Put one nail in each of the four tubes.
3. Set up the tubes as follows:

Tube 1 Put in some dry calcium oxide, and close the mouth with cotton wool. Calcium oxide absorbs moisture from the air.
Tube 2 Add water which has been boiled and then allowed to cool in a sealed container. Cover the nail entirely, and then add a layer of oil on top of the water.
Tube 3 Leave the tube open to the air.
Tube 4 Add water so that the nail is covered, and leave this open to the air.

This means that each nail is exposed to different conditions. These are (1) dry air (2) water only (3) air and moisture and (4) water and air.

4. Examine the nails each day for about a week. Record your observations in a table. Make sure that you record accurately whether there are any changes on the surface of the nails.

In tube 1, with dry air only, there should be little or no rusting. This means that **air by itself** is not enough to cause rusting. In tube 2, the nail was in water only. Water which has been boiled contains little or no air because the air has been driven off by boiling. Again, there should be little or no rusting. This means that **water alone** is not enough to cause rusting. The nails in tubes 3 and 4 will have rusted, although there may be more rust on the nail in tube 4. In both cases, the nail is exposed to both air and water. Remember that there is water vapour in the atmosphere. We know that water and oxygen in the air are needed for the iron to rust. We can write a word equation to try to show what happens:

$$\text{iron (s) + oxygen (g) + water (l)} \rightarrow \text{brown iron oxide (s)}$$

This is not very satisfactory, since it does not tell us anything about the oxide of iron. Rust is not a simple compound but we know that water is involved. One way of saying this is to call rust **hydrated iron oxide**. This tells us that the oxide is combined with water in some way. A more satisfactory equation could be:

$$\text{iron (s) + oxygen (g) + water (l)} \rightarrow \text{hydrated iron oxide (s)}$$

Iron objects rust, and need protection from air and water.

Rust on a car

Questions

1. For tube 2, why do you add a layer of oil to the boiled and cooled water?
2. Would you expect the exhaust pipe on a car to corrode quickly? Why?

Are there other ways in which metals corrode?

If you live near the sea, you will have noticed that metal objects corrode easily. These objects are exposed not only to the air, but also to other substances which are carried in the air as part of the fine spray that comes off the sea. Sea spray contains sodium chloride, which is what makes sea water salty. It may also contain fine particles of sand. Similarly, if you live near a bauxite/alumina industrial plant in Jamaica or Guyana, you may notice corrosion problems with metal objects. If you are near a cement factory or oil refinery, you may notice something similar. We obviously cannot investigate all of the possible causes of corrosion of metals. However, we can investigate what happens when one metal, **iron**, reacts with a number of different solutions.

Activity | *Corrosion in different solutions*

In this Activity you investigate the effects of different solutions on the surface of iron.

1. Take six iron nails and clean them as you did in the previous Activity.
2. Put each nail into a separate test-tube or boiling tube.

Tube 1   Leave this open to the atmosphere.

3. Add enough solution to each tube to cover the nails. The solutions to be used are:

Tube 2   Dilute sodium chloride
Tube 3   Dilute sodium hydroxide
Tube 4   Very dilute acid (sulphuric or ethanoic)
Tube 5   Dilute sodium carbonate
Tube 6   Dilute aqueous ammonia

4. Examine the nails each day for about a week and record your observations in a table.

The nail left open to the atmosphere (tube 1) acts as a control (see page 33). You should be able to see that each of the nails has corroded on the surface. You do not need to find out what substance has been formed in each case. If you look at the list of solutions used you will see that there are two alkalis, an acid and two solutions of salts – see pages 220–1. Although it is difficult to tell from the results of one experiment, there is more corrosion of the iron when the solutions are present than when air and water alone are present.

You can now begin to see why we have to be careful about the properties of metals and alloys that are used in the home and in industry. We use water all the time both at home and in industry, and water itself may cause corrosion problems, particularly if it contains dissolved substances.

We have not tried to investigate what happens when the **temperature** is increased. You might expect that this would make metals corrode more quickly. Cooking food involves the use of high temperatures (have you ever checked the temperature of an oven?). Many industrial processes also require heat.

A fractionating column in an oil refinery – see page 188 – works at high temperatures. It has to be made from metals or alloys which will not corrode even at high temperatures.

Corrosion and the sea

Metal objects exposed to the sea for long periods of time need to be protected from corrosion. The sea contains many corroding salts, such as sodium and magnesium chlorides.

Ships and small boats are exposed to the sea and the air all the time, so metal objects on them will tend to corrode quickly unless they are protected. Ships and boats also have to be made from materials strong enough to stand up to winds and waves. Some small boats are made of wood, but most larger ships are made from iron and steel. They move through the sea, which can be thought of as a dilute solution of sodium chloride, with other salts added.

Corrosion such as rusting starts at the surface of a metal object, so one of the ways of trying to prevent corrosion is to cover the surface of the metal. The outer surface of the hull of a ship will be protected by paint, which has to be renewed regularly. Other methods are described on pages 214–15.

An ocean liner is painted with special paints to prevent corrosion

Stainless steel

This is an alloy which will not corrode easily. It is a steel containing chromium, nickel and iron. There are many different stainless steels, widely used in industry and for objects in the home. They are more expensive than ordinary steels.

Questions

3. Box girder bridges look like this in cross-section. Why do you think bags of a drying agent such as silica gel are placed inside the boxes?
4. Why do surgeons use instruments made of stainless steel when they operate?
5. If you had to send a car by sea from one country to another, how would you try to protect the bodywork of the car?

How can we try to prevent iron from rusting?

What can we do about air and water?

Rust is not only unsightly; it may also be dangerous. You know that both air and water are needed for iron to rust. Rusting always starts on the surface of metal objects, so it may be possible to prevent this process of corrosion by stopping air and water from reaching the surface. There are several possible ways of doing this and you can investigate some of them now. In each case, the surface of a nail is covered and you can compare how it rusts with a nail which has not been covered at all.

Activity | Can covering prevent rusting?

1. Put four test-tubes in a test-tube rack. Pour water into each test-tube so that the nails will be completely covered.
2. Clean four nails carefully, as in the previous Activities. Put one nail in **Tube 1**. This acts as the **control**.
3. Prepare the other three nails as follows:

Tube 2 Dip the nail in oil paint and let it dry.
Tube 3 Dip the nail in a thick oil or cover it with grease.
Tube 4 Cover the surface of the nail with a metal such as nickel by **electroplating** it with a solution containing a nickel salt. Your teacher will help you.

4. Examine each nail every day for 7 to 10 days. Record your observations in a table.

In each case, except for the control nail in tube 1, the iron surface of the nail has been protected from both air and water. Can you see any differences between the nails in tubes 2, 3 and 4? The nail in tube 1 will have rusted a little, but the other nails should not have rusted at all. This tells us that preventing air and water from getting to the surface of iron objects may stop them rusting. You should note the fact that we have used only four small iron objects at room temperature for a short period of time. We expect our iron cooking pots to last for a long time, and these are used at high temperatures. If your school laboratory has a water bath, you could repeat the last Activity but keep all the tubes at 60 °C. You will need a control in which you keep a set of nails at room temperature for the same period of time. What would you expect to happen?

Most chemical reactions proceed faster at higher temperatures, and the rate at which iron rusts is also faster at higher temperatures. For this reason, the coverings on the surfaces of the nails might not be so effective at higher temperatures.

In industry, we may not want to have a covering on the surface of a metal container unless we are sure that this covering is very secure. It will need to be checked regularly.

We use iron and steel in many different ways. Can you make a list of at least 10 uses? Iron and steel used in the construction of large objects such as ships and buildings may be exposed to all sorts of weather conditions. These will include wind and rain, each of which may carry corrosive substances found in the atmosphere.

To test the effectiveness of the surface coverings on your nails, investigate what happens:

- when they are covered and left out in the weather for a long period of time
- when they are covered and then left in solutions such as those used in the Activitiy on page 213.

Activity | How good are surface coverings?

1. Prepare eight nails (two sets of four) in the same way as in the previous Activity.
2. Put one set of nails outside in the school grounds (where they can be checked regularly). Examine the nails every other day for up to a month. Record your observations. Suggest which coverings seem most effective.
3. Get into groups and let each choose one of the following solutions: dilute sodium chloride, aqueous ammonia, dilute sodium hydroxide, dilute sulphuric acid. Place the remaining four nails in test-tubes, each containing the solution you have chosen.
4. Examine the nails every day for 7 to 10 days. Record your observations. Which of the coverings is most effective against the solution you have chosen? Compare your results with those obtained by other groups using different solutions.

These Activities may help you to understand that there is no one method which is best for stopping iron from rusting. Iron and steel objects are used in many different ways and we have to choose the best method for preventing rust for a particular object and the way in which it is used.

Painting A bridge may be protected by painting it. This costs money since the paint is worn away by the action of wind and rain.

Greasing In a car engine or on a bicycle, iron and steel parts may be protected by oil or grease. You may have seen advertisements proclaiming the value of various motor oils. Looking after a machine involves maintenance, which costs money – but much less money than a complete breakdown!

Can we stop rusting in other ways?

One of the ways in which we can try to prevent iron from rusting is to stop air and water from getting to the surface of the iron. We can use paint or oil or grease to protect the iron from rusting.

Galvanised iron There is another important way of protecting the surface of iron. You may be familiar with 'zinc sheets' or 'galvanise' used for roofs and fences.

Zinc sheets are usually described as being made of two metals, iron and zinc. In fact, it would be more correct to say that they are made from steel and zinc. Their correct name is **galvanised iron sheets**. The steel is coated with zinc, and the zinc will corrode before the iron, because it is more reactive than the iron.

Making galvanised iron The sheet steel is often dirty and rusty when it reaches the factory, and has to be cleaned before it can be used. It is placed in a bath of dilute acid, which removes most of the rust. It is then washed in water, before being put in more acid to remove the last traces of rust. Then it is washed again and dried.

The cleaned sheets are placed in a bath of molten ammonium chloride, which forms a layer on the surface of the metal. The ammonium chloride layer reacts when the steel sheets are passed into molten lead. Hydrogen chloride is formed and this removes any remaining traces of impurity on the surface of the steel.

The cleaned steel sheets are then passed into molten zinc, and an alloy of zinc and iron forms on the surface of the steel. A coating of zinc forms above the alloy layer. Finally, the sheets are passed through molten ammonium chloride again, before being cooled and washed with water. Galvanised iron sheets are very widely used in many parts of the world, including the Caribbean.

Stainless steel Stainless steel, which contains both nickel and chromium, is an alloy. Its surface does not usually need to be protected because it does not rust in the same way as cast iron.

Sacrificial anodes on the guard surrounding the propellor of a ship

Sacrificial anodes Ships are protected from rusting in a number of different ways. One method is the use of sacrificial anodes. These are pieces of a metal, such as zinc, which is more reactive than the metal from which the hull is made – usually steel. The anodes are attached to the hull and corrode before the metal of the hull does. They have to be replaced as they are worn away. In other words, they are **sacrificed** to protect the hull. A similar method is used to protect electrical cables laid underground.

Questions

1. In cold countries, salt is spread on the roads when there is snow or ice. Why is this done? How could cars be protected from the salt?
2. How would you protect a machete or cutlass (which is used regularly) from rusting?

Questions

3. Try to find out how galvanised iron reacts with (a) water (b) dilute acid and (c) dilute alkali.
4. Why would you expect stainless steel to be more expensive than cast iron?
5. What would you expect to happen if the surface of a galvanised iron sheet is bent or broken?

What happens when we try to dissolve things?

The importance of water

Water is a very important solvent. Many of the reactions that take place in our bodies use water as a solvent (see page 138). However, water is not just a solvent – it may also be involved as a chemical in a reaction. Such reactions include the building up and breaking down of molecules (page 72) and photosynthesis (pages 66–7). When water is formed during the oxidation of food, heat is evolved (see pages 104–5). If 350 g of water is formed each day during respiration, over 9 000 kJ of energy is evolved. This energy is used in the body.

Activity | Does everything dissolve easily in water?

Your teacher will give you some different substances.
1. Take a small quantity of one of the substances and put it in a test-tube.
2. Add a few cm^3 of distilled water, and shake the tube. If the substance does not seem to dissolve, add a little more water and shake again.
3. If the substance still does not seem to dissolve, warm the tube gently, but do not allow the water to boil.
4. Record whether the substance is soluble or insoluble in cold and warm water.
5. Try to filter the contents of the test-tube.
6. Record your observations in a table.

Substance	Cold water	Hot water
Sodium chloride	soluble	soluble
Cane sugar		

You should find that only insoluble substances leave a **residue** on the filter paper. The soluble substances form solutions, and these pass through filter paper.

At this point you may need to be reminded of the meaning of some words. When a solid is dissolved in a liquid, the solid is called the **solute** and the liquid is called the **solvent**. If you dissolve blue copper (II) sulphate in water, the copper sulphate is the solute and the water is the solvent. A blue **solution** of copper (II) sulphate is formed.

It is also possible to have **solid solutions**. Many alloys contain solid solutions in which one metal is dissolved in another. When a liquid solution is formed it is possible to see through the solution: it is **transparent**. For example, copper sulphate solution is blue but it is possible to see through it. However, in some cases, we do not get a solution but a suspension.

Activity | Making a suspension

1. Add dilute hydrochloric acid to sodium thiosulphate solution and allow the mixture to stand.
2. Can you see through the substance in the tube? You should find that there are yellow particles (of sulphur) in the test-tube. Here we have a suspension because the particles do not settle, but remain suspended in the liquid.
3. Now try to filter this suspension of sulphur. Most of the solid particles remain on the filter paper.

There are also **colloidal solutions** (colloids) in which the particles are smaller than those in a suspension, but bigger than the solute particles in a true solution. Examples of colloidal solutions include starch solution, milk and egg albumen.

Activity | Solutions and suspensions at home

1. List some of the common liquids used at home.
2. Try to divide the list into two groups: (a) solutions and (b) suspensions.
3. Record your observations in a table.

> ## Questions
>
> 1. Define the terms (a) solute (b) solvent and (c) suspension.
> 2. Jelly is an example of a colloidal solution. What is (a) the solute and (b) the solvent?
> 3. 'Solutions are examples of mixtures'. Is this statement correct?

Water will dissolve many solids to form aqueous solutions. However, there are many solids which do not dissolve in water.

Activity | Can we find solvents other than water?

1. Use the same substances as those you tested with water. Work in a fume cupboard if possible. Put a small amount of the first substance in a test-tube and add a few cm^3 of a liquid such as propanone, methylated spirits or 1,1,1-trichloroethane. Use great care handling these liquids.
2. Test the solubility of each solid in each liquid when cold.
3. Do not heat a test-tube directly. The liquids are flammable.
4. Record your results for each solvent in the same way as for water.

The solvents you have investigated are called **organic solvents**. They are very different from water. Substances which dissolve in water do not necessarily dissolve in these organic solvents. Generally, organic solvents will dissolve other organic substances. Carbon is the important element in organic substances – see pages 24–5. If you use plastic apparatus in your laboratory, such as measuring cylinders and beakers, take care that the organic solvents do not dissolve them.

Some organic solvents are widely used in industry. **Propanone** is used for dissolving fats and resins, and **tetrachloromethane gas** (acetylene) is stored under pressure in cylinders and used for welding.

Questions

4. What type of solvent would you use for cleaning paint brushes?
5. Carbon tetrachloride (tetrachloromethane) was once used in fire extinguishers. Find out why it is no longer used in this way.
6. Is nail polish likely to contain water as a solvent? What should be an important property of the solvent present in the nail polish?

It is possible to make a colloidal solution by mixing a liquid and a liquid. The solution formed is called an **emulsion**. (You will see tins of paint with this word on them.)

Activity | Making an emulsion

1. Add a little kerosene to water in a test-tube. You should find that the two liquids do not mix.
2. Put a rubber bung in the mouth of the tube to close it and then shake it vigorously. Record what you see.
3. Now allow the test-tube to stand for some time. Do the contents remain the same?
4. Shake the test-tube again, and then add some soap solution. Allow the test-tube to stand.
5. Record your observations.

The soap solution helps to stabilise the emulsion formed by kerosene and water. This property is used for purposes such as cleaning clothes.

6. Repeat the experiment, starting with some clear oil and some ethanol. Add some water to the clear solution, and then shake the test-tube.

7. Record your observations. Allow the test-tube to stand. You should see that the emulsion begins to separate out.
8. Instead of adding water, add soap solution to the emulsion and allow it to stand. Record your observations. As before, the soap solution helps to stabilise the emulsion.

Solvents and cleaning

Solvents may be used for removing stains from clothes. To test this you will need to have different pieces of old cloth, each of which has a different stain. For example, use different kinds of dried foods, oil and drinks.

Activity | Can we use solvents to remove stains?

1. Test each organic solvent on each of the different stains.
2. Take great care when handling the solvents, since propanone and methylated spirits will burn very easily, while 1,1,1-trichloroethane is poisonous. Try not to get any solvent on your skin. If you do, wash it off with large amounts of water.
3. Record your observations in a table.

You should find that different solvents dissolve different stains. If you are going to remove a stain you need to know what caused the stain, and then choose the correct solvent. If you are going to use large amounts of a solvent – in a dry cleaning establishment, for example – you will need to consider a number of factors, including the effectiveness of the solvent on a range of stains, its flammability, toxicity (whether it is poisonous) and cost. Propanone is too expensive for use at home, so we usually use cheaper solvents, such as methylated spirits and turpentine.

Dry cleaning Cleaning clothes with soap and water is not always the best approach. If you check the label of a garment you will usually find the instructions for cleaning the item. For example, the label may say 'Dry clean only'. Labels on other items may allow you to 'Machine wash at 40 °C'. You should not try to wash clothes with soap and water if the label tells you to 'Dry clean only'. The solvents used by dry cleaners are usually organic and are chosen for their effectiveness and affordability.

Questions

7. Would water be a good solvent for grass stains on clothes?
8. Which solvents are commonly used at gas stations?
9. Why is warm, soapy water more effective in removing grease from dishes than cold water?

What do we mean by hard and soft water?

How can we test the hardness of water?

We know that water is a very important solvent – see page 216. However, it will not dissolve everything. Cold water is not much use if you are confronted with a pile of greasy dinner plates! It is the combination of water and a detergent which makes for easier and more effective washing and cleaning. When water and a detergent are mixed and then shaken, you expect to see a lather formed.

In many parts of the Caribbean, water is obtained from sources where the rock is coral or limestone. It is difficult to obtain a lather with such water. We need to investigate why water from some sources forms lather easily, while water from other sources does not. (Water sources were discussed on page 140.)

Activity | Different types of water

1. Collect water from as many different sources as you can. Your samples should include: tap water, rain water, distilled water, sea water and river water. You could also try to obtain samples of water which has been polluted in some way. If this is not possible, make up a sample of dirty water yourself.
2. Put about 5 cm^3 of each sample into separate test-tubes.
3. Add a few drops of soap solution to each water sample. Make sure you add the same amount in each case.
4. Close the mouth of each test-tube with a rubber bung and shake the tube vigorously for a short time.
5. If little or no lather is formed, add a few more drops of soap solution and shake again.
6. Record how much soap solution is used to form a lather. You can use a table like the one shown below.

Water	Drops of soap solution added	Lathers easily	Forms scum
Tap			
Rain			
Distilled			
River			
Sea			
Polluted			

Allow each test-tube to stand. Observe what happens to the lather. You should find that some of the water samples, such as rain and distilled water, form lather easily. When the test-tube is allowed to stand, the lather remains for some time. Other water samples, such as the polluted water, hardly form a lather, and if a lather develops, it does not last long. If there is little or no lather you will see **scum** forming.

Water which forms a lather easily is called **soft** water. Water which will form a lather only with difficulty is called **hard water**.

What makes water hard? In some Caribbean countries, such as Barbados, the water that comes out of the tap is hard. Tap water has usually been treated at a **water treatment plant** – see pages 140–1. This water comes from sources such as rivers, streams or underground wells. (There are very few rivers in Barbados. By contrast, there are many on the island of Dominica, which has about a quarter of the population of Barbados.)

These different sources of water contain different substances in solution. Water is a very good solvent, as we have seen, and water that passes through rocks or soil or sand will dissolve soluble substances it meets on its way. If there is a strong current, for example in a fast-flowing river, solid particles will be picked up as well. Water needs to be treated before it can be distributed so that it does not carry substances which might be dangerous to health.

There are two types of hardness of water.

Temporary hardness Temporary hardness is caused by the presence of the substance calcium hydrogen carbonate. We know that there is carbon dioxide in the atmosphere. Carbon dioxide can dissolve in water and form a substance called carbonic acid. When rain falls, it may contain some dissolved carbon dioxide – i.e. carbonic acid. If this rain falls in an area where there is limestone (a form of calcium carbonate), some of the calcium carbonate may react with the carbonic acid to form the **soluble** substance calcium hydrogen carbonate. (Calcium carbonate is very insoluble.) The soluble substance will be carried along in the water of rivers and streams. Treatment at a water treatment plant does not remove calcium hydrogen carbonate.

This explains why tap water obtained from limestone areas is hard – it does not form a lather easily, but produces scum instead. When water such as this is boiled, **insoluble** calcium carbonate is formed and then deposited – see the photograph on the next page. The build-up of such deposits affects the efficiency of kettles, because the heating element has to heat the layer of solid as well as the water.

Permanent hardness Permanent hardness is caused by the presence of either calcium or magnesium sulphate. As with calcium hydrogen carbonate, these can be picked up by water travelling over rocks or through soils.

Questions

1. What is the difference between hard and soft water?
2. Would a hospital face any problems if it had to use hard water?
3. How would you make a sample of temporarily hard water soft without using chemicals?

'Scale' or 'fur' inside a kettle

Is hard water a problem? If we have to use water which is temporarily hard, we may end up with problems such as the one illustrated in the photograph. As the layer of deposit (often called 'scale' or 'fur') builds up, more energy has to be supplied to heat the same amount of water. Some people wash their kettles out with vinegar (which is a dilute acid) as a way of removing the solid layer.

You should be able to tell whether the water you use at home is hard or soft. If it is hard, you can use a water softener, but this costs money. Hard water makes washing more difficult, and more detergent has to be used to obtain a lather.

Activity | How can we soften hard water?

1. Put about 5 cm³ of soft water in a test-tube, add a few drops of soap solution and shake for about 20 seconds. Leave the test-tube to stand, and use this as a control.
2. Use the samples of hard water provided. One will be permanently hard and the other temporarily hard. Take about 25 cm³ of each and boil them in separate 50 cm³ beakers.
3. Allow the hard water samples to cool and then take 5 cm³ of each sample. Test each sample separately with a few drops of soap solution, as in step 1.
4. Record your observations in a table. Compare the control sample (tube 1) with the hard water samples. You should find that boiling makes temporarily hard water soft but has no effect on permanently hard water. You could also investigate whether distilling water makes it soft.

Activity | Can chemicals soften water?

1. Take separate samples of both kinds of hard water, and put 5 cm³ of each in separate test-tubes.
2. Add a small quantity of washing soda to each sample, close the test-tube and shake vigorously.
3. Compare the properties of water which has been treated with washing soda with untreated hard water. Use equal amounts of soap solution. Record your observations.

Washing soda consists mainly of the substance sodium carbonate. When this is added to hard water, it will react with some of the calcium ions present to form the insoluble substance, calcium carbonate. This is precipitated out. The hard water then becomes soft because the substance causing the hardness has been removed. A simple word equation for the reaction is:

calcium sulphate (aq) + sodium carbonate (aq) →
calcium carbonate (s) + sodium sulphate (aq)

Washing soda will remove both temporary and permanent hardness. Boiling will remove temporary hardness only.

One of the most effective ways of softening hard water is to use an **ion-exchange resin**. Ion-exchange resins contain large particles which have sodium ions weakly bound to them. When hard water is passed through an ion-exchange resin, the calcium ions in the water replace the sodium ions in the resin and stay attached to the resin. The water which comes out of the resin contains sodium ions rather than calcium ions.

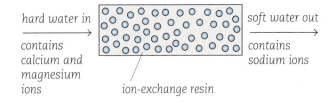

How an ion-exchanger works

The sodium ions are gradually removed from the resin until eventually it becomes 'exhausted', in other words there are no sodium ions left. A concentrated solution of sodium chloride is passed through the resin so that the sodium ions are replaced in the resin. Once recharged, the resin can be used again for water softening.

Although hard water seems to present a number of practical problems, it is not always a bad thing. There is some evidence that heart disease is more prevalent in areas with soft water.

Questions

4. Why is distilled water used in science laboratories and in car batteries?
5. What are the advantages of using an ion-exchange resin for softening water?

Why are acids and alkalis important?

What do we mean by the terms acid and alkali?

You have already used acids and alkalis in your science courses. Solutions of acids such as sulphuric and hydrochloric acids are common in the laboratory. Alkalis such as sodium hydroxide and aqueous ammonia are common too. You may have seen the terms **base** and **alkali**. Alkalis are solutions of soluble bases. Some insoluble bases will be discussed later.

Activity | Using litmus paper

1. Put about 5 cm^3 of dilute sulphuric acid into a test-tube. Test this with small pieces of red and blue litmus paper. Record your observations.
2. Repeat this experiment with different solutions, such as dilute hydrochloric acid, aqueous ammonia and sodium hydroxide solution. Record your observations in each case.

You will notice that the acids turn blue litmus red, while the alkalis turn red litmus blue. Litmus is an example of an **acid-base indicator**. These indicators change according to the acidity or alkalinity of the solution tested.

Activity | Acid-base indicators

1. Repeat the first Activity using the same solutions but other acid-base indicators. Choose two or three from the following: methyl orange, universal indicator, phenolphthalein, bromothymol blue.
2. Record your observations. You should find that the colour of the indicators changes according to the acidity or alkalinity of the solution.
3. Repeat the Activity, using other solutions to see what effect these have on the indicators. You may need to take care when selecting these solutions. If a solution is coloured, this may interfere with what you see.

There are other common indicators. Earlier in your science course, you may have used indicators derived from flower petals such as *Hibiscus*.

The pH scale

It is often important to know whether solutions are acidic or alkaline. For example, if you accidentally spill a liquid on your skin, you need to know what to do as a safety precaution. Concentrated sulphuric acid reacts very strongly with water, and this is very different from the behaviour of dilute ethanoic acid. Solid sodium hydroxide (caustic soda) will harm your skin very quickly but it is possible to wash off dilute sodium hydroxide solution with large amounts of water. For these and other reasons we need to know *how* acidic or alkaline a substance is.

We measure the acidity or alkalinity of a solution with the pH scale, which you may have used before. This scale consists of a set of numbers from 0 to 14. On the scale, 7 is the point of neutrality: a solution of pH 7 is neither acidic nor alkaline. This is the pH of pure water. Solutions with a pH of *less* than 7 are acidic. Acidity increases as the pH goes down from 7 towards 0. A solution of pH 2 is more acidic than a solution of pH 6, for example. Solutions with a pH *greater* than 7 are alkaline. Alkalinity increases as the pH increases from 7 to 14 – thus a solution with pH 13 is more alkaline than a solution of pH 9. We can summarise this information like this:

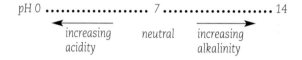

You can estimate the pH of a solution by using universal indicator (UI). If you add a few drops of UI to the solution, you can compare the colour of the indicator as it appears in that solution with a colour chart showing the entire range of colours of UI. One of the most common alkalis, as you know, is sodium hydroxide solution. This is formed in the laboratory by the (explosive) reaction of sodium and water, or by the reaction of sodium oxide and water. A base is a substance which reacts with an acid to form a salt and water only:

acid + base → salt + water

The oxides of metals such as copper and iron are insoluble in water. These are bases, but they do not form alkalis. Magnesium oxide is only slightly soluble in water.

Questions

1. Define (a) an acid (b) an alkali and (c) a base.
2. What is the nature of solutions with the following values of pH: 1, 2, 4, 7, 9 and 12?

Activity | What happens when acids and alkalis react?

1. As a rough guide, take 25 cm^3 of dilute sulphuric acid (measured using a measuring cylinder) and put it in a 100 cm^3 beaker. Add a few drops of a suitable indicator. Record the colour of the indicator.
2. Add dilute sodium hydroxide solution *slowly* from another measuring cylinder, mixing the solutions carefully. Stop adding alkali as soon as there is a *permanent* colour change. Measure the volume of alkali added.

3 You will have found that, however careful you are, it is not possible to measure accurately the amount of alkali added. You can now try a more accurate method.
4 Measure out carefully 5 cm³ of the acid into a test-tube. Add a drop of indicator. Now add the alkali a drop at a time, using a dropping pipette. Count the number of drops of alkali used up to the point where there is a *permanent* colour change. Record your results. Now repeat the experiment to confirm that you get a similar result. Make sure that you wash the dropping pipette thoroughly with distilled water after use.

There is an even more accurate method for this experiment. To measure the volumes of acid and alkali accurately, use a **pipette** and **burette**. When an acid and an alkali react so that the reaction is complete – in other words, there is no acid or alkali remaining – the reaction is called **neutralisation**.

acid + alkali → salt and water

In this experiment, the word equation is:

sulphuric acid (aq) + sodium hydroxide (aq) → sodium sulphate (aq) + water (l)

Sodium sulphate is a salt. If sodium hydroxide reacts with hydrochloric acid, the salt formed is called sodium chloride.

Activity | What happens after neutralisation?
1 Repeat the experiment, starting with 50 cm³ of the acid, and add just sufficient alkali to neutralise the acid completely. Check the solution with red and blue litmus.
2 Put the neutral solution in a large evaporating basin and heat it over a beaker containing boiling water (a water bath).
3 As the liquid in the evaporating basin is driven off, what do you notice has formed in the basin? This is one way of obtaining soluble salts from their solution.

Question
3 Write word equations for the reactions of (a) dilute sodium hydroxide solution and dilute nitric acid and (b) copper oxide and dilute sulphuric acid.

What do some common chemicals contain?
Everything that you use at home or in school contains chemical substances. Even your body is a mass of chemical substances! The most common chemical is water. Pure water is neutral and has a pH of 7. Given all the chemicals you use every day, how do we know whether they are acids or alkalis?

Activity | Are common chemicals acidic or alkaline?
1 Collect as many samples as possible of common chemicals. Be careful about choosing coloured chemicals. Why? Try to dissolve any solids in water.
2 Test a small sample of each chemical or solution with a few drops of universal indicator.
3 Record your observations in a table. Note whether the chemical is acidic or alkaline.
4 Read the label on the chemical. Try to identify the substance making the chemical acidic or alkaline.

Questions
4 Acids and alkalis are potentially dangerous. If you got some aqueous ammonia (a) on your skin and (b) in your eye, what should you do?
5 Why is it important to know whether a substance is acidic or alkaline?
6 Should you store any acidic or alkaline substances at home? If your answer is yes, what precautions should you take?

Activity | How can we try to remove food stains?
You probably know that it is quite difficult to remove food stains from clothes, particularly if they are allowed to dry. These stains may contain acids or alkalis. Tea and coffee contain tannic acid, and citrus fruits contain citric acid.
1 Try to obtain samples of old cloth and then stain them with different substances, such as tea, coffee and fruit juice. Each stain should be kept separate from the others.
2 Try to remove these stains, starting with the simplest method – using cold water. Use warm water if cold does not work. Finally, try warm soapy water if warm water alone does not work. You may find that some stains are difficult to remove. In these cases, keep the stained cloth.
3 Record your results in a table.
4 Now try removing these difficult stains by using a *mild* alkali such as a solution of sodium hydrogen carbonate or a substance such as borax. You may find that the more difficult stains are removed by one of these substances.

This is another example of a neutralisation reaction. Acidic food stains can be removed by using a mild alkali.

Questions
7 Would you use concentrated sulphuric acid or concentrated sodium hydroxide solution to remove food stains? Why?
8 Could you use an acid or an alkali to remove a grease stain?

Household chemicals

Water: a very common household chemical

Water is a very common substance and it is crucial to our existence (see pages 138 and 140; and pages 146–7). It has a very wide range of uses which depend upon its unique properties.

Some of these uses of water depend on its properties as a solvent, while others may depend on its lack of reactivity. (Remember that pure water has a pH of 7, i.e. it is neutral – see page 220.)

Activity | Water in the home
1. Make a list of the common uses of water in your home.
2. Try to divide these uses into groups according to the property of water which is important for each use.

As you will probably have found, water is used in many different ways, although its most important property is probably as a solvent. You might like to think about some of the substances we use every day which contain water. For example, milk consists mostly of water. It contains fat, milk sugar and proteins as well, but there is more water in milk than anything else. So if you drink tea or coffee with milk, you are drinking mostly water. If you have a stomach upset, you may take a common medicine such as Milk of Magnesia. This does not contain milk, but is a suspension of magnesium hydroxide in water. Most of the common hot sauces used for flavouring food contain a lot of water.

All of the examples quoted so far have been liquids. However, water can be found chemically combined in solid substances, such as blue copper(II) sulphate. Copper sulphate which does not contain water is white anhydrous copper(II) sulphate. Substances such as washing soda and Andrew's (liver) salts contain water in their crystals.

Look at the table below. You may be surprised at the amount of water present in common food items.

Percentage composition

Food	Water	Carbohydrate	Fat	Protein	Other
Cabbage	94	4	–	1	1
Milk	87.5	5	4	3.5	–
Eggs	78	–	10	12	–
Fresh fish	77	–	3	18	2
Meat	63	–	18	18	1
Dried fish	18	–	10	63	9
Maize (corn)	10	78	1	8	3
Soya beans	9	34	18	34	5

You may also wish to think about the implications for your diet of some of the percentage figures. Meat and soya beans have a relatively high fat content, as do eggs.

Questions
1. Which property of water is the most important for most of its common household uses?
2. How important is the boiling point of water for its uses in cooking?
3. List some foods which are cooked without water.

Safety notes If you look back to pages 164–6, you will realise how important it is to handle chemicals safely. For example, if you use a flammable solvent such as propanone, there must not be a flame close by because the vapour of propanone is volatile, and can catch fire easily. If you have to heat a small quantity of propanone, you have to use a water bath. Also, your teacher may have demonstrated some of the properties of the metal, sodium. This is highly reactive even with water, and must be handled very carefully.

There is now a much greater emphasis on safety in the laboratory, and your school may have a policy about the wearing of laboratory coats and safety glasses when potentially dangerous substances are being handled. This policy may also apply to the heating of chemicals in general.

The laboratory is not the only place where you will handle chemicals. You will also handle them at home. It makes sense to apply the same safety considerations at home too. In the laboratory, you or your teacher will know about the properties of the chemicals you handle, and the bottles containing them should have suitable safety or hazard warnings. The labels on chemicals used in the home should tell you which substances they contain, but this may not be enough. You have to know about the properties of these substances. This is more important, perhaps, for something like a cleaning liquid, such as bleach, or a pesticide, than it is for toothpaste. You can use a toothpaste knowing that it has to be safe for use in your mouth – but bleach and pesticides are not!

Questions
4. Do you think that every substance should carry a detailed label giving a complete list of all the chemicals it contains?
5. Old names are still often used for chemicals. Find out the modern names for (a) carbolic acid and (b) muriate of potash.
6. Processed foods often contain preservatives (page 77). Should these be recorded on food labels?

Part 1

Activity | What do some common chemicals contain?

1. Collect labels from or samples of up to 10 common household chemicals. It is too expensive or possibly dangerous to collect all the actual chemicals, so collect their labels or copy the information from the labels. *Do not remove any labels from chemicals still in use.* What do you think is the reason for this advice? Do not limit yourself to chemicals used inside the home – there are many chemicals used in the garden, for example.
2. If you cannot collect many labels, try to get the information you need by reading carefully the labels of chemicals in use in your home or on sale in a shop. If you do this in a shop, make sure that you get permission from the owner/manager first. Some of the substances you might look for include soap powder, detergents, disinfectants, baking powder, fertilisers and pesticides. However, there are many others to choose from.
3. Try to find out whether the information given on the labels about the chemicals contained in the substances is enough to tell you whether they are acidic, alkaline or neutral.
4. Check to see whether there are any hazard warning labels (see pages 164–5) on any of the containers. For example, there are some very dangerous pesticides still available, such as Paraquat and Malathion: these should be handled with very great care. Wash your hands after handling these containers. You may find that some labels are not very helpful, so in these cases you have to obtain advice or information from others about how to handle the substances safely.
5. Try to find out whether the government in your country has a policy about the labelling of materials sold to the public.
   You may need to contact a department or ministry responsible for consumer affairs. Many newspapers in the region now have regular features on consumer affairs and you could contact the journalist(s) responsible to find out more about matters such as labelling and food safety.
6. Discuss what you have found.

How can we use household materials properly?

You know by now that chemicals – even water – have to be handled with care. Some common substances can be dangerous unless they are used in the proper way. An example is disinfectants, which are very important for destroying bacteria in bathrooms and kitchens: you might find it very uncomfortable to swallow even a small amount of a disinfectant, even if it is a mouthwash!

Normally, you should never taste or swallow chemicals. There are obvious exceptions to this, such as drugs prescribed for the treatment of an illness, or common medicines such as Milk of Magnesia.

Substances such as plastics (page 204–5) can be dangerous too. Some plastics give off poisonous fumes when heated and plastic bags can be dangerous because of their electrostatic properties. There are now often warning labels on plastic bags to ensure that these are kept away from young children because a bag placed over the head can cause a child to suffocate. Young children want to explore their environment, and will often try to touch and swallow things which are potentially dangerous. To protect them, such items have to be kept out of their way. As they grow older, they have to be taught how to handle household materials properly.

There are three basic questions we should ask about any household material in common use:
- What chemicals does the material contain – and what are their properties?
- Where and how is the material stored?
- What is the material used for?

This means that we are applying the same sort of thinking about safety in the home as we do in the laboratory.

Activity | Using household chemicals safely

1. Make a list of between 5 and 10 common household chemicals.
2. For each chemical, suggest ways in which it might be dangerous if not properly used.
3. Suggest precautions (if any) that should be taken when storing each of the substances listed.

Questions

7. Many people now use vitamin supplements to add to their diet. How important is it that these should be labelled properly?
8. If a label stated that a substance contained caustic soda, what precautions should you take in handling it?
9. Do you think *all* the substances present are listed on the labels of household cleaners? Why?
10. There is now often a 'use by' date on labels for processed foods. Would it be dangerous to use such food after this date?
11. Do you read the labels on food items? If not, how do you know what you are eating?
12. Would you taste a food item without knowing what it was or how it was prepared?

Household chemicals

Handling other household materials safely

Although we have looked at a range of household chemicals in the preceding pages, there are several more materials, used either in the house or outside it (such as in the garden), and it is useful to know something about their properties. (See also pages 164–7.)

Activity | Fuels
1. Make a list of any fuels, such as kerosene or charcoal, used in your home.
2. List the use(s) of each fuel.
3. How are the fuels stored? Should you take any special precautions in storing them?

Activity | Lubricants
1. Make a list of any lubricants used inside the home or for machines such as bicycles, sewing machines or lawn-mowers.
2. What does each of these lubricants contain?
3. Do you need to take any precautions when using them?

Activity | Insecticides
In our homes, we want to make sure that we keep the insect population to a reasonable minimum – although we may feel that there should be **no** mosquitoes or cockroaches!

Insecticides can help, although we have to take care with their use, particularly if there are small children, pregnant women or people with respiratory problems living in the home.

1. List all of the insecticides used in your home.
2. Examine the labelling on one or more of these materials. What are the 'active ingredients'?
3. Does the labelling contain any warnings about use by children?

Activity | Fertilisers, insecticides and herbicides
These are chemicals which may be used in a garden or on plants kept in the house. Fertilisers encourage growth, while insecticides and herbicides are used to keep down insect and plant pests respectively. (See also page 155.)

1. Make three lists, one each for the fertilisers, insecticides and herbicides used by you or your parents.
2. Examine the labelling on at least one example of each type of material.
3. What are the important ingredients in these examples? (Check, for example, the nitrogen content of a fertiliser.)
4. Are there any hazard labels on the insecticides or herbicides? How should these materials be stored and used?

Questions
1. There are some solvents and glues which are addictive when sniffed. Do you think that the sale of such materials should be banned? Alternatively, should there be warnings about the dangers of misuse of such materials?
2. If you found an old metal container with a liquid in it but no label on it, what would you do with it?

Activity | Handling food preservatives, food additives and food supplements
There are many methods of preserving food, starting with the very simple, such as salting and drying. (Dried, salted cod was one of the first exports to the Caribbean from Canada, and is now part of the staple diet for many people.) The hot sauces common in the Caribbean are based on hot peppers (such as the 'Scotch Bonnet' in Jamaica), but contain preservatives (see page 77) such as vegetable oil and vinegar.

1. Examine the labels on a range of processed food items.
2. Make a list of the **preservative(s)** contained in each item. You may need to get advice from your parents and your teacher about this since preservatives may not be identified as such.
3. Try to find out about the properties of at least one substance that makes it a good preservative.
4. Try to find out which methods of preserving food are used by your parents or relatives.
5. Make a separate list of the food **additives** (see page 77) in each item.
6. Do we need such additives? Why or why not?

Preserving food is big business, and there have been many debates about the effect of some preservatives. One such is sulphur dioxide, used in preserving dried fruit. Apricots preserved with sulphur dioxide retain their colour relatively well, while those preserved in other ways look quite dark. Appearance of food can be improved by using chemicals such as alar (which makes apples look shiny) but it is doubtful whether these chemicals are beneficial in any other way.

Questions
3. How would you try to preserve fruit such as pineapple or banana?
4. Does improving the appearance of fruit or vegetables make them a better buy?

Part 2

Activity | Food supplements
You may want to refer back to pages 78–9. You already know about the constituents of a balanced diet, all of which you can obtain by choosing the right balance of foods. However, many people choose to take food supplements as well. The most common of these are vitamins.
1. Make a list of the vitamins in your local pharmacy.
2. If there is any literature available, find out about the purpose of taking any **one** of these vitamins.
3. Could you obtain the appropriate daily intake of this vitamin by choosing foods correctly, rather than taking a vitamin supplement?
4. There are many other food supplements, and you may have seen advertisements praising the virtues of materials such as echinacea or folic acid. Make a list of five common food supplements on sale in your local pharmacy.
5. Try to find out the purpose of taking at least one of these supplements.

Questions
5. A very famous scientist of the 20th century proposed that very large doses of vitamin C would provide immunity from the common cold. How would you test this hypothesis?
6. Do you take any food supplements on a regular basis? Is there a good reason for this?

How can we find out what common materials contain?
Handling materials safely requires knowledge about their properties. Some of this knowledge may be obtained from sources such as container labels. These may tell you about the substances in the material and how much of each substance is present.

The task of determining accurately how much of each constituent there is is called **analysis** and it is carried out by scientists called **analysts**.

The analyst may need to find out about the elements and/or compounds present in a sample of a material. For example, if we started with two samples of copper(II) sulphate, one white and the other blue, we should find that both samples contain copper and substances from the sulphate group. The white sample will contain no other elements or groups, while the blue sample will contain hydrogen and oxygen in the same ratio as in the compound water.

There are a number of simple tests used by scientists to identify particular elements or groups, and three of these are described here.

Activity | Testing for copper
1. Put about 5 cm^3 of blue copper sulphate solution in a boiling tube. Add dilute sodium hydroxide solution a few drops at a time until there is no further change.
2. Now add aqueous ammonia until there is no further change.
3. Record your observations.

When sodium hydroxide solution is added to a solution of a copper salt, a pale blue precipitate is formed. This precipitate dissolves in aqueous ammonia with the formation of a deep blue solution. This is a good test for the presence of copper.

Activity | Testing for iron
1. Put about 5 cm^3 of green iron sulphate solution in a boiling tube.
2. Add dilute sodium hydroxide solution a few drops at a time until there is no further change.
3. Repeat steps 1 and 2, using aqueous ammonia instead of dilute sodium hydroxide solution.
4. Record your observations.

When either dilute sodium hydroxide solution or aqueous ammonia is added to a solution of an iron salt, a green precipitate (iron hydroxide) is formed. This is a good test for the presence of iron.

Activity | Testing for chlorides
1. Put about 5 cm^3 of sodium chloride solution in a test-tube and add an equal quantity of dilute nitric acid.
2. Add a few drops of silver nitrate solution. A white precipitate (of silver chloride) shows that chloride is present.

The very simple tests which you have carried out are just a starting point. Analysts require a great deal of training but are important, since they can provide the government and the public with information about the food we eat and the materials we use.

Questions
7. You are provided with two substances, one black and one blue. How would you try to show that each substance contained copper?
8. Examine the labels on a selection of processed food items. Do these tell you what you want to know about the contents?
9. You are provided with a white substance which is said to be either sodium carbonate or calcium carbonate. What tests would you perform to show which it is?

How do machines make work easier?

Activity | What is the law of levers?

1. Suspend a metre rule at the 50 cm mark as shown in the diagram. Balance it with pieces of plasticine if necessary.
2. Make hooks of bent wire and use them to hang masses on the rule.

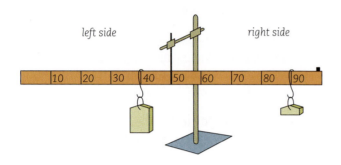

3. Place a 40 g mass on one hook about 10 cm from the centre (50 cm mark). Place a 10 g mass on the other hook and slide it along so that it balances the 40 g mass, keeping the rule horizontal.
4. Record the masses and distance from the pivot (balancing point) in a table like this.

Left side (LS)			Right side (RS)		
Mass m_1	Distance d_1	$m_1 \times d_1$	Mass m_2	Distance d_2	$m_2 \times d_2$
40	10	400	10	40	400

5. Use several different masses at different distances on one side and balance the rule by placing other different masses on the other side.
6. What do you observe? How is it that a small mass can balance a much larger mass?

When the rule is balanced, i.e. horizontal: mass (m_1) × distance (d_1) on the left side = mass (m_2) × distance (d_2) on the right side. This is called the *law of levers*. From the law this formula is derived:

$$m_1 d_1 = m_2 d_2$$

therefore $\quad m_1 = \frac{m_2 d_2}{d_1}$

or $\quad d_1 = \frac{m_2 d_2}{m_1}$

You can find m_2 or d_2 by changing the subject of the formula and by substituting. So, if one mass or distance is unknown, it can be calculated from the other three known values.

Example

$m_1 = 40$ g $\quad d_1 = ?$ $\quad m_2 = 10$ g $\quad d_2 = 20$ cm

$$m_1 d_1 = m_2 d_2$$

$$d_1 = \frac{m_2 d_2}{m_1}$$

$$d_1 = \frac{10 \times 20}{40} = 5 \text{ cm}$$

Levers

From the Activity, you see how a small mass can balance a larger mass. Small forces can balance, resist or overcome larger forces. The key is to alter the distance through which the force is acting. We can do this using levers. There are three main types of lever and many different tools and instruments are based on them. Levers are grouped into orders or classes.

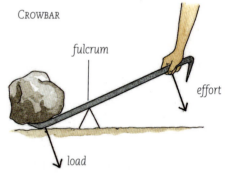

CROWBAR

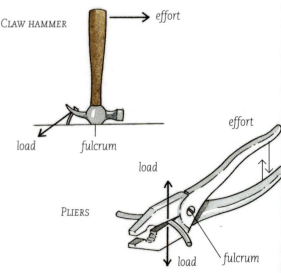

CLAW HAMMER

PLIERS

Class 1 levers – these have the pivot or fulcrum between the effort (force) and the load. This kind of lever can be used as a force multiplier or as a force reducer to move an object or load

Part 1

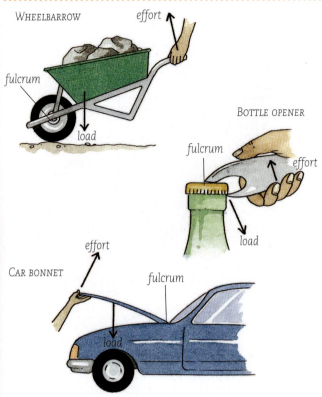

Class 2 levers – these have the fulcrum at one end, the effort at the other end and the load inbetween. This class of lever is a force multiplier because the load is always nearer to the fulcrum than the effort

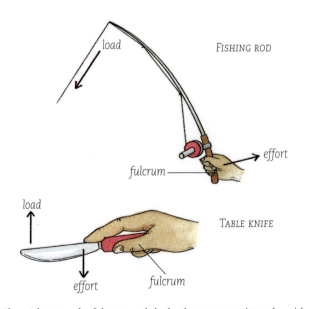

Class 3 levers – the fulcrum and the load are at opposite ends, with the effort inbetween. The effort is always nearer the fulcrum than the load. The lever is therefore a force reducer and never a force multiplier

Activity | The human arm

1. Look at the diagram of the human arm. See where the biceps muscle is attached. This is where force (or effort) is applied to bend the arm. The elbow is the pivot or fulcrum where the arm bends. The load is suspended from the hand.

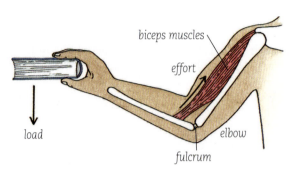

2. Measure the distances *effort to fulcrum* and *load to fulcrum* on your arm in millimetres. If the load exerts a force of 50 N, what is the force exerted by the effort? (Omit forces due to the weight of the arm.)
3. Discuss your observations in class.

This Activity shows you how a great force must be applied by the muscle of the arm to overcome the force exerted by the load. The biceps muscle must therefore be very powerful in order to lift a modest weight.

Do levers make work easier?

You saw previously that on a balanced rule, the *mass × distance on one side* is equal to *mass × distance on the other side*. The mass is acted upon by gravity to produce a force (known as weight) so we can more correctly say that *force × distance on the left side* is equal to *force × distance on the other*.

If the force is measured in newtons and the distance in metres then the work done by the lever is in *joules*, since force (F) × distance (d) = work (w). Since F × d on both sides must balance, the work done, or energy expended, on one side must be the same as on the other side. So, if the force available is small you can use it to overcome a larger force by selecting the right distance for it to act through. Thus, by using a lever, it is possible to do work which would otherwise not be possible.

Questions

1. From the expression $m_1 d_1 = m_2 d_2$, express each term in relation to the others.
2. Give other examples of Class 1, 2 and 3 levers not mentioned in the text.

How do machines make work easier?

Besides levers, there are other machines which can make work easier. These are shown below. They include the inclined plane, the wedge, the screw, and the pulley (see page 230). These machines make work easier by allowing a load to be moved by a small effort. This relationship is described mathematically as:

$$\frac{load}{effort} = \text{mechanical advantage (M.A.)}$$

Example What is the M.A. of a machine which exerts an effort of 200 N and overcomes a load of 1 000 N? *Answer*

$$M.A. = \frac{load}{effort} = \frac{1\,000}{200} = 5$$

The larger the mechanical advantage the easier the machine makes doing work.

The inclined plane In the figure below, two men are loading a truck using the inclined plane. Using the inclined plane is considerably easier than lifting the 200 kg box by hand directly into the truck through the distance (height) h. Note that the same amount of work is done (ignoring the friction of the rollers) when the box is pushed through the longer distance s, of the inclined plane. For the inclined plane, the mechanical advantage is the length of the inclined plane (s), divided by its vertical height (h).

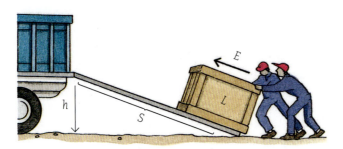

The wedge The wedge is also an inclined plane. In the figure in the next column it is being used in the form of a chisel to split a log (E = effort, L = load). The thickness of the wedge is an important consideration, is it not? The principle of the wedge can also be applied to the cutting edges of knives, swords and even razor blades. Can you explain this connection?

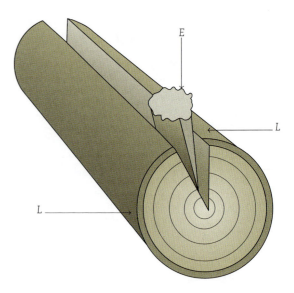

The screw The screw has many uses. In the figure below, it is used as a car jack, which is essentially a long screw carrying a nut. The nut is joined to a steel bar, which supports the load of the car when the jack is in use. When the long nut is turned, for example through one revolution, using the short steel bar (called a **tommy bar**), the car is raised a small distance. This is equal to the *pitch of* the screw, and is not usually more than 2 or 3mm in height. The same principles of calculation apply here as for the inclined plane (above). A relatively small force is applied over a long circular distance to raise a very large weight through a very small distance. This process is repeated several times until the car has reached sufficient height off the ground.

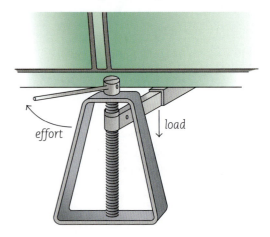

A car jack in use

Part 2

The wheel and axle Various applications of the wheel and axle are shown below. In each case, effort is applied through a relatively large circular motion (wheel) to overcome a load, which moves through a proportionally smaller distance. The mechanical advantage can be calculated from the equation:

$$M.A. = \frac{load}{effort} \quad \text{or} \quad M.A. = \frac{radius\ of\ wheel}{radius\ of\ axle}$$

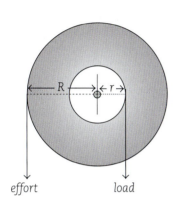

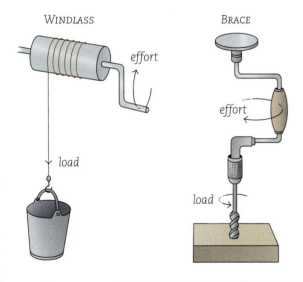

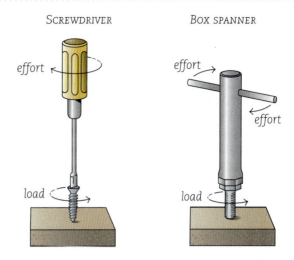

SCREWDRIVER   BOX SPANNER

Gear wheels Wheels with cuts in their outer rim are called **gears**. When the teeth of two gears are meshed together, one gear can turn the other. If each of the gear wheels were connected to a shaft, then one gear could supply the effort while the other could be the load, thus arriving at the same arrangement as the wheel and axle. Consider the two meshed gear wheels in the figure below. Which of the gaers would you choose to apply effort to in order to get the best mechanical advantage?

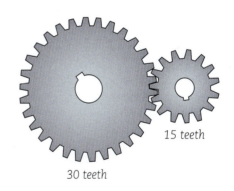

Questions

1. Describe how you would work out the mechanical advantage of a pair of wire cutters.
2. In the wheel and axle system, can the turning circle of the effort be too large? Discuss.
3. In the wheel and axle system, can the turning circle of the load be too small? Explain.
4. In the two meshed gear wheels above, what is the mechanical advantage of the system when (a) turning power (effort) is supplied by the larger wheel and (b) turning power (effort) is supplied by the smaller wheel?
5. In the two meshed gears above, when the smaller gear makes 10 revolutions, how many will the larger gear make?

How efficient are machines?

Activity | How do we use pulleys?

Make your own pulley like the one in **picture 1** (below). Use empty cotton reels or spools to make the pulley and then try these experiments.

1. Raise a mass of 1 kg by pulling on a spring balance as shown in picture 1. Use a spring balance calibrated in newtons. When the mass is not moving it is balanced by the pull on the spring balance (i.e. in equilibrium). What is the force in newtons? What is the force when the mass is just moving upwards?

   If a spring balance is not available, select a suitable rubber band and draw evenly spaced marks on it. Measure the distance between two marks, or the length of the whole scale, before and after force is applied. However, this would need to be calibrated later.

2. Suspend the mass in another way, as shown in **picture 2** (below). Now, what is the force on the spring balance **A**? Is it approximately half that in picture 1? What is the force exerted on the other string **B**? What is the force on string **C**?

   The force due to gravity, which pulls a mass of 1 kg, is 10 N. Therefore, the pull on string **C** is 10 N because the force is shared between strings **A** and **B**. The pull on each of these two strings is 5 N (excluding the pull due to the mass of the pulley).

What force exerted by pulling on string **B** can cause the mass to rise? Since the pull on string **B** is 5 N when equilibrium is reached, it follows that a pull of less than 5 N would cause the mass to fall and a pull of only slightly more than 5 N would cause the mass to rise. What is the mechanical advantage of the system?

$$\text{M.A.} = \frac{\text{load}}{\text{effort}} = \frac{10\text{ N}}{5\text{ N}} = 2$$

Consider **picture 3**. The load is supported by four ropes on two sets of pulleys, each rope taking only a quarter of the total load. Hence, if the pull on the rope E is only slightly more than a quarter of this load (strain), the load is pulled upwards. This arrangement (as shown in picture 3) is called the **block and tackle**. Here, the mechanical advantage (theoretical) is 4 and for each unit of length that the load moves the winch (effort) rope moves four times further.

How is efficiency measured?

Consider picture 3. Does the rope itself have mass? Is energy required to turn the pulley? Does the lower set of pulleys have mass? Even without lifting any load, energy is still required to operate the system of pulleys. If the load is very small, a large proportion of the energy put in goes into moving the pulley and the rope, and into overcoming friction. When the load is large, a much smaller proportion of the input energy goes into the operation of the pulleys and a larger proportion goes into moving the load, i.e. there is a greater work output.

$$\text{Efficiency (\%)} = \frac{\text{work output}}{\text{work input}} \times 100$$

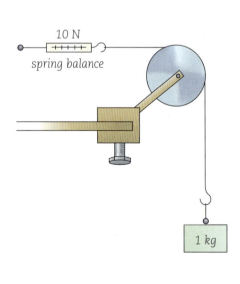

Picture 1

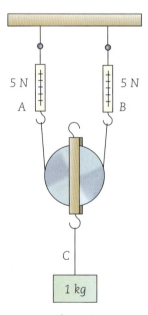

Picture 2

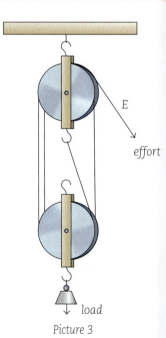

Picture 3

Work = force × distance (see page 227)

$$\text{Efficiency} = \frac{\text{load} \times \text{distance load moves}}{\text{effort} \times \text{distance effort moves}}$$

For levers and other machines, efficiency is calculated in the same way. This graph (below) shows how load and efficiency are related in pulleys.

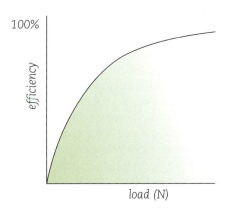

You probably think that the best machine to use must be the most efficient one. But you must also consider the question of convenience. For example, an inclined plane might be the best machine for loading a truck (picture 4), but a pulley may be best for getting loads to the top of a building. Like the lever, the pulley can be used to change the direction along which a force is acting. A motor used with a pulley can raise a load to the top of a building (see picture 5).

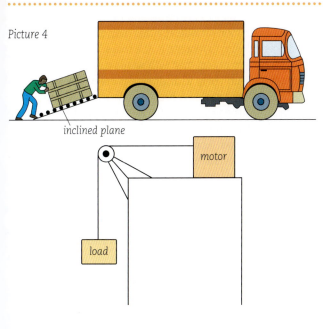

Picture 4

Picture 5

Block and tackle systems on cranes used to load ships.

Can machines be dangerous?

Machines, both simple and complex, must be properly maintained if they are to be used safely and efficiently. Harm may be caused by improper use, bad selection and through failing to take the necessary **safety precautions**. For example, in simple machines, too weak a cable in a pulley system may cause an accident and too thin or narrow an inclined plane may fail to stop a car from rolling back. With more complex machines, for example those found in your school's Home Economics and Industrial Arts Departments, expert advice is needed on maintenance and proper use. However, while you may learn to use these machines safely and carry out simple maintenance, rectifying faults and carrying out repairs should be left to those who are trained and qualified to do this work.

Questions

1. How can you tell at a glance what the mechanical advantage of a pulley system is?
2. Describe how you would test experimentally the efficiency of a pulley system.
3. Give examples of simple machines being used more for convenience (such as for changing the direction of the applied force) than for mechanical advantage.
4. Discuss any special precautions you must take in the operation of a complex machine you have learnt to use (for example, a sewing machine).
5. Can a machine be 100% efficient?

What is soil?

Activity | What is soil made of?

Collect a spadeful of garden soil, preferably from under a tree where there is a layer of decaying leaves.

1. Spread a handful of the soil onto some newspaper. Use a hand lens to examine and identify: pieces of decayed organic materials made from dead leaves etc., small soil particles and larger pieces of gravel.
2. Pick out some soil particles and spread a *very* thin layer on a glass slide. Look at them using the low power of a microscope and describe their size and shape.
3. Put 50 cm^3 of the soil into a 100 cm^3 measuring cylinder and add 50 cm^3 of water. Hold your hand over the top of the measuring cylinder and shake the contents together. Allow until the next day for the soil to settle.
4. The different parts (**components**) of the soil will settle into different layers depending on the size and weight of the particles. This is called the **sedimentation method** for examining soils. Look at this picture and identify the components of the soil. Use a ruler to measure the height of each component. *Do not* use the readings on the side of your measuring cylinder.

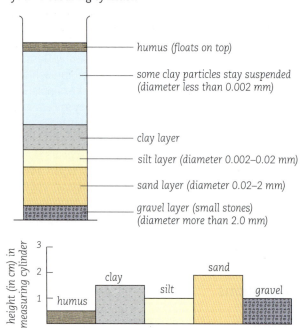

The soil components can be classified as:
(a) *Solids* Soil particles of different kinds (clay, silt, sand and gravel) derived from rocks, and organic matter (humus) derived from decayed organisms.
(b) *Liquids* Water and dissolved mineral nutrients.
(c) *Gases* Air in the spaces between soil particles.

How is soil formed?

The development of soil takes many thousands of years.

The uppermost layer of soil (**topsoil**) which contains the **humus** (see opposite page) and is most useful for growing plants such as crops, may be less than 50 cm deep. Below this is the **subsoil** which contains more mineral nutrients and supports the roots of trees. Further down still are the rocks from which the soil particles have been derived.

Several factors are important in the breakdown or the **weathering** of rocks to form soil.

1. **Mechanical factors** Mechanical factors break down the rocks without changing their chemical composition (compare mechanical digestion, page 84). Water, as rain, rivers, and lakes, wears away at rock surfaces and breaks away small pieces of rock. These rock fragments grind against each other and the rock surface. Water may be trapped in cracks in the rocks. The water expands as it freezes and breaks apart the rocks still further.

   The wind can blow rock fragments and break them down when they hit against rock surfaces. Sea breezes contain particles of sand and salt which also wear away at the rocks. On rocky shores the waves make rocks and stones hit against each other so that they break into smaller pieces and become smooth.

   On a hot day, rocks will expand, and at night, when it is cooler, they will contract. The rocks will cool and this causes cracks to form in the rocks. This is especially true in desert areas where the days are hot and the nights cold.

2. **Chemical factors** Chemical factors break down the rocks by chemical reactions (compare chemical digestion, page 84). New substances are formed. Water may contain dissolved carbon dioxide and this may dissolve minerals such as calcium carbonate in the rocks. The oxygen in the air, on its own or together with water, can oxidise certain compounds and help to disintegrate the rock.

   Thus rocks are slowly broken down, and the rock particles are carried by water and wind to be deposited and form the mineral part of the soil.

3. **Biological factors** Biological factors cover the effects that plants and animals have on the rocks. The first plants to grow on a bare rock surface are algae and lichens which can grow into the rock surface by producing acids that dissolve away the rock. Then these organisms decay and form a thin layer in which other plants can grow.

   Burrowing animals such as earthworms and termites eat their way into soft rocks and form channels into which air and water can run. Together with soil bacteria they help to decay the dead animal and plant materials to produce humus, the organic part of the soil.

Activity | Looking around you

Over a period of a week look around the school grounds and surrounding areas on your way to school.

1. Find a tarmac or stone path that has plants growing on it. How can the plants live in these places? What effect are the plants having on the path?
2. Look for crevices in a stone wall. What do you find there? What effect are the plants having?
3. Look for a stone wall or rocks near a tap or where water has been running. What effect has water had?
4. Find a place where there is an exposed surface down through the rocks, for example where rocks have been cut away to make a road. Identify the topsoil, subsoil and underlying rocks.

Activity | Different kinds of soils

1. Collect samples of soils from a variety of places around the school. Label them A, B, C, etc. and keep a record of where you found them. Copy the table below.

Soil	Where it is found	Humus	Clay	Sand	Soil type
A					
B					
C					

2. Fill different measuring cylinders with equal amounts of water. Add the different soil samples to different measuring cylinders. Shake each sample. Let them settle overnight.
3. Measure the height of the humus, clay and sand and complete the table.
4. From your table pick out:
   (a) A *sandy soil*: lot of sand, hardly any clay, little humus.
   (b) A *clay soil*: lot of clay, little sand, little humus.
   (c) A *loam soil*: sand, clay and humus (about half sand, third clay, a sixth humus).
5. Complete the last column of your table.

Comparisons of particles in sandy, clay and loam soils

Sandy	Clay	Loam
Mostly large sand particles	Mostly small clay particles	Mixture of sand and clay
Large air spaces	Small air spaces	Medium air spaces

Functions of different soil components

Humus Humus is the decayed remains of plants and animals. Compost added to the soil increases the amount of humus. Humus contains mineral salts, released by bacteria, which are needed for plant growth. It makes small soil particles stick together to make larger soil crumbs. Humus absorbs water, so the soil keeps its water and dissolved minerals better. How to find the amount of humus in different soils is described on page 234.

Clay particles These are the smallest soil particles. They are so small and light that many of them stay suspended when soils are mixed with water. In soil, the clay particles hold onto water and so tend to make the soil sticky (heavy) and waterlogged, without very much air.

Sand particles Sand particles are larger than clay particles, so they do not pack together so tightly. The spaces allow water to drain through very quickly. Sand is heavier than clay and forms a lower layer during sedimentation. A sandy soil may be dry, and easily blown away.

Air Plant roots and other living things in the soil (pages 242–3) need oxygen. Good soil has spaces between the particles and crumbs, which contain air. Air is also needed by bacteria which decay materials to form humus. How to find the amount of air in different kinds of soil is described on page 235.

Water Soil particles are surrounded by a film of water which is taken up by the plant roots. If the spaces between soil particles are too large (sandy soil), water drains through too quickly and the soil is dry. If the spaces are too small (clay soil) water fills up the spaces and there is no space for air. How to measure water drainage and water retention are described on page 235.

Questions

1. How are components of the soil classified?
2. Why do the soil particles form different layers in the sedimentation method?
3. Describe how mechanical, chemical and biological factors help in the formation of the soil.
4. What can you learn about soil formation by looking at different places around you?
5. How would you use the presence of soil particles to distinguish between sandy soil, clay soil and loam soil?
6. Choose samples of sandy, clay and loam soils and put them in similar pots. Plant 10 bean or maize seeds in each pot. Water the seeds. Compare the plants' growth after two weeks and describe any differences that you have observed. Can you explain these differences?

How do soils differ?

Soils differ in the amounts of the different kinds of soil particles they contain (page 233) and in the amounts of humus, air and water in the soil.

For the Activities on these pages you will need samples of sandy, clay and loam soils. Look at the table you prepared on page 233 and go and collect a sample of each kind of soil.

Activity | Observation of soils

1. What do the three kinds of soil look like? (For example, colour, particles visible.)
2. What do they each feel like when they are (a) dry and (b) wet? (For example, warm, cold, sticky, gritty.)
3. Use some of the wet soils and see if you can roll each sort out into a long worm.
4. Write your results in a table like this.

	Sand	Clay	Loam
Appearance			
Feel (when dry)			
Feel (when wet)			
Can it be rolled?			

5. Describe how you could distinguish between sandy, clay and loam soils on the basis of the observations you made above.
6. Use these methods to identify an unknown soil, given to you by your teacher.

Activity | Testing different soils for pH

The pH is a measure of acidity or alkalinity. A pH value of less than 7 is acidic, pH 7 is neutral and a pH of more than 7 is alkaline (see page 220). Different soils have different pH values.

1. Put a little of the soil to be tested in a test-tube and mix it with barium sulphate powder and distilled water.
2. Put your thumb over the end of the test-tube and shake the contents vigorously. Allow them to settle. (Barium sulphate is added to clump or **flocculate** the clay particles so that they become larger and sink. This leaves a clear solution to test for pH.)
3. Add a few drops of universal indicator solution or dip a piece of universal indicator paper into the test-tube. Then compare the colour of the solution or paper with the colour chart provided with the indicator.
4. Is the soil acidic, alkaline or neutral?
5. Test other soil samples.

Activity | Measuring the humus in the soil

First method
1. Shake a sample of soil with water and let it settle (sedimentation method, page 232).
2. Measure the depth of the humus (for example 0.5 cm) and the height of the soil altogether (for example 5 cm).
3. The percentage of humus is then:

$$\frac{0.5}{5} \times 100 = 10\%$$

4. Repeat with equal amounts of other kinds of soil.

Second method Different groups of students should do this Activity with different kinds of soil.
1. Measure 20 g of a loam soil into a crucible or tin can.
2. Put the soil in an oven at 40 °C for 30 minutes, or out in the sun for 2 hours. (This will drive off the water.)
3. Let the soil cool down and then find the mass of the dry soil and crucible (or tin). (This is the **original mass**.)
4. Heat the crucible or tin strongly with a Bunsen flame for about 20 minutes. (This will make the humus burn and it will be driven off, see below).

5. Let the remaining soil cool down and then find the mass of the soil and crucible (or tin).
6. Continue heating, cooling and finding the mass until the mass is constant. (This is the **final mass**.)
7. The percentage by mass of humus is then:

$$\frac{\text{original mass} - \text{final mass}}{\text{original mass}} \times 100$$

Questions

1. Which kind of soil has the most humus?
2. How do your results compare when using the two methods? Which do you think is more accurate? Why?
3. In the second methods why are (a) step 2 and (b) step 6 so important?

Activity | Measuring water drainage and water retention

Water drainage is the amount of water which comes out (drains out) from the soil. **Water retention** is the amount of water which is kept back (retained) by the soil.

1. Collect three filter funnels and put cotton wool in the neck of each one.
2. Measure out 30 cm³ of sandy, loam and clay soil.
3. Put a soil sample into each funnel. Rest the funnels in the necks of three measuring cylinders. Label each one.
4. Gently pour 100 cm³ of water into *each* sample of soil.

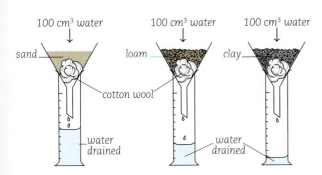

5. The water which drains through is caught in the measuring cylinders. Copy the table below and enter these 'water drained' figures in the *first* column.

	Water drained (cm³)	Water retained (cm³)
Sand		
Loam		
Clay		

6. Now, you added 100 cm³ of water to the soil. If 40 cm³ drained through, then 100 − 40 cm³ = 60 cm³ has been retained. Work out the amount of water retained by each soil sample and enter these figures in the *second* column.

Activity | Measuring the air in the soil

1. Put 50 cm³ of a sample of soil into a 100 cm³ measuring cylinder.
2. Add 50 cm³ of water.
3. Stir the soil and water. Observe the bubbles of air escaping from the air spaces between the soil particles (see below).

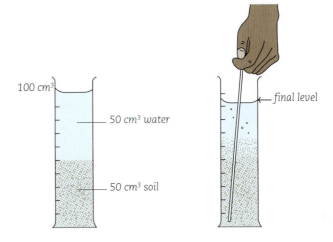

4. The original volume of soil plus water in the measuring cylinder was 100 cm³. Find the final volume of soil plus water.
   How much air has escaped (original volume − final volume)?
5. The percentage by volume of air is then:

$$\frac{\text{volume of air which has escaped}}{\text{original volume (50 cm}^3\text{) of soil}} \times 100$$

6. At the beginning of the Activity there was air between the soil particles. What is there now in these air spaces? Would this soil be suitable for plants to live in? Explain your answer.
7. Repeat this Activity with the same amount of different kinds of soil. Which soil contains the most air?

Questions

4. (a) Which soil has the greatest water drainage (least retention)?
   (b) What effect will this have on the soil, and on the plants living in the soil?
5. (a) Which soil has the greatest water retention (least water drainage)?
   (b) What effect will this have on the soil, and on the plants living in the soil?
6. (a) Which soil has an in-between value for water drainage and water retention?
   (b) What effect will this have on the soil, and on the plants living in the soil?
7. You have observed sandy, clay and loam soils. Try to explain the reason for as many of the observations as you can.
8. Plants need water *and* air from the soil. Which of these things is likely to be in short supply in (a) a sandy soil and (b) a clay soil? Explain your answers.
9. In what ways is loam a better soil than either a sandy or a clay soil?

How can we use our knowledge about soil?

Land covers about 30% of the surface of the Earth. We use this land for housing and industries, to grow crops and rear our animals, and for roads and other purposes.

The underlying rocks are also important, for example for the mining of bauxite in Jamaica and Guyana, or deposits of petroleum in Trinidad and Barbados. The igneous rocks of the volcanic islands of the Eastern Caribbean (page 314) impart their characteristics to the soils found there.

On these pages we will see how we can improve clay and sandy soils for agriculture, and how we can use rocks and soils for road-building. On pages 238–9 we will examine plant growth in a variety of soils, and investigate good farming practices. On pages 240–1 we will see how we can conserve the soil.

Clay soils
Characteristics
- Mostly small clay particles
- Small air spaces
- Water drains poorly (a lot is retained)
- Soil stays wet (becomes water-logged)
- Little air
- Tend to be acidic

Improvement **Add lime** Lime is calcium oxide. It is very useful to add lime to clay soils because the lime makes the clay particles clump together or **flocculate** to form larger particles. Lime is alkaline and so helps to neutralise the acidity of clay soil. It also adds calcium to the soil which is needed for plant growth.

Add sand Sand particles are bigger than clay particles and if they are mixed into clay soil they make the air spaces larger. This means that water will drain through better, that the soil will not be so wet, and that there will be more air for the plants.

Add compost We can prepare compost (page 157), which decays to humus in the soil. Humus improves the texture of clay soils because it makes the particles stick together to form larger soil crumbs. Humus also absorbs water so the soil becomes less water-logged. It also contains minerals which are slowly released for plant growth.

Add chemical fertiliser If chemical fertilisers are added to clay soils, particles of the fertilisers will stay between the soil particles and may be too concentrated.

It is therefore best to wait until the soil has been improved with lime, sand and humus before adding chemical fertilisers. Chemical fertilisers do not improve the texture of the soil, but they do release their chemicals faster than humus does.

Loam soils
Loam soils are a mixture of sand and clay particles, so the air spaces are of medium size. Because of this the water drains properly (without water-logging) but enough is retained for plant growth. Loam soils have good amounts of air and water. This is important for healthy growth of plants and animals.

Loam soils are the best soils for plant growth, that is they are the most fertile. Plants and animals have lived in the soil, and then died and decayed. The soil contains humus from the decay of plants and animals.

Loam soils can be improved by adding extra humus (or compost). The humus releases mineral nutrients slowly into the ground as these are used up by the plants. Chemical fertilisers, such as NPK (containing nitrogen, phosphorus and potassium) provide extra nutrients. The improvements we make to clay and sandy soils are meant to make them more like loam soils.

Sandy soils
Characteristics
- Mostly large sand particles
- Large air spaces
- Water drains easily (little is retained)
- Soil becomes dry
- A lot of air
- Tend to be neutral (can be saline)

Improvement **Reduce salinity** Sandy soils near to the sea tend to have a lot of salt in them. This comes from sea water which has drained into them, and from salt spray. Only certain plants, like coconuts, are adapted to live in these conditions. If sea sand is to be used in the garden it should first be washed to remove any salt.

Add clay Clay particles are smaller than sand particles, and if they are mixed into sandy soil they fit into the air spaces, so making the spaces smaller. This means that water will not drain out so quickly, and so the soil will not become so dry and there will be more water for plant growth.

Add compost Compost decays into humus. Humus improves the texture of sandy soils because it holds on to water which would otherwise have drained out of the soil. This makes it a better medium for plant growth. It also contains minerals which are slowly released into the soil for plant growth.

Add chemical fertilisers If chemical fertilisers are added to sandy soils they will just be washed out by the water which drains through. It is better to wait until the soil has been improved with clay and humus before adding fertilisers. The smaller air spaces and additional humus of the improved soil will allow the chemicals to work better.

Activity | *Investigating soils for road-making*

1. Prepare two sandy, two clay and two loam soil samples in six plastic containers.
2. Put one set of each soil type outside in the sun for 2 hours to dry the soil.
3. Add water to the other set.
4. Fetch six masses of 100 g and put one on top of each soil sample.
5. Describe how well each soil supports the mass (load).

Questions

1. (a) Which soil supports the load the best? (b) Which characteristic of the soil may account for this?
2. Which soil supports the load the worst? (That is, the load will sink into the soil.)
3. What difference does wetting the soils make to (a) sandy (b) clay and (c) loam soils?
4. Which kind of soil would be best for road-making?

(a) Soil road in very dry weather

(b) Soil road in very wet weather

Questions

5. Make use of the photograph and picture above to describe the main problems of soil roads in (a) very dry and (b) very rainy weather.
6. Laterite roads are roads made of soil which has weathered, and have a hard top surface. Compared to tarmac roads, what would be (a) their advantages and (b) their disadvantages?

Tarmac roads

Main roads have a covering of **tarmac**. This is a mixture of **tar** and **macadam**. The tar is runny when hot and fills up the spaces in the road before it sets into a hard surface. Macadam consists of hard stones broken down into small pieces. They help to make the final surface very hard and able to withstand the passage of heavy vehicles.

Before the tarmac is added, the **road base** is prepared. This consists of broken-down rocks which have small stones added in between. Heavy machinery is used to flatten this to make a firm foundation.

If the tarmac wears thin, cracks or melts, then water may run in, and make a **pothole**.

Tarmac roads are needed to support the weight of many vehicles. Notice that the road surface is, however, breaking up in some places

Questions

7. What kind of weather makes a pothole get bigger fastest? Why do you think this is?
8. Look at the sides exposed in a pothole. Draw it. Identify the tarmac, layer of small stones and larger rocks used in building the road.
9. Why are potholes dangerous?
10. Broken-up limestone is often put into potholes. Why? Does it solve the problem? Why?
11. What is the best way to deal with potholes?
12. A farmer has a very sandy soil. What advice would you give him for improving its fertility?
13. Is it *always* useful to add (a) humus and (b) chemical fertilisers to *all* kinds of soil? Why?
14. What difficulties would you have in driving a car along a sandy beach?
15. What modifications do heavy-duty cross-country land-rover vehicles have? Why are these necessary?
16. 'It is better to have a good laterite road than a tarmac road full of potholes.' Comment on this statement.

How do we get the best out of the soil?

Soil is important because we use it for growing crops and for rearing animals. Its physical and chemical properties determine how useful and fertile it will be.

Physical properties Physical properties describe the size of the soil particles which make up the soil. Look back to pages 233–6 and then complete this table.

Physical property	Sand	Clay	Loam
Size of particles			
Size of air spaces			
Amount of water retention			
Amount of air			

Chemical properties Chemical properties describe the chemicals in the soil. They either assist or harm plant growth. We can find the pH of the soil (page 234). On the whole, plants prefer neutral soils or ones which are slightly acidic. We may need to adjust the soil pH. If the soil is too acidic we add lime. If the soil is too alkaline we add ammonium sulphate. Some plant preferences are shown below.

pH range	Plants which grow best
4.5 – 6.0	Potato
5.5 – 7.0	Tomato, carrot
6.0 – 7.0	Lettuce, onion
6.0 – 7.5	Cabbage, grass

Salinity is the amount of salt in the soil. Sandy soil near the sea tends to be saline. Only a few plants are adapted to grow in this soil, for example coconuts.

Nutrients are the mineral salts in the soil, containing, for example, nitrates, phosphates and potassium. Decaying material, humus and fertilisers add nutrients which can be used for plant growth.

Activity | Comparing plant growth in different soils

You are going to use different kinds of soils to grow your seedlings. The soils will be watered with different liquids and you will compare the resulting growth of the seedlings.

1. Use six half-gallon plastic containers such as those used for ice cream. Make holes in the bottoms and put in a layer of stones to allow for drainage (see top of next column).

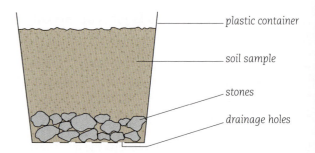

2. Set up the following containers and label them A to F like this:

Type of soil		Water with
A	Sandy soil (washed to remove salt)	Rain water
B	Clay soil	
C	Loam soil	
D	Loam soil	Sea water
E	Loam soil	Dilute acid
F	Loam soil	Dilute alkali

3. Put five similar seedlings into the soil in each container. Water them with the same volume of the correct liquid every other day.
4. Measure the height of the seedlings after three weeks and describe their general appearance.

Questions

1. How would you predict which plants would grow best on (a) sandy (b) clay and (c) loam soils? Give reasons.
2. How would you predict which plants would grow best on (a) acidic (b) alkaline (c) saline and (d) nutrient-rich soil? Give reasons.
3. How well did the seedlings grow in containers A to F of the Activity? Account for each of your findings. What appears to be the best conditions for plant growth?
4. List the physical and chemical characteristics of (a) clay and (b) sandy soil. How can each one be improved for better plant growth?
5. (a) Look at the ideas for good farming practice shown opposite. Describe how each suggestion will result in improved crop yields with minimum depletion of soil nutrients.
(b) If you are able to talk with farmers in your area, discuss these points with them. Find out which of the actions they carry out, and which ones they think would cost too much, take too much time or not be very effective.

Good farming practice

Do rotate crops
On a certain plot of land it is best to grow different crops each year and to include a year of fallow (page 240). But this may be difficult if you are growing sugar cane or bananas.

Why? Different crops take up different nutrients from the soil. Leguminous plants (beans and peas) should be included in the rotation. Disease organisms left in the soil could attack the same crop the next year, but not a different crop.

Do make compost
Compost is partly decayed plant and animal remains.
Farmyard manure can be added to compost heaps (page 157) and it will be decayed by bacteria so that it will be harmless to crops and to humans.

Why? Compost returns nutrients to the soil. It further decays to humus which improves the texture of the soil because it binds small soil particles together as soil crumbs.

Do mulch
Mulch is dried grass. It must be dried in the sun first to kill it, so that it will not start growing again. Mulch is put on the ground as a layer around plants.

Why? Mulch stops water from evaporating from the soil and so less water is needed for the plants. It also stops weeds from growing, and it can decay slowly into humus.

Do

Do plough back
Plant remains should usually be ploughed back into the soil. The exceptions are plants which are troublesome weeds, such as nutgrass, or which are infected in some way. These should be removed and burned.

Why? Nutrients from the old plants are returned to the soil. This is especially important in the case of beans and peas which have nodules on their roots which contain nitrates (page 243).

Do remove weeds
Before you plant your crop, the ground should be cleared of weeds. While the crop is growing the weeds will also have to be removed by hand or with the careful use of chemical sprays (herbicides).

Why? Weeds compete with the crop you are growing, both in taking up space and in using nutrients from the soil which the crop needs.

Do irrigate
Irrigation means watering plant roots as directly as possible. Underground pipes are the best method but these are expensive to install. Irrigation canals can also be useful. Do not water at midday when water will evaporate and not be used by the plants. Water should go to the plant roots which need it (trickle irrigation).

Why? Irrigation makes the best use of the water available.

Don't

Do not burn off plants
Do not burn off plants unless they are infected or are troublesome weeds. Plants should be returned to the soil so that the nutrients are not lost. Burning sugar cane makes harvesting easier, but it destroys some of the sugar.

Why? Burning the ground breaks down the humus and affects the soil structure. The ash from burning can return some minerals to the soil, but burning usually does more harm than good.

Do not use raw manure
Raw manure is animal faeces. It may contain disease organisms like pathogenic bacteria and eggs of parasitic worms. Raw manure should be added to compost heaps to be made harmless.

Why? Disease organisms may infect the humans who work in the fields or may infect the crop plants. Using urine and faeces directly on the fields can be dangerous.

Do not remove ground cover
If a field is not being used for a crop (e.g. in the fallow year of crop rotation) it should still have grass or other plants growing on it (e.g. for feeding cattle).

Why? Ground cover helps to reduce the soil erosion (page 240). This is especially important on sloping ground where uncovered soil can easily be washed away by the rain.

How can we conserve the soil?

On the previous pages (232–9) we looked at how we can describe different kinds of soils and the ways in which we can improve them. Now we will see how we can keep (**conserve**) the soil in good condition.

Conserving the soil

Crop rotation This means growing different crops each year on the same plot of land, for example:

Year	Crop	Reasons
1	Root crop (e.g. yams or cassava)	Different crops take different nutrients from soil. Not affected by same diseases
2	Cereal crop (e.g. maize or rice)	
3	Legumes (e.g. beans or peas)	Nodules add nutrients to the soil
4	No crop: fallow (e.g. grass for cattle)	Ground cover stops soil erosion

Prevent soil erosion Land, such as fallow land, should always have a covering of plants, otherwise soil will be washed away (soil erosion).

Activity | The importance of ground cover

1. Collect two shallow Styrofoam packing trays like those used for meat or vegetables.
2. Fill one tray with tightly packed garden soil.
3. Cut a piece of turf (soil and its grass cover) from the garden which will fit tightly into the second tray.

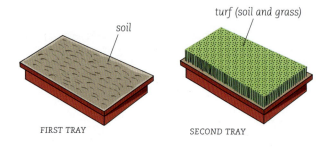

FIRST TRAY — soil
SECOND TRAY — turf (soil and grass)

4. You are going to let water run onto the two trays. Lay the trays flat in a sink. Arrange them in such a way that you can collect any water that runs off from the trays.
5. Turn on the tap at a steady rate, and for a measured 2 minutes run water first onto the first tray, then the second tray as each lies flat in the sink.
6. Repeat step 5, but with each tray tilted at an angle of about 20°.

Questions

1. (a) How much water ran off and was collected in each of the four cases? Show the results in a table.
   (b) How much soil was washed off? (Compare the muddiness of the water.)
2. How does (a) the plant cover and (b) the slope of the trays affect the amount of soil washed off?
3. In which of the following situations would you expect the most soil erosion, and why?
   (a) Sloping ground with plant cover.
   (b) Flat ground with plant cover.
   (c) Sloping ground with no plant cover.

Activity | Looking for soil erosion

Look around near your school or your home for examples of soil erosion. It may not be as bad as this.

Severe soil erosion which can occur when land is without cover. In the dry season earth is blown away, and in the wet season it is washed away

1. Find an example of soil erosion (a) on flat land and (b) on a slope. Describe what you see.
2. In each case try to suggest what has caused the erosion.
3. What use are plant roots (of small plants and big trees) in helping to stop soil erosion?
4. What use are plant leaves in helping to stop soil erosion?
5. From your experience, is soil erosion worst when the rains are heaviest? Is it the length of time that the rain falls, or the strength with which it falls that is most important? How can you test your suggestion?

How can we prevent soil erosion?

Keeping the ground covered with plants is one way in which we can prevent soil erosion. What else can we do?

Activity | Contouring

Contouring means planting crops along the contours of the land, that is *around* a hill and *not* up and down the hill. This Activity will demonstrate what difference contouring makes.

1. Your teacher will stick strips of blotting paper onto the blackboard in these two different patterns.

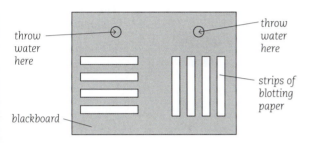

2. Starting from the same place, an equal amount of water (10 cm^3) is thrown above each set of strips.
3. Compare the amount of water which drips down the blackboard.
4. The strips of paper represent rows of plants in the soil on the side of a hill:
   (a) represents rows planted along the contours
   (b) represents rows planted up and down the hill.

Questions

4. From your Activity, which arrangement of plants do you think would reduce soil erosion?
5. In the photograph below identify the rows of plants. Have these been planted along the contours?

Terracing

Terracing means building up flat surfaces of soil around a hillside. Terracing is necessary on sloping ground where erosion is a problem. Large rocks and boulders are used. A wall is built up and the space between the wall and the slope is filled in with soil. This is repeated up the hillside. Notice that there are now level surfaces, instead of the sloping surface that was there before.

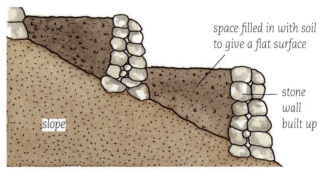

How a terrace is built up (cross section)

Terraces built on the side of a hill

Questions

6. What do you understand by the following:
   (a) crop rotation (b) soil erosion (c) contouring (d) terracing?
7. (a) What are the advantages of crop rotation? Give reasons for the inclusion of each of the plants grown in a typical four-year cycle of crop rotation.
   (b) What are some of the problems with crop rotation that a farmer might have?
8. What conditions lead to soil erosion?
9. List three ways in which soil erosion can be prevented, and describe one of them in detail.
10. What would be the effects of:
    (a) deforestation (removal of trees for timber without replanting seedlings to replace them)?
    (b) soil washed from uncovered land accumulating in nearby rivers?

What organisms live in the soil?

On page 144 we looked at some of the food chains we find in aquatic habitats. Now we will look at organisms that live in a terrestrial habitat in the soil and in leaf litter.

Activity | Collecting leaf-litter animals
1. Choose a place under a tree where there are lots of decayed leaves (leaf litter).
2. Dig up a spadeful of leaf litter and carry it inside in a plastic bag.
3. Spread your sample on a newspaper, taking care to catch anything that moves. Use a large spoon or forceps, as some animals may bite or sting.
4. Put the animals into separate glass bottles.
5. Take out the animals one by one and identify as many as you can using the diagram below.
6. When you have finished, *return the organisms* to where you found them.
7. From the information below, and anything you may have observed, try to identify whether the animals are first order consumer (herbivores), second order consumers (carnivores), or omnivores, scavengers or decomposers (see also page 144).

Activity | Collecting soil animals
1. Make small holes, 2 mm in diameter, in the bottom of a half-gallon plastic or tin container.
2. Remove leaf litter from an area of ground, and dig up a spadeful of the soil underneath. Fill the container two-thirds with soil.
3. Support the container above a beaker of water.
4. Arrange a light above the container. The light and heat will drive soil animals out of the soil and into the water (below).

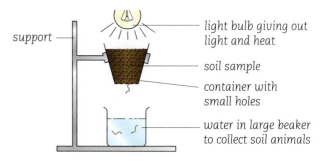

5. Identify any animals, using the diagram below.

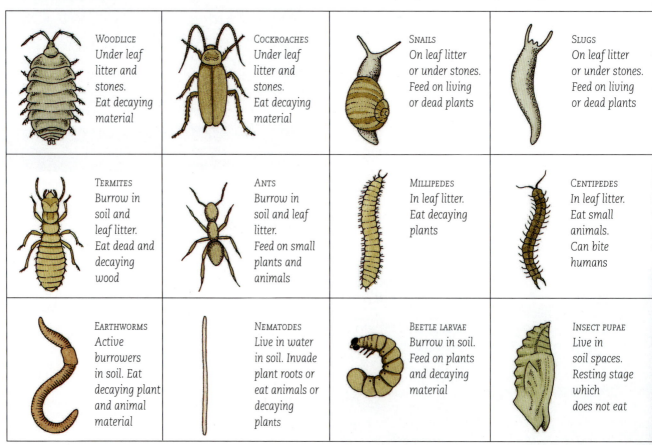

Soil animals

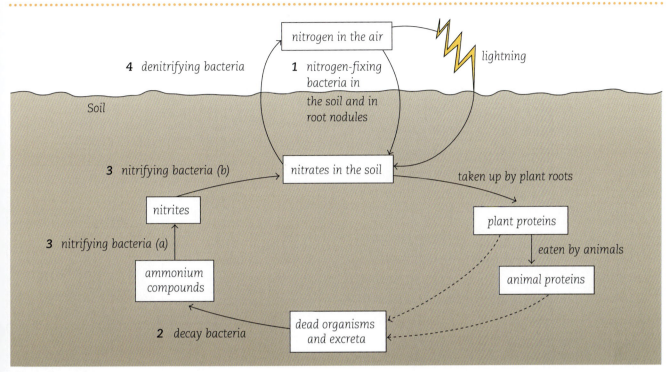

The nitrogen cycle

Important soil animals

Earthworms Earthworms burrow into the soil taking decaying leaves into their burrows. They mix up the soil layers. The burrows they make also mean that air gets down into the soil. Earthworms eat soil, or rather they take earth into their bodies, digest any juicy bits of decaying material, break up the soil and then pass it out as their faeces (worm casts).

Earthworms are not very common in the hard, dry soils in many parts of the Caribbean. They are also rare in waterlogged soils.

Termites Termites eat dead and decaying wood. They soften the wood with their digestive juices and then chew it. Single-celled organisms (protists) in their guts actually digest the cellulose in the wood. Some termites make nests underground and so mix up the soil.

Ants Ants make burrows in the soil. They feed on small animal and plant remains in the soil and they soften and break the soil into fine particles.

Important soil bacteria

The soil you collected in the Activities will contain plant roots and the animals you identified. But most of the living organisms in the soil are soil bacteria which you cannot see without a microscope. These bacteria play a role in decaying dead material so as to produce nitrates which can be recycled. This process is called the **nitrogen cycle** (above).

There are four main groups of bacteria living in the soil which are involved in the nitrogen cycle (labelled 1 to 4 on the diagram above).

1 Nitrogen-fixing bacteria These bacteria live freely in the soil or in the bumps (**nodules**) on roots of legumes such as beans and peas. These bacteria convert nitrogen in the air into nitrates in the soil.

2 Decay bacteria Decay bacteria act on dead organisms and their wastes (excreta and faeces), to make ammonium compounds.

3 Nitrifying bacteria These bacteria are of two types which (a) convert ammonium compounds to nitrites, and (b) convert nitrites to nitrates which can be taken up by plant roots.

4 Denitrifying bacteria Denitrifying bacteria undo the good work of the other bacteria. They are common in waterlogged soils because they can work without oxygen. They convert nitrates back into nitrogen in the air, so these nitrates cannot be used by plants.

Questions

1. In what ways do (a) plant roots and (b) two named soil animals affect the soil?
2. In what ways are the amount of nitrates in the soil (a) decreased and (b) increased?
3. Describe the steps in the nitrogen cycle.
4. How can knowledge about the nitrogen cycle help farmers to keep their soil fertile?

What organisms live round about us?

On pages 242–3 we looked at the organisms we might find in the soil and leaf litter. We can classify these organisms as shown below.

Organism	Food	Feeding level
Plants	Make their own	Producers
Woodlice	Decaying material	Consumer (scavenger)
Cockroaches	Decaying material	Consumer (scavenger)
Snails	Living or dead plants	First order consumer
Slugs	Living or dead plants	First order consumer
Termites	Wood	First order consumer
Ants	Plants and animals	Consumer (omnivore)
Millipedes	Dead or decaying plants	First order consumer
Centipedes	Animals	Second order consumer
Earthworms	Decaying material	Consumer (scavenger)
Nematodes	Living plants or animals	Consumer (parasite)
Beetle larvae	Plants, decaying materials	Consumer (omnivore)
Fungi Decay bacteria	Break down dead organisms and wastes	Decomposers

Let us remind ourselves how the organisms at different feeding levels are interrelated.

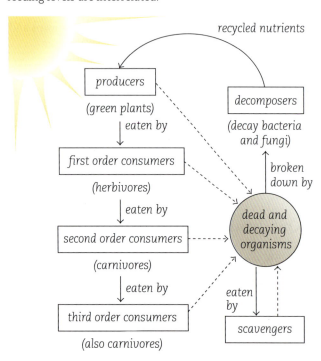

Activity | Collecting land organisms

1 Things you will need.
- Scissors or pruning shears to cut pieces from plants.
- Large spoon or trowel to collect soil.
- Knife or penknife to scrape mosses and snails from rocks and stones.
- Long stick to knock organisms out of trees and bushes.
- Animal traps and nets for catching organisms.

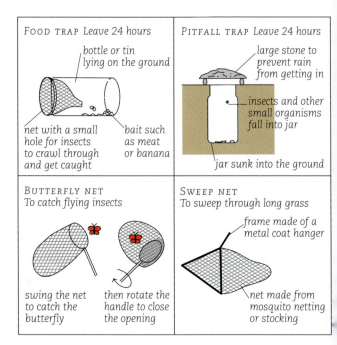

- Thick gardening gloves for handling organisms with spines, and those that bite.
- Notebook and pencil to record where organisms are collected and what they are feeding on.
- Plastic bags and bread ties, tins and jars for transporting organisms.
- Paper tape and pen for writing labels for containers.
- Newspaper or plastic bowls for sorting animals and examining soil.
- Forceps to handle small animals or ones that bite.
- Hand lens to examine small organisms.

2 In small groups, collect organisms in particular places in the school grounds. Try and identify them by using the diagrams on pages 13–15 and page 242.
3 Record what the animal was eating when you found it, such as a butterfly feeding on nectar. Also collect a piece of the vegetation on which an animal was found.
4 Some animals are hard to find because they hide, or are coloured in a similar way to their surroundings (**camouflaged**). Record any examples of camouflage.

Food chains and food webs

This table shows some of the organisms you might observe, together with what eats them.

Organisms that are eaten	Organisms that eat them
Grass	Cows and grasshoppers
Cabbage	Humans and slugs
Plant leaves	Slugs and caterpillars
Decaying plants	Earthworms
Flower (nectar)	Butterflies and humming birds
Cows	Humans
Grasshoppers	Praying mantises and toads
Slugs, caterpillars, earthworms and butterflies	Small birds
Praying mantises	Lizards
Toads, small birds and humming birds	Large birds

Food chains show which organisms feed on which other organisms. An example of a food chain from the table would be:

Grass → Grasshopper → Praying mantis → Lizard
(producer) (first order consumer) (second order consumer) (third order consumer)

Food webs show interrelated food chains. Here is a partly completed food web made from the table.

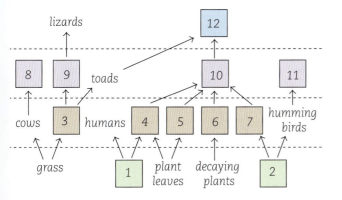

Questions

1. Find, and write out, the other nine food chains from the table above. Each one must start with a producer.
2. Use your information to fill in the spaces 1 to 12 in the incomplete food web.

Ecology

The study of living organisms in their natural surroundings is called **ecology**. There are several other important terms which we use.

Groups of living organisms A **population** is a group of organisms of the same kind or species in a particular place. For example, the grasshoppers in a field would be a population.

A **community** is a number of populations living together. For example, all the different populations of organisms living on the school grounds is a community.

The place where organisms live The habitat is the *place* where the populations and communities are living. For example, the school grounds is a habitat.

We can divide habitats into *aquatic* habitats such as fresh water, sea water and estuaries, and *terrestrial* habitats which are on land and in the soil.

The conditions surrounding organisms The *conditions* that are found in a habitat are called the **environment**. The environment includes the climatic conditions, such as temperature and amount of rain, the soil conditions and the other living organisms. For example, in its habitat a caterpillar will need to find its food, and also to escape from its enemies. Organisms need to be well adapted to their environment in order to survive.

Everything together The living *organisms* in a particular *habitat* affected by the *environment* are called an **ecosystem**. For example, all the organisms which live in the school grounds and the environment which affects them are called an ecosystem.

The ecosystem is in balance; this is sometimes called 'the balance of nature'. The food chains and food webs which are found in an ecosystem tend to keep the numbers of animals and plants constant. For example, if the numbers of grasshoppers increased in a particular year, some of them might die because they could not find enough grass to eat. Others would be eaten by praying mantises and toads, and so the numbers would return to their usual level.

Questions

3. What do you understand by each of the following terms? Give an example of each: (a) producer (b) first order consumer (c) second order consumer (d) decomposer (e) herbivore (f) carnivore (g) omnivore (h) scavenger (i) parasite (j) population (k) community (l) habitat (m) ecosystem.
4. Why do food webs always start with producers?
5. In the food web, name (a) an omnivore (b) five herbivores and (c) three second order consumers.
6. How do organisms (a) at the same level and (b) at different levels in a food web affect each other?

How are materials recycled in nature?

Living organisms need certain basic materials for them to live and grow. They need carbon and nitrogen compounds, oxygen and water. These materials must be continually recycled in nature.

Activity | The roles of photosynthesis and respiration

1. You will be given some sodium hydrogen carbonate solution (carbon dioxide (CO_2) indicator solution). This changes colour as follows.

Colour	Amount of carbon dioxide (CO_2) dissolved
Reddish	Same as in the air
Purple	Less than in the air
Yellow	More than in the air

2. Set up four boiling tubes like this:

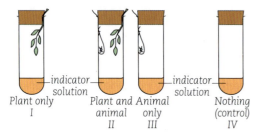

Plant only I · Plant and animal II · Animal only III · Nothing (control) IV

3. Leave them for about 2 hours. Then observe what has happened to the indicator solutions. Explain your results.

Questions

1. The table below gives the results of a similar experiment. Match the results to the correct boiling tubes I–IV shown above.

Column 1	Column 2	Column 3	Column 4
Indicator stays reddish	Indicator goes purple	Indicator stays reddish	Indicator goes yellow
CO_2 same as in air	Less CO_2 than in air	CO_2 same as in air	More CO_2 than in air
No living organisms present	Plant has taken up more CO_2 than it has produced	Plant takes up CO_2 produced by plant and animal	Animal has produced CO_2 in respiration

2. Explain how (a) photosynthesis and (b) respiration affect the amount of carbon dioxide in the air.
3. What would you expect the results in each of the boiling tubes to be if you had left them in the dark?

Activity | The roles of evaporation, condensation and feeding

1. In class, set up an aquarium in a large glass container. First put in sand and stones, and then fill it two-thirds with rain water.
   If you use sand from the beach, first wash it to remove the salt. Add the rainwater a little at a time.
2. Add some plants (producers) such as pondweed, and some animals (consumers) such as tadpoles, mosquito larvae and pupae, water snails, and an aquatic beetle larva.
3. Close the top with a sheet of glass (below). Leave it for a week.

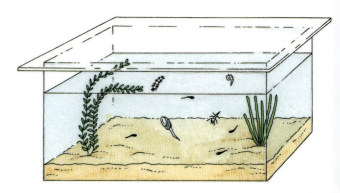

Questions

4. Observe the glass surfaces above the water. Can you see little drops of water? Where might these have come from? What processes produce them? Why are these processes important?
5. Do the animals stay alive? What do they eat? How does this supply them with carbon and nitrogen compounds?
6. Describe three food chains in the aquarium. Make a diagram showing the food web (page 245).
7. Can you see any bubbles of gas coming from the plants? What might these bubbles be? What are the plants doing when it is light?
8. The animals and plants need oxygen for respiration. Where is the oxygen, and how is it renewed as it is used up?

Recycling materials

In these two Activities we have only been able to look at a few of the processes involved in the recycling of materials. Other information will be found for the nitrogen cycle on page 243, the water cycle on page 139 and the cycling of oxygen and carbon dioxide on page 108.

The diagrams opposite summarise the processes by which nitrogen, water, oxygen and carbon dioxide are recycled in nature.

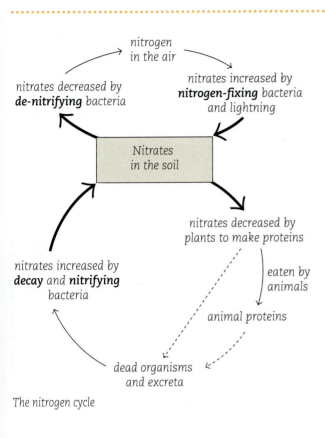

The nitrogen cycle

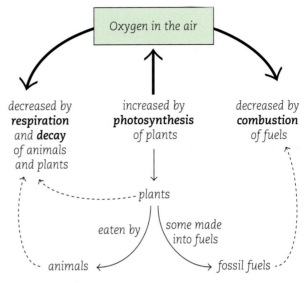

The oxygen cycle

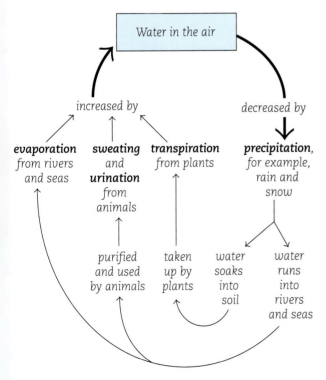

The water cycle

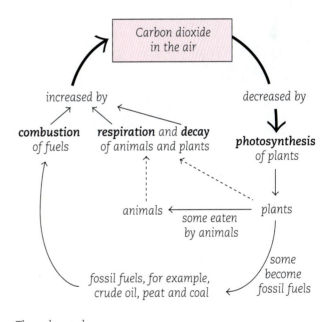

The carbon cycle

Questions

9. Describe how (a) nitrogen (b) water (c) oxygen and (d) carbon dioxide are recycled in nature. For each cycle list the processes which (i) increase and (ii) decrease each substance. Which of these processes are influenced by human activity?
10. How are the oxygen and carbon cycles interrelated?
11. Why are the nitrogen, water, oxygen and carbon dioxide cycles important?

Pollution and conservation

Pollution
Pollution is the accumulation of waste products such as sewage or poisonous chemicals, or excess energy such as noise or heat. These **pollutants** upset the balance in the environment. They can cause annoyance, ill health and destruction of buildings and living things.

Activity | *Examples of pollution*
1. Three examples of pollution are shown in the photos on this page: land pollution from poor disposal of waste materials, air pollution from a cement factory, and water pollution from detergents.
2. Find other examples of pollution in your surroundings. For each one say whether it affects land, water or air. Describe the problems which might be caused and give suggestions as to what might be done about each one.

Here are some examples of pollution.

Pollutants from burning fossil fuels Soot (carbon) blackens leaves and buildings and contributes to respiratory diseases. Carbon dioxide and oxides of sulphur and nitrogen form part of acid rain (page 193) damaging buildings and plants. To reduce the problems, we need to extract pollutants from factory smoke, and use smokeless fuels.

Pollutants from cars Car exhausts (see page 193) can produce carbon monoxide (which in excessive amounts can cause death), and lead (which can cause brain damage). To reduce these problems we should make sure our cars are properly tuned, use lead-free gasoline, and fit special converters.

Noise pollution This may come from mechanical drills, jets or loud music. Noise pollution can cause annoyance, or, if sustained over time at a high level, can cause damage to your hearing or could even cause your eardrums to rupture. Levels should be controlled as much as possible, and workers exposed to noise pollution should wear ear-plugs.

Pesticides (page 155) Many of these chemicals can cause ill health to the people who spray them. Also, by being passed up the food chains, they may cause damage to a wide number of other animals including ourselves. Pesticides should be used carefully and harmful ones should be banned.

Radioactivity This is produced by nuclear power stations and nuclear bombs. Radioactivity can cause increased rates of leukaemia. It is essential to have good safety precautions in place and to dispose of radioactive wastes very carefully (see also page 21).

Accumulated waste This can provide a living place for pests and could cause the spread of disease (pages 150–5). Sewage (human faeces and urine) may get into waterways where any harmful organisms (page 152) may continue their life cycles in different organisms before re-infecting other humans. Good hygiene and sanitary measures, and recycling of wastes can reduce the problems.

Fertilisers Excessive amounts of fertilisers in waterways change conditions and so poison fish. If possible, we should rather use small amounts of fertiliser and supplement this with compost.

Crude oil This can come from tankers and oil rigs. It ruins beaches and kills fish and sea birds. It is important to clean up oil spills and to try to prevent them from happening.

Heavy metals In Japan, for example, mercury from industrial wastes turned up in sea fish and caused disease in humans when these fish were eaten. Chemical wastes should be removed from waste water before it leaves the factories.

Conservation
To conserve means 'to preserve and keep in balance'. In relation to our environment it means we should try to reduce the production of pollutants and the effects they cause, as well as trying to improve features of our surroundings.

Here are some examples of conservation.

1 **Reduce pollution** Pollutants can be reduced, but methods for doing so are costly. Individuals, companies or governments have to pay. Steps can be taken to reduce pollution:
- Instead of fossil fuels use smokeless fuels, solar energy (page 195) or non-polluting sources.
- Put filters and alkaline solutions in factories to remove soot and neutralise acid-making oxides.
- Keep cars tuned, fit converters and use lead-free gasoline.
- Improve public transport and reduce number of cars.
- Replace harmful pesticides with safer ones.
- Improve hygiene and sanitary measures.

2 **Recycle materials** Natural resources (raw materials) used to make household items are scarce. We need to recycle.
- Non-biodegradable materials (not broken down by bacteria) can, for example, be re-used (plastic containers) or collected and recycled (metal, plastics and paper).
- Biodegradable materials (decayed by bacteria) such as household scraps, can be composted or used to make biogas (page 157). Sewage (human faeces and urine) can be treated to make clean water or fertiliser.

3 **Conserve soil and water** The problem of soil erosion, and ways of conserving soil are discussed on pages 239–41. We should avoid over-working the soil as this can lead to it being washed or blown away. We can grow high-yielding crops, use compost, and put water directly to the roots to increase productivity. Water reserves in the soil might be used more quickly than they can be replaced by rainfall. This is partly caused by removing rain forests and also by the increase in the population which increases water use. We should use water wisely and recycle and conserve it as much as possible.

4 **Preserve the balance of nature** Living things are all interconnected. If we use pesticides, introduce species to a country, over-fish oceans, or pollute our environment, we will feel the effects later. We need to be aware of outcomes.

5 **Protect endangered species** Humans have a special responsibility to maintain the numbers of wild animals and plants rather than hunting or trapping them, or removing them for profit. If a population of plants or animals becomes small it is in danger of becoming extinct. Countries can pass laws prohibiting, for example, the capture of sea turtles, manatees and certain parrots. They can also prohibit the collection of certain rare plants from the wild. Over-fishing also threatens certain species.

Groups of wild animals are sometimes kept and bred in zoos. This may help, but it is difficult to return captive animals to the wild as they may not be able to hunt for their own food. Setting up conservancy areas is better – if they can be kept clear of predators and poachers.

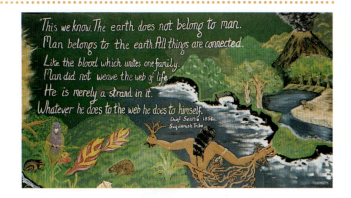

6 **Conserve natural beauty** There are often problems with the use we wish to make of the land. For example, companies may want to cut down rain forests to sell the timber, but this means the people who live there would have to be moved. Or an area near the shore may be mined for peat, resulting in a loss of wild life.

Land is increasingly needed for housing and for growing crops and rearing animals to support a growing population. But if areas of natural beauty can be maintained, they can be used for recreation and tourism, which can bring in foreign exchange. Governments should try to keep a balance between economic development and conservation.

Activity | Practising conservation

1 Choose one of the methods of conservation and describe how it could be implemented in a factory, farm or workplace in your community.
2 Choose a different method of conservation and describe how you and your friends could try to implement it in your school or community.
3 Imagine that a new housing estate is to be built in an area of great natural beauty in your country. What are the advantages and disadvantages? What possible problems might there be? How could they be reduced?
4 Design a poster with advice on helping the environment. Discuss the ideas with guardians and friends, and see how many suggestions you can carry out.

Questions

1 What is meant by (a) pollution and (b) conservation?
2 What reasons might be given by (a) the government (b) a factory and (c) an individual for not practising conservation activities? For each group, discuss some arguments you would use to stress the importance of conservation for future generations.
3 A factory needs a new cleaning system. Who should pay: (a) the government (b) the factory owners or (c) the people who buy the products? Give reasons for your answer.

Why do we need exercise?

Activity | How fit are you?

1. Choose a step which is about 20 cm high. Step onto it with one foot, then bring up the other foot. Then step down onto the floor one foot at a time. Ask a friend to time you. Repeat this up-and-down process at the rate of once every 3 seconds.
   WARNING Do not continue the exercise if you feel dizzy or very much out of breath.

2. Stop after 3 minutes. Wait for *exactly 1 minute*. Then find your pulse (page 94) and count your heart rate for 15 seconds. Compare your reading to these.

Pulse (heart beats in 15 seconds)		Fitness rating
Men	Women	
Below 18	Below 20	Excellent
18–20	20–22	Good
21–25	23–28	Average
Above 25	Above 28	Poor

We may not be able to develop our muscles like this athlete, but we can exercise to be healthy

What are the effects of exercise?

Exercise is any physical movement that is done vigorously enough to make you breathe deeply and become warm and sweaty. So exercise is not restricted to sport but can also include digging in the garden and walking briskly. You can choose which kind of exercise you prefer.

Exercise has various effects.

Lungs During exercise you breathe more deeply to get more oxygen into your lungs (pages 98–9). Your lungs will become stronger and the alveoli open up more. This means that your lung capacity increases, so there is more oxygen in your body for the use of your muscles. Your muscles work more effectively and are less likely to get cramp or become easily fatigued.

Heart and circulation During exercise the heart beats harder and more quickly to pump blood to the muscles. Over a period of time it becomes more efficient and so you are less likely to have a heart attack (page 97).

Exercise also widens the arteries. This means that if there are any deposits (**atheroma**) laid down in the arteries (page 96) these are now less likely to cause a blockage.

The strong heart beat and additional movements of the leg muscles means that blood flows more readily back to the heart through the veins. This reduces the likelihood of developing varicose veins.

Muscles During exercise the muscles work more and need more oxygen. As they are used regularly they become 'toned up' and ready for further exercise. After a new exercise you may feel 'aches and pains' which show that your muscles have been strained. Start with short periods of exercising and slowly increase the amount of time.

Joints Exercise helps to keep joints moving easily (supple). You can bend, for example, without feeling stiffness. This helps to guard against developing arthritis. Using the joints also keeps the ligaments and muscles healthy.

Mass Regular exercise combined with an energy-controlled diet can lead to loss of mass. This is especially important if you are too heavy (page 47) as this puts an extra strain on the heart. Different kinds of exercise use up different amounts of energy (see table opposite).

Overall functioning Overall functioning of all the body systems is improved with exercise. For example, you digest food more effectively and have more regular bowel movements.

The functioning of the kidneys improves, and by the increase in perspiration you get rid of more wastes.

Feeling of well-being Exercise is enjoyable. Exercising in the fresh air makes you feel better and often makes you sleep better so that you will wake up more refreshed. The social pleasures of shared exercise and of team games are also important.

Which exercise(s) should you choose?

There are several points to bear in mind.

1. Do you want an exercise that produces a useful outcome, for example, gardening? Or one that is for relaxation, such as walking?
2. Do you want an exercise you can do on your own, like swimming; or with a companion, for example table tennis; or with a team, for example cricket or football?
3. Do you want a dynamic exercise where you move your whole body, for example swimming or jogging? Or a static exercise where you develop some of your muscles as in weightlifting?
4. Are you fit enough to take on strenuous exercise such as jogging which may be a strain on your heart and muscles? Or would it be better at first to do something less strenuous, like brisk walking?
5. Do you have enough money to buy any necessary equipment (page 260) for the exercise, or could you share or borrow equipment?
6. Do you have sufficient time to spend on your chosen exercises to improve and feel a sense of achievement?
7. Are the exercises you have chosen interesting for someone of your age? Do you have enough variety of different kinds of exercise so you will not be bored?

When you exercise you need a supply of energy which mainly comes from the carbohydrates and fats which you eat (page 74). Different exercises use up different amounts of energy. They also have different effects on improving the heart, lungs, joints and muscles.

The fitness values of common exercises

Exercise	Joules used in 20 min	Improvements in		
		heart & lungs	joints	muscles
Walking	240	*	*	*
Housework	360	*	**	**
Gardening	360	*	**	**
Brisk walk	400	**	*	**
Gymnastics	560	**	****	**
Digging	560	**	****	**
Easy jogging	640	***	*	**
Tennis	640	***	***	**
Disco	640	***	****	*
Squash	800	***	***	**
Brisk jogging	840	****	**	**
Cycling	880	****	**	***
Swimming	960	****	****	****

Negligible *   Fair **   Good ***   Excellent ****

What are the possible problems with exercise?

Heart attack If you are very unfit and suddenly start strenuous exercise, such as prolonged jogging, there may be problems if the coronary arteries have been narrowed by atheroma (page 96). The heavy exercise puts a sudden strain on the heart which may not get enough oxygen. This could cause angina or a heart attack.

Before beginning strenuous exercise, people are advised to consult a doctor who will check their blood pressure (page 96) and their general health.

Cramps and 'stitch' You should not take strenuous exercise soon after a heavy meal. This is because as food is digested it is transported away from the small intestine by the blood. If the blood is required for this purpose, then there may be an insufficient blood supply to the muscles of the legs and this can cause cramp. If, for example, you are swimming at this time, it might mean that you could drown.

Exercising too soon after a meal can also produce a pain in the abdomen called a 'stitch'.

Effect of the sun In the Caribbean the effect of the sun is usually beneficial and leads to a feeling of well-being. But if we exercise excessively in the hot sun and do not drink enough we may lose too much liquid because of sweating to keep cool. This could lead to dehydration, and heat exhaustion or heatstroke (page 171).

Too much exposure to the sun can also cause sunburn, with blisters and peeling skin. The ultraviolet rays (page 320) in sunlight are also associated with the development of skin cancer especially in fair skinned people and albinos. Bright sunlight can also affect our eyes and we must never look directly at the sun (page 133).

Questions

1. (a) If people are unfit what does this mean about their various body processes and organs?
   (b) In what ways will exercise help them to become more healthy?
2. Here are four kinds of exercise:
   (a) long-distance running (b) gymnastics (c) cricket (d) sprinting.
   Which of the people listed below do you think would enjoy and be likely to do well in each one?
   (i) A person who enjoys team games.
   (ii) A small, light person with a lot of stamina.
   (iii) A person with strong leg muscles and supple joints.
   (iv) A person with strong leg muscles and large lung capacity.
3. What are some of the problems you should think about before you begin strenuous exercise out-of-doors? How could you deal with these problems?

How can we train our bodies?

Activity | How do training programmes compare?
1. Examine the training programmes of a sprinter and a long distance runner (you may need the help of your sports teacher). See how their training programmes compare.
2. Do both emphasise physical fitness? Which athlete concentrates on increasing muscular strength? Which athlete concentrates on increasing stamina? Do the two types of athlete look different?

What is the effect of training?
The reasons for training are to enable an athlete to become physically fit so that he or she can stand up to the rigours of performance without injury and also to improve his or her performance over the opposition. If strength is required, the training will include strengthening exercises. If stamina is required, prolonged exercises are performed. Weight training or some form of resistance training is used by almost every athlete and sportsperson today. An increase in strength, stamina, coordination and efficiency of muscle action, as well as improvement of blood circulation, strengthening of the heart and improved oxygen absorption, are all likely outcomes of proper training.

All parts of the body are linked to the **central nervous system (C.N.S.)**. Thus, all activities are coordinated. Muscles work together rather than against each other. To bend or straighten an arm some muscle must contract while another must relax. Signals must be sent from the brain or spinal cord to the muscles, instructing them when to contract or relax.

In even the simplest movement, dozens of muscle movements are controlled by the C.N.S. (see page 122). Movements that occur without conscious awareness are either **reflex actions** or **conditioned reflex actions**. Reflex actions occur naturally. Examples are the knee jerk or eye blink reflexes. In contrast, conditioned reflexes are learned by constant repetition and practice. By training, athletes can practise sequences of conditioned responses so that they are performed swiftly and smoothly.

During training, many changes may occur in the tissues. This is the body's way of adapting to the conditions that are brought to bear on it. Some tissues are built up, for example muscle, and some are broken down, for example fat. For this to happen, an adequate diet is important. Exercise may increase the demand for certain foods, for example sugars and vitamins, and may also increase the efficiency with which they are taken up from the food.

Training in a modern gym

Do athletes need to train under special conditions?
Physical activity causes the body's temperature to rise. If athletes are not prepared for this, they may become heat exhausted. Training to cope with higher body temperatures or high environmental temperatures is usually part of an athlete's training programme.

In 1968, the Olympic Games were held in Mexico City, which is more than 2 000 m above sea level. People who normally live at sea level found it difficult to breathe at this altitude. People who always live at high altitude become adapted to these conditions by increasing their lung capacity and increasing their number of red blood cells, so taking up oxygen more efficiently. Many athletes preparing for the Mexico Olympics decided to train at similar high altitudes, so as to train their bodies for the Olympic conditions.

Blood boosting is a novel, though illegal, method of overcoming the conditions at high altitude. A doctor withdraws some blood from the athlete. The body soon manufactures more red cells. Just before the competition, the red cells from the previously withdrawn blood are re-injected. The overall effect is that there are extra red blood cells so that the ability of the athlete to use oxygen efficiently is increased. The athlete is then able to perform better, particularly if a lot of stamina is required. Instead of blood boosting, some athletes have opted for the illegal substance EPO, which increases the number of red blood cells.

One of the main ideals of athletics is suggested in this quotation: *mens sana in corpore sano* – a sound mind in a sound body. Athletes try to gain mastery of their bodies by training and developing them so that the body and mind become a harmonious and expressive whole. The habitual use of drugs is a threat to these ideals, chiefly because of the way they affect the nervous system, interfere with perception and produce harmful side-effects.

Some common drugs

Alcohol Alcohol (ethanol) is the drug found in popular alcoholic drinks. It has many effects on the body both in the short term and in the long term. When taken in small quantities (for example, 500 cm^3 of beer over 2 hours) it can easily be metabolised in the liver and used as a source of energy like sugars or fats. In larger quantities, alcohol passes directly into the bloodstream. It circulates around the body and affects the thinking and learning areas of the brain that control activities. Activities that require alertness and concentration, for example driving, are carried out less effectively.

The parts of the brain that control the functioning of the kidneys and blood vessels are also affected. The blood vessels near the skin expand and the person feels warmer. This is an illusion: actually, the body temperature falls. In cold weather, this could lead to a dangerous lowering of the body temperature.

Also, the water balance in the body may be upset, leading to dehydration. If the water balance is upset, it could affect the athlete's cooling system and the effect on brain function could affect judgement. Alcohol impairs muscle coordination and this could reduce an athlete's performance. All of these effects could lead to a serious drop in athletic performance.

Phenobarbitone Phenobarbitone is one of the most widely used of the family of drugs called **barbiturates**. These have been used as sedatives to encourage sleep and to counteract the effect of over-stimulation from other drugs. Barbiturates are depressants, which affect the central nervous system, peripheral nerves, muscles (including heart muscle) and various tissues. Under the influence of this drug, the athlete is hardly in a mental or physical state to compete.

Amphetamines Amphetamines are drugs similar to adrenalin (which is produced by the human body). They have been used to treat an uncontrollable desire to sleep and as a stimulant to counteract the effect of depressant drugs such as barbiturates. Amphetamines produce a stimulating effect. However, when this effect wears off the user is left in a state of depression. Amphetamines have been used (unwisely) by athletes to increase their performance but in so doing they take their bodies beyond safe limits of physical exertion. Unfortunately in some cases this proves to be fatal.

Caffeine The stimulant caffeine, which is found in tea, coffee and cola, is beneficial in small quantities. In large quantities, it is counted among the performance-enhancing drugs with harmful side-effects and is banned by the International Olympic Committee.

Nicotine This is the active ingredient (drug) in tobacco. Again, it affects the central nervous system, first exciting then depressing it. The drug itself causes changes in moods which may have significant effects on athletic performance. However, the process of obtaining the drug by smoking may also have harmful effects. Cigarette smoke contains, in addition to nicotine, many irritants which affect lung tissue. These irritants stimulate cells in the lung to produce an excess of mucus which then interferes with gaseous exchange, particularly the uptake of oxygen from inspired air. Athletic achievements that call for a sustained effort and a prolonged demand for oxygen are severely hampered.

Should athletes use drugs?

Most athletes and sports organisations are not in favour of using drugs to improve performance. It is claimed that drugs are harmful to the body in the long run and the assistance they give to athletic performance is only temporary and may be dangerous. But while this statement was accepted in the past, many athletes today question it. They claim that certain drugs if used wisely, under medical supervision, have no harmful side-effects. **Anabolic steroids** mimic natural hormones to stimulate muscle building in the body, as do certain forms of exercise and weight training. So the question may well be: why should one form of modifying the body be rejected in favour of the other? What do *you* think?

Questions

1. Do some research to find out what happened at the Olympic Games held in Mexico City in 1968. Why did this happen? Tell your class what you found out.
2. Is blood boosting wrong, or is it in keeping with 'fair play' for athletes?
3. What body modifications take place during athletic training?
4. How can different drugs affect muscular coordination?
5. 'Athletes who smoke tend to be sprinters rather than distance runners.' Discuss.

What kinds of objects float?

Object	Length (l)	Breadth (b)	Height (h)	Volume (l × b × h)	Mass (m)	Density (m/v)	Does it float?
Wooden block							
Polystyrene block							
Aluminium block							

Activity | Does density affect buoyancy?

1. Collect a number of regularly shaped blocks of different materials, both metals and non-metals, for example, aluminium, wood, candle wax and polystyrene.
2. Measure their length, breadth and height in centimeters. Weigh each to obtain the mass in grams.
3. Copy the table above and enter your results. Calculate density from the formula:

$$\text{density} = \frac{\text{mass}}{\text{volume}}$$

4. Put each block in water. Does it float?

Some objects have complicated irregular shapes and their volumes cannot be calculated easily from simple measurements. If the object is small, the volume may be found by displacement in a measuring cylinder – see (a) below. If the object is large, a Eureka can may be used – see (b) below.

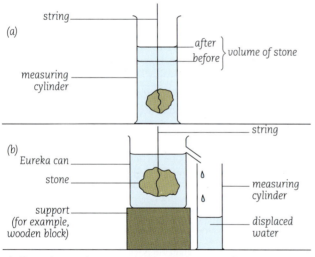

Finding volume using a measuring cylinder or Eureka can:
1. Find the mass of the object using a balance.
2. Partly fill the measuring cylinder with water; take the reading.
3. Lower the stone into the measuring cylinder until it is completely immersed.
4. Then take the new reading. The volume of the stone is the difference between the two readings.

This method of finding volume will reveal whether or not an object floats. For irregular objects that float, you need to modify this method slightly. What would you do?

It is also possible to find the density of liquids. First, find the mass of 50 cm³. Calculate the density using the formula: density = mass/volume. Find the density of a number of liquids including sea water, tap water and cooking oil.

Write out a list of substances in order of density. The table below may include some of the materials that you have investigated. Do the figures agree with yours?

Substance	Density g/cm³	Substance	Density g/cm³
Methylated spirits	0.8	Zinc	7.1
White spirit	0.85	Steel (iron)	7.8
Ice	0.92	Brass (varies)	8.5
Tap water	1.0	Copper	8.9
Glass (varies)	2.6	Lead	11.3
Sand	2.6	Mercury	13.6
Aluminium	2.7	Gold	19.3

Your experiments and these results show that substances with a density less than 1 g/cm³ will float in tap water and substances with a density greater than 1 g/cm³ will sink.

Do sea water and tap water have the same densities? If you worked carefully you would have found that sea water has a density greater than 1 g/cm³. This explains why it is easier to swim in the sea than in fresh water (which has about the same density as tap water). The greater density of salt water is due to the dissolved salt. In the Dead Sea (Israel) where the amount of salt in sea water is approximately 30%, it is almost impossible for a swimmer to sink.

Does temperature affect density?

You learnt how convection currents were formed in liquids on page 114. The key to their formation lies in the relative differences in density between hot/warm water and cold water. Water decreases in density when heated and increases in density when cooled, reaching a maximum at 4 °C.

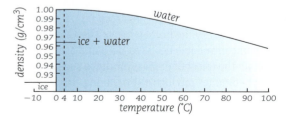

Variation of water density with temperature

Because fresh water and sea water have different densities, and temperature also affects density, ships will have different degrees of buoyancy depending on whether or not they are in fresh water or sea water, or whether it is summer or winter.

The **Plimsoll line** on the hull of ships shows how far down a ship is in the water so it is possible to judge the load that can be carried safely. This was first proposed in the British Merchant Shipping Act in 1896 by Samuel Plimsoll M.P.

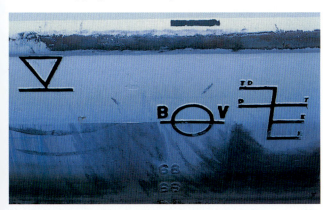

The Plimsoll line is a circular disc 30.48 cm (12 in) in diameter with a line 45.7 cm (18 in) drawn through its centre. The upper edge of the straight line is the maximum summer safe load line. The lines to the right indicate the maximum load lines for different seasons and circumstances. The vertical line is always 53.3 cm (21 in) from the centre of the disc, towards the ship's stern

Activity | Can iron float?
1. Get a clean razor blade (**be careful**) and a bowl of clean water. Allow the water to settle.
2. Balance the razor blade on the middle finger of your right hand if you are right handed. Then lower the razor blade gently onto the surface of the water. If your hand is steady and the water is still the razor blade should remain on the surface.
3. Repeat the procedure but this time use a needle. Carefully lower the needle onto the surface of the water. If this is difficult, carefully wipe a film of Vaseline over the needle and the razor blade. This prevents water from wetting the objects. Notice how the needle and the razor blade distort the surface when floating on it.
4. Put a drop of detergent in the water while the razor blade and needle are floating. Observe the effect. Can you get the objects to float again in the same water?

What is surface tension?
The surface layer of liquids behaves like a thin skin. The force that holds these molecules together is called **surface tension**. If an object's mass is distributed over a wide area the surface tension forces will be able to support it.

Many insects, for example pond skaters and mosquito larvae, use surface tension forces for support on the surface of water. If the forces are weakened, for example by dropping detergent or oil into water, the forces will no longer support these insects and they will sink.

How do iron ships float?
If you place a lump of iron into water, no matter how carefully, it will sink because, as you know, iron is much denser than water. When people first had the idea of building ships from iron, many people laughed at the suggestion because they expected that ships made of iron would sink.

Activity | Investigating flotation
1. Take the block of wood that you found the density of in the previous Activity on page 249. Float it in the full Eureka can and collect the water displaced. What is the volume of water displaced? What is the mass of water displaced?
2. Repeat the experiment using strong salt water (for example, 20% strength) in the Eureka can. What is the volume of salt water displaced? What is the mass of salt water displaced? Are they the same as for fresh water?

Experiments like this show that the mass of liquid displaced is the same for all liquids, for example fresh water and salt water. Is this what you found?

This leads to the **law of flotation**, which states that *a floating body displaces its own weight of the fluid in which it floats*. This means that a boat or a ship, like a piece of wood, when placed in water will sink until the weight of water displaced is just equal to its own weight. So overall the ship, including the cargo and all air spaces in it, must weigh less than the same volume of water. Because of all the air spaces in a ship, the total weight is less than that of an equivalent volume of water – thus it floats.

All bodies in water are acted upon by an upward force causing them to weigh less in water than in air. This upward force is called an **upthrust**. The less dense the material, the greater the upthrust. This principle has been applied to the manufacture of water safety devices. Most of these involve inflated tubes or containers of some sort. Some life rafts are made from expanded polystyrene foam or polyurethane foam, which contains trapped air bubbles.

Questions
1. How does density affect buoyancy?
2. How is it that iron ships float?
3. Explain how a ship, in the course of one journey, can have different buoyancies without loading or off-loading cargo.

How can we survive under water?

You might have seen deep-sea divers who dive in search of conch shells and so on. How do they survive? It can be shown that water contains dissolved air. Large bodies of water, which support life contain dissolved air. This gives fish a supply of oxygen, which they are able to obtain by means of gills (see page 143).

But humans breathe air and are well adapted for life on land. When we wish to enter the strange and fascinating world under the water we may find it fraught with dangers that we must overcome if we are to come out again unharmed.

Have you ever seen scuba-divers? They use **s**elf-contained **u**nderwater **b**reathing **a**pparatus (the first letters of each word form the word **scuba**). Although scuba-diving is very popular in many parts of the world, the underwater environment can make it a dangerous sport. What are some of the most important dangers you can think of?

Is pressure a problem?

The diagram below shows water escaping from three similar holes at different heights in a container. The water from the bottom hole travels furthest while water from the top hole travels the least distance. This shows how pressure, which forces the water through the holes, is greatest when the distance from the water surface to the hole is greatest.

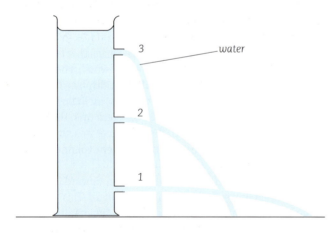

Activity | Making a Cartesian diver
Use a hollow glass figure or a small dropping pipette.
1. Take sufficient water into the glass tubing so that it can just float in water.
2. Place it in a glass jar completely full of water with a well-fitting stopper. Place the stopper in the container, gradually pressing on it.
3. If the density of the diver is well adjusted, it should sink when a little pressure is exerted by pushing on the stopper and it should float when the pressure is released. Can you explain why the diver moves up and down?

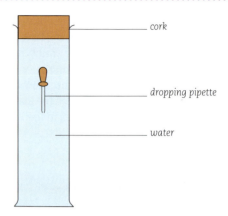

The air bubble in the Cartesian diver was compressed, thus reducing the buoyancy. In the same way, the buoyancy of a scuba-diver is affected by water pressure and the expansion of air. At the surface, the diver breathes air at atmospheric pressure. When the diver goes deeper, this air is compressed. When the diver breathes in air at depth, the air will expand to increased volume when the diver rises to the surface.

The aqualung

The aqualung – the regulator has a mouthpiece attached for breathing and a reduction valve which enables air to be delivered to the diver at safe pressures

Activity | Gases in a fizzy drink
Buy a carbonated soft drink. Gently remove the cap. Observe how bubbles form in the liquid. Can you explain this?

High pressure increases the solubility of gases in liquids. The drink in the bottle or can is under pressure. Once the stopper (cap) has been removed, the pressure is reduced and the gas (carbon dioxide in this case) comes out of solution from the drink because it is less soluble at the new lower pressure.

Similarly, as a diver travels deeper and water pressure increases, more and more gases dissolve in the diver's blood and tissue fluids. This may lead to 'the bends' (see next page) when the diver returns to the surface.

Does diving affect hearing?

Sound travels much faster under water than it does in air. A diver will become acutely aware of the many sounds made by marine life. But the diver's ears may be damaged by the pressure under water if precautions are not taken.

The diagram below represents the **middle ear** (see page 269). The membrane at **A** represents the eardrum. The plugged tube at **B** represents the Eustachian tube. At sea level, before the diver descends, the pressure inside and outside the middle ear is balanced. When the diver descends, the increased pressure due to sea water pushes on the eardrum and stretches it. If the diver 'clears' his or her ears then the pressure is relieved and the danger is over. By holding their nose and blowing, the diver can force air from the throat through the Eustachian tube in the middle ear. This air is then at the same pressure as the surrounding water, so the pressure on the eardrum is relieved.

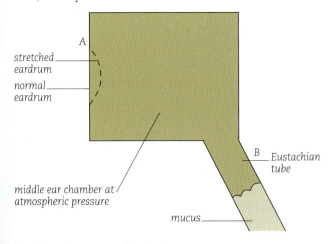

The effect of pressure on the middle ear

Can air under pressure be poisonous?

Nitrogen Since nitrogen is the major component of air (78%), most dissolved air in the body is nitrogen. Nitrogen can have anaesthetic effects on the brain. These effects become most noticeable at depths of 30 m and over. The effects, known as **nitrogen narcosis**, are lack of coordination, confused thinking and the disappearance of self-preservation instincts. At a depth of 60 m, the effect of nitrogen narcosis becomes very serious indeed.

Oxygen If pure oxygen is used for a diver's gas supply instead of ordinary air, dangerous effects develop in a depth as low as 8 m. Symptoms include coughing, vomiting, convulsions and unconsciousness. If the air or gas mixture contains 20% oxygen, **oxygen poisoning** can occur at depths of 80 m and beyond.

In professional deep diving, these problems of gas poisoning have to be overcome. This is done by reducing the proportion of oxygen and replacing nitrogen by the gas **helium**, which is not toxic.

How can divers avoid the bends?

The bends refers to the cripplingly painful sensations that many of the early scuba-divers experienced. Today we know that they are caused by air bubbles cutting off circulation in various parts of the body. The pains indicate oxygen-starved and dying tissues. Precautions necessary to avoid the bends include: adequate preparation before dives, learning to dive within safe limits of time and depth, not returning to the surface too quickly, and carrying out safety procedures.

Decompression Getting rid of dissolved air is called **decompression**. Decompression involves slow ascents to predetermined depths and remaining there for a certain time to allow bubbles of dissolved air to be released harmlessly in the lungs. If divers have to return to the surface quickly they must enter a **recompression chamber**, otherwise their health remains at risk. Inside a recompression chamber is a closed compartment where the air pressure can be controlled to a safe level for the diver. When a diver is being treated in a recompression chamber, compressed air corresponding to the dive conditions is used. This makes the air bubbles in the body, which are causing the bends, re-dissolve. The pressure in the chamber is then released slowly and safely. There are still too few recompression chambers in the Caribbean for the amount of activity, both amateur and professional. So, if divers get the bends they have to be treated in hospital.

What causes 'burst lung'?

This is a serious condition, which can end in death. As the name suggests the lungs can rupture, causing loss of blood, collapse of the lungs and death. Burst lung can be caused by a too rapid ascent from depth without taking the precaution of breathing out heavily to release lung pressure. A slow ascent, breathing naturally, helps to avoid this usually fatal occurrence.

Scuba-diving can be a very enjoyable and safe sport but mistakes can be crippling and fatal. Adequate training with a reputable club or professional body is vital before any scuba-diving is attempted.

Questions

1. Why is deep diving more dangerous than shallow diving?
2. What is the function of the aqualung?
3. Why would it be dangerous to dive with a blocked Eustachian tube?
4. What precautions must a diver take to avoid the bends?

How are objects affected by moving through air?

What is air?

The nature of the substance air was a puzzle for thousands of years. The early scientists placed **air** in a special group along with **earth**, **fire** and **water** and called them the **elements**. These four elements were thought to be the fundamental substances that made up the universe. Today we know that air is a mixture of gases and **gas** is one of the states of matter along with **liquid** and **solid**. It is made up of particles called **molecules** (see page 18) that are far apart from each other and move at high speeds. As objects move through air, they collide with the molecules. These tend to resist movement, so that objects slow down unless energy is constantly supplied to keep them moving. Does the size and shape of objects affect their movement through air?

Activity | How do different objects fall?
1. Drop a flat sheet of paper. Does it fall straight down?
2. Drop a ball or stone from the same height. Does it fall the same way? Fold the paper into different shapes and try it again. Does it behave in the same way?
3. If you screw the paper up into a ball, it will fall in the same way as any other ball.

You might conclude that it must be the way the shape of the paper acts on the air that makes all the difference. Of course, it is not easy for you to get rid of the air. But your teacher can remove most of the air from a thick glass jar using a vacuum pump. Your teacher can show you that steel balls or glass marbles and flat pieces of paper will fall at the same rate when all or most of the air in the jar has been removed. (See also page 297.)

Can sports players use air resistance?

In competitive sports, players can make use of **air resistance**. For example, polishing a cricket ball on one side is a technique which some bowlers use to get the ball to swing or change direction when bowled as in the picture. This polishing makes the ball smoother and less likely to meet air resistance on the smooth side than the other rough side. The rough side creates more resistance, called **drag** (page 297), in the air when the ball is bowled. The effect is that the ball follows a curved path, drifting towards the rough side. Of course, the ball must be gripped and bowled correctly. The position of the seam, the speed of the ball and the amount of spin imparted at delivery have all been shown to be critical factors in producing the **reversed swing** effect so much admired in the Pakistan pace attack.

Spinning of the ball, for example a football, can alter the flight path, causing it to swing due to air pressure differences caused by the spin. You might have seen this in the 'banana kick' made famous by Brazilian footballers.

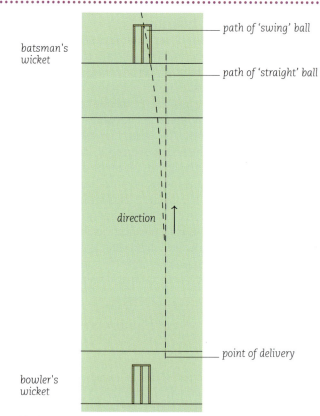

Polishing or spinning a cricket ball can make it swing when bowled

How can winds affect sports playing?

Wind is air in motion. This air movement can have important effects. For example, without wind there can be no sailing. Good sailors are ever mindful of the direction and strength of the wind; so are kite fliers. Sailing boats cannot sail directly against the wind since it would blow them backwards. However, they may sail *close* to the wind. Kites make use of the wind to lift them. The tail is important to keep the kite stable in some designs. In high wind, long tails are necessary and the angle of slope of a kite also has to be smaller than in lower winds. Games like table tennis and badminton are very sensitive to winds, so they are normally played indoors.

Does wind affect athletic performance? If a sprinter runs with the wind this may lead to an improved time. When this occurs, the time is recorded as being 'wind assisted' and is usually disallowed as an official (record) time.

Activity | How far can you throw?

Throw a cricket ball underarm with your elbow straight. What two variables must you control to get the best throw, i.e. greatest distance?

It is important not only to regulate the speed of arm movement but also to adjust the angle of the arm before the ball is finally released.

Activity | What is the best angle for throwing something into the air?

1. Construct a catapult using either a sawn-off syringe or another type of improvised gun, which can fire pellets at different angles. Ensure that the firepower will be the same every time by using an elastic band (see below).

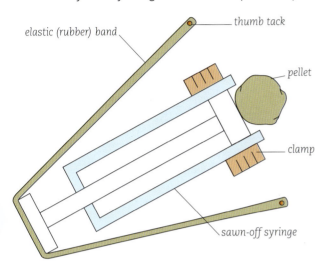

2. Start with the catapult horizontal and fire the pellet. Use a metre rule to measure the distance at which it falls. Increase the angle in steps of 10° and repeat. Record your results in a table.
3. Determine, to the nearest 5°, the angle of fire at which a pellet is thrown furthest.
   This is the angle that throwers aim for when trying to cover the maximum distance.

Apart from the hammer (44°) human throwers do not throw their furthest around a 45° angle, but throw better at lower angles, because more power is produced. For shot put, the best putting angle is 40°, for javelin the best angle is 30°, and for the discus, 30°. For long jump, in effect a throwing event involving the whole body, the best angle is again low, at 28°.

Putting the shot. Measure the putting angle

What about the high jump and pole-vault? Ideally, the best angle to achieve maximum height is 90°. But a high jumper or pole-vaulter, using a 90° angle, would knock away the bar he or she was trying to jump over or fall back down onto it. For the high jump a forward motion is also needed, so that when jumpers make the vertical jump near the bar, the forward momentum carries them in a steep arc over the bar. The pole-vaulter makes a fast approach to the bar and uses a strong, flexible fibreglass pole to make a steep arc over the bar.

Questions

1. The shot must be put at the best angle for the maximum distance to be achieved. Is the best angle here 45°? Is it the same for (a) the javelin (b) the discus and (c) the hammer?
2. Look at this sequence for javelin throwing. Use a protractor to measure the angle of the javelin with the horizontal in the final step.
3. What effect does air have on falling objects?
4. Explain how cricket bowlers use shine to 'move' the ball.
5. What is the best throwing angle used by a mechanical thrower? How is it found?
6. Why does a pole-vaulter or high jumper need a run-up?
7. Why do the best throwing angles used in sport differ from the best angles used by mechanical throwers?

Javelin throwing

What do you expect from sports equipment?

What is your sports equipment made from?
There are many kinds of sports and most of them require special equipment. For any sport, the equipment available covers a range of prices and quality. Whether you buy or make your own sports gear you need to know that it is worth the money or the time invested in it.

The **durability** of sports equipment will depend on the way it is used, for example, table tennis balls cannot withstand being stepped on and footballs cannot withstand being kicked into barbed wire. It is essential that you learn the correct way to use your equipment. Most of the knowledge is specific for the sporting activity and is acquired through proper instruction.

Activity | Looking at sports equipment
If you look at some sports equipment, you will observe that it is made from familiar materials whose properties are known or can be investigated. By knowing these properties you can arrive at a fairly good idea of what your sports equipment can withstand. You will then have a better understanding of how to take care of it and get more use from the equipment.
1. Examine some sports equipment.
2. Copy the table below, adding more columns to the list if necessary.
3. What does the table tell you about the properties (characteristics) of different materials? How do they compare for strength and elasticity? What standard of comparison would you use?

Materials which form a major part	Sports equipment	
	Football	Cricket bat
Wood		✔
Rubber		✔
Plastic		
Cotton		
Steel/iron		
Aluminium		
Leather	✔	
Other		

Activity | How do materials respond to stretching?
1. Hang up cords or wires of different materials one at a time as in the diagram. Make and place a wire pointer somewhere on the material.

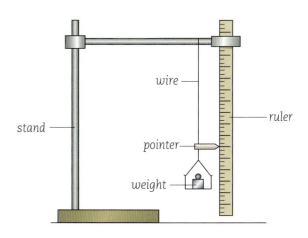

2. Load the string with 200 g or more at a time, and measure the extension produced.
3. Plot the results on a graph of load (weight added) against extension. Compare the shapes of the graphs for different materials.

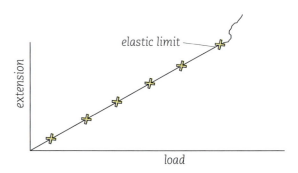

Graph of load against extension for a metal wire, for example copper. How does it compare to the materials you have tested?

As weights are added, the material stretches. The value of the load (force) at which the material fails to return to its original length is called the **elastic limit**.

In sports equipment, strength is also important. Usually, **strength** of material must be considered along with size, for example, ropes used in climbing or mountaineering must be thick enough to give a proper handgrip. The **weight**, **durability** and **cost** are also important.

Which material is stronger?
To compare the strength of two materials exactly the **mass per unit length** is important. For example, if steel wire is compared with nylon then a metre of nylon string should be compared with a metre of wire of the same mass. Of course, because steel is denser than nylon, the steel wire will be much thinner.

Activity | Finding the strength of a multi-stranded rope

1. Select one strand from the rope. Hang weights on as in the previous Activity. Find the breaking point and therefore the maximum load the strand can hold.
2. Find the average breaking point of several strands.
3. Count the number of strands in a rope. Calculate the strength of the rope from this formula:

strength = average load on one strand × number of strands

What care is needed for your sports equipment?

Unless properly treated, natural materials like cotton and leather can be attacked and destroyed by micro-organisms such as mildew. Cotton must not be kept moist over long periods. Leather must not be allowed to become waterlogged. The wood of a cricket bat must be oiled. Ultraviolet light from the sun can have a harmful effect on plastics. Clear plastic will eventually become cloudy and then brittle. Rubber tends to crack when it has been dried in the sun. Sea water will severely affect leather and speed up corrosion in metals, iron in particular (see page 213). Sea salt left behind when sea water dries, will also attack rubber. The majority of these bad effects will take place so gradually as to be unnoticed.

The good sports dealer will tell you how to care for and maintain your equipment. From the start, you must establish a proper routine in preparing equipment before use, using it carefully and maintaining it afterwards.

Activity | Why are there different playing surfaces?

Think of some popular sports which need specially prepared and laid-out playing areas. How do you know whether or not they are suitable for a particular sport or activity? Visit a sporting site and discuss this problem with the groundstaff or sports supervisor. Examine playing surfaces for some of these sports: cricket, football, hockey, netball, basketball, table tennis, lawn tennis, hard court tennis, squash, roller skating and athletics.

Make a table like this for your comments. Discuss your findings in class.

A large number of sports can be termed **ball games**. In most of these, the ball makes contact with the playing surface, therefore **bounce** is important. Because the players are also moving, in some sports over the same surfaces as the ball, **friction** is also important.

How does bounce affect play?

In games like tennis and squash, the higher the ball bounces the longer the player has to hit the ball, because it is airborne for a longer time. On a grass tennis court the grass absorbs some of the energy from the ball and the bounce is less than on a concrete surface. Players have to make special allowance for this. In squash, bounce is reduced by using a soft ball, which absorbs a higher proportion of the energy when the ball hits the hard playing surface. In cricket, a very hard pitch will cause high bounce, which can make life difficult for batsmen. The most disliked playing surfaces have uneven bounce because the movement of the ball after bouncing is difficult to anticipate.

How can you increase friction?

All ball-game players choose suitable footwear to increase frictional contact between the ground and themselves. In this way, they can make quick changes in speed and direction without sliding or falling over. These movements are very necessary for a good performance in a game. Careful choice of footwear is therefore important.

In field sports and athletics, boots and shoes have spikes on the soles to prevent slipping. On hard wooden or concrete floors rubber soled shoes are used because they are best able to grip the floor and offer resistance. However, if too much grit and sand are present the advantage of rubber is counteracted.

Questions

1. Where can you find information on the proper use and care of sports equipment?
2. Why might the same type of equipment, for example tennis racquets from different manufacturers, need different treatment and maintenance?
3. What makes a playing surface (a) fast or (b) slow? What type of cricket pitch favours (a) fast bowlers and (b) slow bowlers?

Sport	Cricket	Football	Tennis	Table tennis	Badminton
Location	Outdoors		Outdoors	Indoors	Indoors
Dimensions					
Surface composition	Turf	Concrete/asphalt	Plywood		Wood/concrete
Surface texture		Hard	Hard		

How can we find out about our surroundings?

Living organisms have to interact with their surroundings to survive. Even the simplest living things have means of detecting food, of looking for suitable conditions for growth and of avoiding adverse conditions in their surroundings. Living organisms must have a system for gathering information and responding or they will remain passively at the mercy of whatever conditions they find. In humans and higher mammals where life has developed past mere existence to sophisticated levels, **senses** have been developed to gather useful information.

What sense organs can you see being used in this photograph?

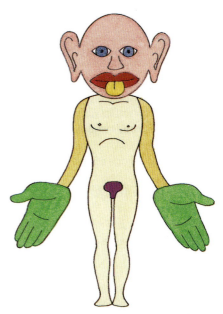

Simple representation of sensory neuronal connections

How important is sight?

The five senses, sight, hearing, touch, smell and taste, are well known. Of these, sight is by far the most important for the vast majority of higher animals, humans in particular.

The sense organs concerned with sight are called the eyes (pages 278–9). But these cannot work without light. Fortunately, there is usually natural light available. Even so, we have learnt how to produce light so that our eyes can be used when there is no natural light. To show how important sight is, look at the diagram showing the proportions of brain cells related to vision and hearing. A lot of the brain is concerned with processing information from the eye. (See also page 125.)

Light is a type of energy radiation (page 320). Different objects behave in different ways when light shines on them by absorbing some parts of the light, while reflecting other parts. Different objects appear distinct from each other. Using sight, many different living organisms can 'see' their environment. But light operates only in straight lines. If objects are hidden behind others, they may not be seen.

How important is hearing?

Sound has some advantages over light. Sound travels in straight lines, but sound vibrations can penetrate matter much more easily than visible light. A sheet of black paper or aluminium foil will effectively block the brightest light but will not block sound (except very high-pitched sounds).

Very few animals produce light, but many produce sound. For example, most vertebrates have vocal cords but they also produce sounds by breathing, by brushing against objects and by moving about. Organs of hearing (ears) can easily pick up these sounds (pages 268–9) and even tell which direction they are coming from. In humans, a much smaller part of the brain is involved in the processing of sound than light (see diagram above and page 125).

In bats, sound and hearing are very important. Most bats rely on hearing echoes to receive information, rather than on sight. They can successfully hunt their insect prey at night, in total darkness, by sending out a sound signal, which is reflected from the object to their ears. By listening to the echoed signal, the bat can detect the distance and shape of the objects nearby. Radar (see page 320) works on the same principle but uses radio waves instead of sound waves.

Are smell and taste important?

Smell was one of the first senses to evolve. It involves closer contact with an object than either sight or hearing because molecules of some part of the object must interact with some part of the cell membrane in the sense organ. In humans, the senses of smell and taste work together but only give very limited information about the environment. However, in some animals the senses of smell and taste are very well developed, with large parts of the brain processing information for these organs. Some fish living in murky water, where visibility is very limited, for example the dogfish (a species of shark), probably have smell and taste as their most advanced senses. Some antelopes can use smell to detect their enemies (lions) when they are well hidden, wild pigs use smell to find food and moths use it to find a mate.

What can touch tell us?
Touch can tell us about textures (hard, soft, rough, smooth) and also hot and cold. These characteristics can only be guessed at from a distance. They cannot be detected with any certainty by sight and hearing.

Are there other senses?
You can tell if you are upright or leaning over: this is because you have an organ of balance in your inner ear (page 269).

The rattlesnake has a special organ for detecting infrared (heat) radiation. It can detect its prey (small mammals) in total darkness and strike with deadly accuracy.

Migratory birds can find their way back home over a distance of several thousand kilometres. This ability is thought to be due to special cells in their beaks that are sensitive to the Earth's magnetic field. Mammals, including humans, are believed to have these cells also.

The electric eel produces an electric field around its body. When another organism swims close to the eel, the electric field is disturbed and the eel senses the other presence.

Other living organisms (and some metal objects) seem to have an electric field surrounding them. This may explain the very strange behaviour often exhibited by some sharks. Killer sharks can locate prey, using various senses – smell, sound and sight. But when the shark moves in for the 'kill' it turns over on its back and closes its eyes. At a very short distance away, it senses the electric field. If metal objects are nearby, electric fields may overlap and the shark is then unable to distinguish between the objects. Many strange metal objects have indeed been recovered from sharks' stomachs.

Can we extend our senses?
Some people say that because humans rely so much on sight, our other senses are underdeveloped. What do you think? Some people have, by training, developed other senses to a very high level. The hearing of blind people can be extremely acute. Professional wine tasters can develop an extremely sensitive sense of smell. But there are limits to how much we can develop our senses: no amount of training will enable us to see the moons of the planet Jupiter. That is beyond the physical limits of the eye.

Such limits are overcome by the use of instruments. By applying scientific laws to make instruments, some of the limitations of the senses can be overcome. By using telescopes, astronomers can see objects far away in space. By using the electron microscope, scientists can 'see' how atoms are arranged in a molecule. With X-rays, doctors can 'see' into the body.

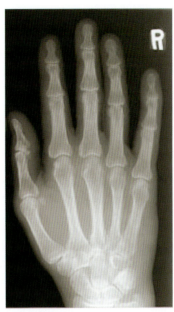

An X-ray photograph of a hand

The amount of information that we can acquire by the senses and extensions of our senses is vast. Information that cannot be processed immediately may be recorded for the future.

Very sophisticated methods of recording information exist; these include books, photographs, records, magnetic tapes and computer memories (chips). But like long-forgotten facts, recorded information is useless if it is not accessible and cannot be retrieved when desired. It is therefore very important to understand how to find things out from these vast quantities of stored information.

Rapid advances in electronic technology have led directly to the manufacture of low-priced computers. Today, not only small businesses but also ordinary individuals own computers. (See pages 28–9.)

Questions
1. Which human sense do you think is most important?
2. How can our senses be extended?
3. Why are our senses limited?

Can waves transfer energy?

Activity | *How are waves produced?*
1. Fix one end of a 5 m length of rope about 1 m above the ground.
2. Hold the free end and shake it up and down in a steady rhythm. Is a wave produced?
3. Increase the rate of movement of your hand. Is the motion of the wave increased? Do you think energy is transmitted along the rope?

Because your hand is holding the rope, its up and down movement is transmitted to the rope, which itself moves up and down, transferring energy from the moving hand along the rope.

If a stone is thrown into a pond, the surface of the pond is disturbed and **water waves** spread out from the point of disturbance to all parts of the pond. Ripples of water spread out in all directions.

Sea waves Sea waves are caused by the wind disturbing the surface. These waves travel from the open sea towards land as barely visible 'swells'. As they travel towards land, they encounter shallower water and rise, breaking and crashing on the shore. Thus the energy picked up from the wind far out to sea is finally lost on the seashore. When there is a hurricane the waves will be much more energetic because of the much higher wind energy provided by the hurricane (page 312).

Think about the effect of sea waves on driftwood and other floating debris. It is obvious that there is very little forward (in the direction of the wave) movement of the water.

In the Caribbean, there are steady easterly trade winds. The east coast of most of the islands constantly receives steady waves. The eastern coastlines are rugged and show obvious signs of continuous **erosion** due to this constant buffeting by the waves.

The total amount of energy stored in sea waves in this way is enormous. In some countries, scientists are experimenting with ways of harnessing wave energy. In some systems, the up–down wave action is used to turn a dynamo, which then generates electricity.

A ripple tank One of the best ways of investigating waves is by using a **ripple tank**. This consists of a tank of water (which may be as little as $\frac{1}{2}$ cm deep), a light bulb or spotlight supported over the centre of the tank in order to cast shadows of the waves on the bottom, a vibrator motor to produce waves and some accessories such as curved and straight bars and bars with gaps.

Activity | *Using a ripple tank*
1. Use a straight bar on the water surface to generate one **plane wave front** at a time.
2. Observe and sketch what happens when the plane wave front hits a barrier. The plane wave front is the **incident wave**. In this case, the incident waves are straight. The resulting waves are the **reflected waves**.
3. Observe and sketch what happens when a plane wave front hits a concave reflector and a convex reflector.
4. How do differently shaped surfaces affect the wave front?

From your experiment with the ripple tank, you can see that waves can be made to change direction by the use of a suitable reflector.

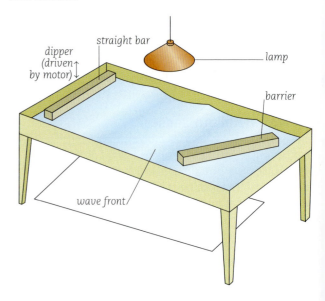

Wave action is used to create compressed air to drive a turbine

Light waves

It can be shown that light energy is transmitted as **wave motion**. Light can be reflected by a suitable surface such as a mirror. Light obeys the same laws of reflection as the waves in a ripple tank (see page 276). Plane mirrors can be used to pick up light from different angles and reflect it to the eye, for example the mirror used by dentists enables them to see behind teeth.

Periscopes You can make a simple periscope from two plane mirrors which are fixed facing one another at an angle of 45° to the line joining them. A periscope enables you to see over the heads of crowds or over the top of an obstacle. Periscopes can be used in submarines but they are more elaborate than the one shown here. Prisms are used instead of mirrors and a telescope is included to extend the range of vision.

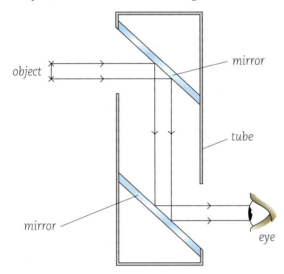

Plane mirrors in a periscope

A plane mirror is sometimes used for a rear-view mirror in motor vehicles, but convex mirrors are better, since they give a wider field of view.

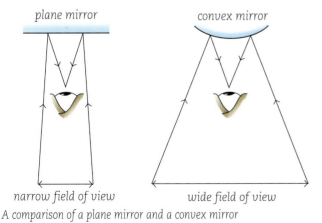

A comparison of a plane mirror and a convex mirror

What is a concave mirror?

A concave mirror will focus rays of light by reflection. In a concave mirror the light travels only as far as the reflecting surface. A convex lens also focuses light, but the light has to pass through the lens.

Where are concave mirrors used? The concave mirror is used for concentrating energy (for example, light). Concave mirrors are used in shaving mirrors, where they give a magnified image; in reflecting telescopes (they are in the largest optical telescope in use); in spotlights and in headlamps, where they are used to produce parallel beams.

The concave surface is also used to concentrate energy such as high-frequency television and radar waves. These **dish aerials** can be seen in most parts of the Caribbean. These aerials pick up and concentrate the signal from a 'stationary' satellite high above the Earth's surface.

They are also used for concentrating heat rays (page 197) and for collecting and concentrating sound waves. However, only high-frequency sound waves can be successfully collected in this way (for example, bird song). Wave motion, both electromagnetic and audio, with wavelengths of about 10 cm and less, is most easily collected by a concave collector.

Dish aerial. This is for receiving radar waves

Questions

1. How would you demonstrate that wave motion transfers energy?
2. How would you show, using water waves and/or a ripple tank, that the angle of incidence is equal to the angle of reflection?
3. Why do the largest telescopes use mirrors rather than lenses for collecting light?

What are sounds?

Activity | How are sounds produced?
1. Take a 30 cm plastic ruler and hold it firmly (or clamp it) at the edge of your desk (or some other flat surface), so that about 20 cm overhangs.
2. With your free hand, pluck the overhanging portion of the ruler and observe what happens. Does the ruler vibrate? Can you see it vibrating?
3. Place your ear directly over it as you pluck the edge once more. Do you hear a sound?
4. Slide another 2 cm onto the desk to make the overhang shorter and pluck the edge again. Does it vibrate faster? Do you hear a sound? Is the note the same as before?
5. Continue sliding the ruler onto the desk to make the overhang shorter. At the same time observe what is happening to the vibrating part of the ruler and the kind of sound produced.

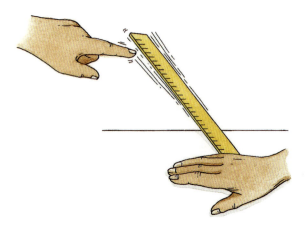

Now, can you explain how this sound is produced? Plucking the overhanging part of the ruler sets it vibrating, causing it to produce a sound. When the ruler is overhanging a long way the vibration is easily seen, although it can hardly be heard. As the length of the vibrating portion is reduced, the notes are heard more distinctly. As the overhang is made shorter and shorter the notes become higher and higher and it is impossible to see the individual vibrations of the tip of the ruler.

Low notes are said to have a **low pitch (low frequency)** and high notes are said to have a **high pitch (high frequency)**. The sounds produced by plucking the tip of the ruler were at various frequencies, corresponding to the frequency of vibration of the ruler. The human ear can pick up sounds of between 20 vibrations per second and 20 thousand vibrations per second. *One vibration per second is known as 1 hertz or 1 Hz for short. One thousand hertz is 1 kilohertz (kHz).*

Sounds can also be produced by vibrating strings, which may be plucked, as in a guitar, or struck, as in a piano. The tightness of the string as well as the size affect the way it vibrates. Which strings on a guitar produce high notes? Which strings produce low notes?

Activity | Sound in air columns
Vibrating columns of air in a flute or other wind instrument can also produce sounds.
1. Set up about eight soft drink bottles all of the same type, and place different amounts of water in them.
2. Blow air across the top of each one in turn, to produce a sound. Are different sounds produced? Do the different sounds relate to the different lengths of the air space in each bottle?
3. Adjust the water levels in the bottles to produce a musical scale. Can you play a tune on the bottles? Tap the bottles gently with the edge of a ruler. What kinds of sounds are produced?

> ### Question
> 1. Steel band music began in the Caribbean and is now popular in many other parts of the world. Watch someone playing a steel pan or attend a steel band concert. How is the sound produced? What is the vibrating surface? How are steel pans made?

How are sounds made louder? The loudness (or volume) of sounds can be increased by supplying more energy to the action that produces the sound. Blowing harder into a flute, plucking a string more vigorously or hitting a drum or steel pan harder will all increase the loudness of the sound.

Why do musical instruments have distinct sounds?
You may have noticed that middle C played on a piano sounds different from middle C on a guitar or steel pan. No matter how well these instruments are tuned, the notes always sound different. These notes are said to differ in quality or 'timbre'. The middle C note has a vibration of 256 Hz. If all instruments produced pure notes, then they would all sound the same. But different instruments produce pure notes with different, extra 'impure' notes along with the main note. These impurities are called **harmonics**.

Harmonics are a characteristic of a particular instrument and are related to the shape, size and material from which it has been made. Harmonics are a desirable feature in musical sounds. One of the most famous of musical instruments is the Stradivarius violin, named after the Italian Antonio Stradivari, who lived in the 17th century. The violin is famous for the special harmonics, which produce a particularly pleasing tone.

Can sound waves be seen?

Sound waves need particles of matter as a **medium** to travel through. They are in fact a back and forth movement of atoms and molecules hitting against each other as the sound signal is transmitted. Therefore, sound cannot travel in a vacuum.

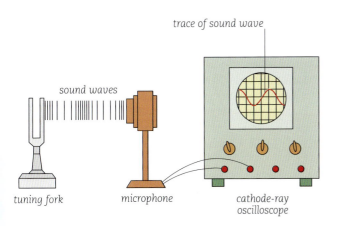

Sound waves can be seen using this apparatus

A microphone can convert sound impulses or waves into electrical signals. As electrical signals, the waves can be displayed (after a certain amount of adjustment), on the screen of an **oscilloscope**. The oscilloscope produces a picture of a sound wave. Sound that is free of harmonics will give a simple picture, a **trace**, on the screen. This trace is called a **sine wave**. Picture (d) below shows a trace with overtones (harmonics).

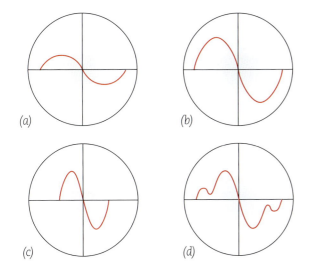

Diagrams (a), (b) and (c) are simple traces. Diagram (d) shows a trace with overtones (harmonics)

The height of the trace on the screen of an oscilloscope indicates the strength of a sound. The height is known as the **amplitude**. The number of peaks on the screen can show the **frequency** of vibration of the sound (if the time scale is known). A wave with a large amplitude carries a lot of energy.

If, as in the picture at the bottom of the previous column, the settings are the same for all of the screens, (a), (b) and (c), then the signal (b) is stronger (has a greater amplitude) than (a) while the signal (c) is at a higher frequency than (a).

Can sound be too intense?

The intensity of a sound is measured in units called **decibels (dB)**. The faintest audible sound is taken as 0 (zero) dB. A decibel is a measure of about the smallest difference in sound intensity that a normal human ear can distinguish. The intensity of a sound wave corresponds closely with the loudness of the sound that we hear.

Noise pollution Sounds and noises (non-musical sounds) at intensities above 120 dB can have serious consequences for humans. Loud noises at night have caused dangerous blood pressure rises in sleepers. A sudden loud noise, such as an explosion, can in extreme cases cause eardrums to rupture and even cause death due to blast pressure. People who work in constantly noisy places such as factories may become deaf, and their ears can become insensitive to lesser sounds. Research carried out in Trinidad on the effects of loud sounds produced in steel bands on the hearing of pan men, supports these findings.

Buildings may also be affected by loud sounds and noises. Strong blast pressure resulting from a nearby explosion can cause damage, broken glass windows and cracks in walls.

High-frequency sounds, outside the range of human hearing (**ultrasound**), may be dangerous. People in the path of ultrasonic waves of high intensity may become confused and depressed and may even lose control over their movements. This effect can last for days.

Low frequency sound, below the threshold of hearing (infrasound) is known to have serious effects on the body. Infrasound in the region of 7 Hz can cause serious internal injury, by making the internal organs vibrate.

Questions

2. Choose a musical instrument. Explain how different frequencies of sound are produced by it.
3. 'No two musical instruments produce the same sound.' Discuss this statement.
4. How may sounds be harmful?

How do our ears work?

Activity | What use are our outer ears?

Our **outer ears** are the flaps of tissue on either side of our head. These collect sound waves from our surroundings and direct them into our ears. This idea was used in the old-fashioned ear-trumpet.

1. Roll up a newspaper to make an ear-trumpet.

rolled up newspaper

Ear-trumpet

2. Put the narrow end into your ear. Do *not* push it right inside. Find out how well you can hear sounds from a certain distance. Try again without the ear-trumpet. Compare your results.
3. Design an experiment to test the hypothesis: 'Students with large outer ears hear better than those with small ones.' What other factors besides large outer ears might be important in hearing?

Activity | What use are the ear canals?

Carefully put clean plugs of cotton wool into your **ear canals**. Do *not* push them in too far. How well can you hear sounds? What effect does the cotton wool have?

Wax is produced in the ear canal to clean and moisten it. Some people produce so much wax that it can block up the canal. This can cause partial deafness and 'buzzing' in the ear. Wax should be removed by a doctor or nurse squeezing warm soapy water into the canal to wash out the wax.

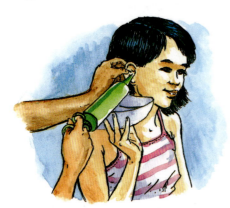

Syringing the ear

What happens in the middle ear?

Look at the diagram on the opposite page.

What use is our eardrum? The **eardrum** separates the ear canal from the middle ear. Think of an African drum or a drum from a steel band. The skin membrane or the metal is hit and so vibrates. This sets up sound waves which travel in the air.

When the sound waves hit against the eardrum they make it vibrate and these vibrations are then passed on to the bones in the middle ear.

The Eustachian tube The eardrum can only vibrate properly if the air pressure is the same on both sides.

Unequal pressure can occur in aeroplanes where we are exposed to changes in cabin pressure when we take off and land. We become slightly deaf because the eardrum is unable to vibrate and does not pass on the sounds. (See also page 257.)

The **Eustachian tube** is a passageway between the **middle ear** and the back of the nose along which air can pass to equalize the pressures. This is when the ears 'pop', and then we can hear properly again.

The ear bones These are the **hammer** (malleus), **anvil** (incus) and **stirrup** (stapes).

The hammer is attached to the eardrum, so when the eardrum vibrates the hammer vibrates too. These vibrations are passed on to the anvil and then to the stirrup which is attached to the oval window which connects with the **inner ear**.

What happens in the inner ear?

Look at the diagram on the opposite page. The inner ear is filled with fluid and contains two main parts: the **semi-circular canals**, which are concerned with balance, and the **cochlea**, which is concerned with hearing.

The stirrup hits against the **oval window** and makes it vibrate. These vibrations make the fluid move and these movements are felt in the cochlea. Compensating movements of the **round window** also occur to keep the pressure constant.

Along the walls of the cochlea there are tiny hairs attached to sensory cells which respond to the vibrations and transform them into nerve impulses. Cells near to the base of the cochlea respond to high-frequency vibrations (high pitch) while those near the tip respond to low-frequency vibrations (low pitch).

The nerve impulses are passed along the auditory nerve to the brain. In the brain there is a hearing centre (page 125) which is concerned with the interpretation of sounds, and so we hear.

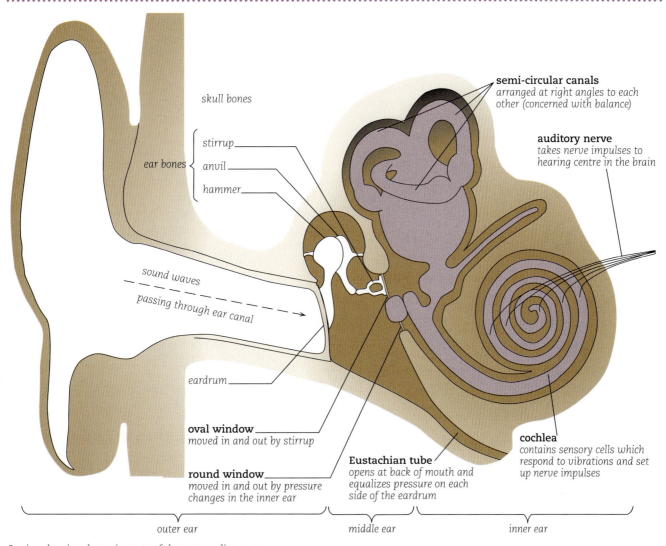

Section showing the main parts of the mammalian ear

How do we keep our balance?

The semi-circular canals in the inner ear are arranged at right angles to each other. Each one detects movement of the head in one particular direction. If you nod your head, shake it or tilt it, the movement will be detected. Fluid moves past sensory cells, and impulses are sent along the auditory nerve to the brain.

The brain also receives information from your eyes and from your muscles as to how your head and body are positioned. All of this information is put together, and messages are sent out from the brain to, for example, your legs or arms to move so you can keep your balance.

If you spin around quickly the fluid in the canals also moves, and may continue moving after you have stopped. You have the impression you are still moving and will feel dizzy.

Questions

1. What are the functions of:
    (a) the outer ear and ear canal?
    (b) the Eustachian tube?
    (c) the round window and the oval window?
    (d) the auditory nerve?
2. Describe the things which happen from the moment that a steel drum is hit to the moment that we hear it and realise it is a steel drum.
3. What is the importance of (a) the ear and (b) the brain in hearing?
4. (a) Describe two causes of temporary deafness.
    (b) Describe two causes of permanent deafness.
5. What are some ways in which we can take care of our ears? (Consider also the question of noise pollution on page 267.)
6. What is a hearing aid? What is its function in the ear? How might different hearing aids work?

interacting with our environment | *gathering information* | *how do our ears work?*

How is sound recorded?

How can we reproduce sound?

You can listen to a speech, and if you are good at shorthand, you can record every word. At a later date, someone could use your shorthand notes to give the exact same speech. Of course, it would not *sound* the same – all the pauses and stresses would not come out exactly as in the original – but the message will get across. A simple tune you hear may be recorded on a music sheet, but could you reproduce the sound?

Today, there are methods of recording whole orchestras and getting near-perfect reproduction. Methods are so good that it is sometimes difficult to tell the difference between a live musician and a recording.

How are sounds recorded?

How is sound recorded on a drum or disc? Think of a tuning fork that is vibrating at a low frequency. A small needle, called a stylus, is attached to one tip of the tuning fork. The needle is held near a rotating drum which is covered in paper blackened with soot. The tuning fork is tapped gently to set it vibrating and then lightly touched to the rotating drum. The fork continues to vibrate and the needle leaves a mark on the paper. This mark, also called a trace, is similar to the sine wave on page 267.

If a tuning fork with a higher note is used, a trace with higher frequency results.

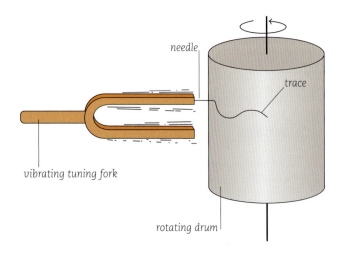

How is a record made?

Imagine that instead of sooty paper a layer of very soft wax is used so that a groove is cut in the wax by the needle. If the wax is now hardened and another needle (attached to a loudspeaker) is passed through the groove at the original speed of the drum, the original sound can be reproduced.

Activity | Looking at records

Use a hand lens to examine some old records (or pieces of old records), 45, 33⅓, and, if possible, 78 rpm (revolutions per minute). How do the grooves look. Are they different in the different types of record?

There are channels in the grooves, which force the stylus to make side-to-side movements when the record is played.

Making a record When a record of any type is being made, first, a lacquer-coated aluminium disc is used: this is the original master. Grooves are cut in a spiral moving from the edge of the disc towards the centre. The original master is used to produce another metal disc called a **master matrix** with raised ridges instead of cut grooves. This master matrix is used only to make other masters, which produce stampers. The stampers are used to stamp out the final records from a block of plastic (vinyl).

Record players Early record players were called **gramophones**. A mechanical clock drive was used to spin the disc, and a stylus attached to a mica diaphragm (loudspeaker) reproduced the sound via a sound box. The gramophone was very limited in the quality and volume of sound it was able to produce.

In a more **modern record player** the principle is the same. However, today electricity is used to enhance and amplify the sound. The stylus is attached to a tiny coil of wire positioned near a magnet. When the stylus follows the movement of the groove, it causes corresponding movements in the coil and makes electrical signals. These signals are amplified (increased in size) to drive the speakers and we hear the sound.

How are sounds recorded magnetically?

The most familiar form of magnetic recording is **tape recording**. Magnetic tape is plastic ribbon, coated on one side with a magnetic powdered material, for example iron(II) oxide. The tape is driven by a motor and runs past and touches the tape head which contains a strong magnet wrapped in a coil of wire. When sound is being recorded, an electrical signal from a microphone is sent to the coil in the head. This creates changes in the strength of the head magnet. As the tape goes past, these changes are recorded on the tape as the tiny magnetised particles move in different directions and remain fixed in their various positions.

To play back the sound, the magnetic tape is moved past the same tape head which then responds to the magnetic fields on the tape. The magnetic fields make (induce) electric currents in the coil in the head and these currents are taken to an amplifier and then a loudspeaker, which increases the sound volume to the desired level.

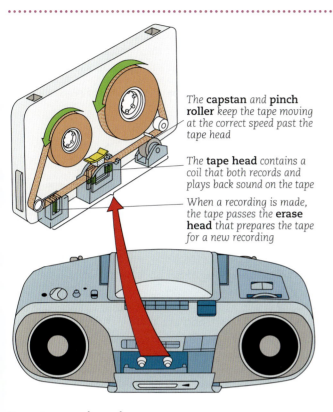

*The **capstan** and **pinch roller** keep the tape moving at the correct speed past the tape head*

*The **tape head** contains a coil that both records and plays back sound on the tape*

*When a recording is made, the tape passes the **erase head** that prepares the tape for a new recording*

How a tape recorder works

Activity | How is sound recorded optically?

Optical recordings of sound are used on cine film. (Some cine films, however, have a magnetic sound-track.)

1. Examine a piece of old 16 mm film with an optical **sound-track**.
2. Identify the picture frames and then the dark, narrow, continuous bands running alongside the picture frames. These are the sound track.

To produce a sound signal, the sound track must have variations. Why? Can you see any variation? There are two types of sound-track variation: (1) *variable density* in which the thickness of the band remains constant but its transparency varies, and (2) *variable width* in which the sound-track varies in width. When a film is being shown, it is run in front of a powerful lamp, which shines through the film and projects it onto a screen. A smaller lamp shines through the sound-track part of the film.

Both types of sound-track produce variations in the brightness of the light passing through them. These variations are detected by a photoelectric cell, which converts them into variations in electric currents. The electric currents are then amplified and finally fed into loudspeakers that convert them into sound waves.

Compact discs

The compact disc is one of the most popular types of recording and is also an optical recording. It uses an aluminium disc, smaller than an ordinary record and spinning much faster at 500 rpm. Each disc is covered by some transparent material and is not much affected by scratches. The sound reproduction is almost completely free of hiss, crackle or any distortion. The sound signal is carried in millions of tiny pits in the disc. A laser beam takes the place of the stylus found in an ordinary record player, and a photoelectric cell reads changes in the laser beam. This information is amplified and then converted back into sound signals by loudspeakers.

A compact disc player and compact discs. The disc is about 10 cm in diameter

Questions

1. Explain how dust can affect the sound quality of a phonograph record.
2. How were electricity and electronics used to improve the gramophone?
3. Why is a laser beam rather than a stylus used to 'read' the new optical discs?
4. How do the playback and recording heads on a tape recorder work?

Can we extend the range of sound?

How much energy is in sound?
Humans probably have the most highly developed vocal apparatus in the animal kingdom. We have vocal chords and muscles which are able to regulate sound to produce speech. The production of speech involves most of the facial muscles, controlling the tongue and breathing. All of these muscular activities use energy. Yet very little energy is converted into the sound waves. The combined shouts of 50 people would not produce more than 1/100 watt of power (page 175), not enough to light even the feeblest torch. Because sound only carries a little energy it does not travel over long distances. Can you think of ways sound *can* travel long distances?

How does the telephone work?
On pages 186–7 you saw how an alternating electric current could produce a wave similar to a sound wave. The important parts of the telephone are the **microphone** and the **receiver** (earphone). The microphone converts sound vibration into alternating electric currents while the earphone converts electric currents back into sounds waves.

The very first telephone was invented by Alexander Graham Bell. The earphone was combined with the microphone. It was an electromagnet, above which was suspended a **magnetic diaphragm**. Look at the picture.

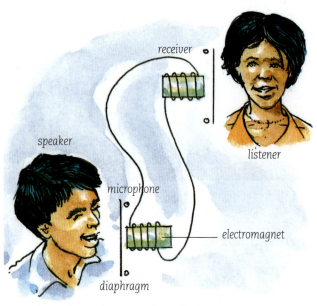

When the speaker makes a sound, the diaphragm vibrates. It is magnetic, so the vibrations produce a varying magnetic field which produces variations in the electric current in the circuit. The current travels to the listener. The variations in the current cause variations in the strength of the magnetic field in the receiver. This affects the magnetic attraction of the diaphragm, causing it to vibrate. These vibrations cause sound waves which are heard by the listener

The modern telephone still uses a receiver which works on the same principle. A separate microphone is used (carbon microphone), which works on a different principle.

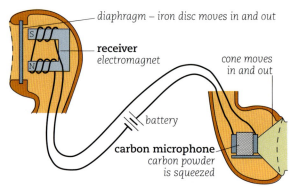

How a modern telephone works

How the carbon microphone works A steady current flows between the carbon blocks and through the granules. When a person speaks, the diaphragm vibrates and the grains are pushed together more or less closely. This makes the electric current flow more or less easily. The current variations are converted into sound waves by the receiver.

When the telephone was developed in 1876, the telegraph, which used the Morse code, was the best method of communication over long distances. Shortly afterwards, in many parts of the world, the telephone became the most popular means of personal communication over long distances.

At first, telephones made use of the electrical cables and other circuitry developed for the telegraph. Today, the telephone is an integral part of a sophisticated communication system that makes use of radio, television and computer systems and shares their technology.

How can radio waves be used?
If you look at the chart of the **electromagnetic spectrum** (page 320), you will see that electromagnetic radiation includes X-rays, visible rays, infrared and radio waves. All these different kinds of radiation have a lot in common. They differ only in frequency (and hence wavelength). They all travel through space at the same speed, which is the speed of light, at 3×10^8 km per second.

Different types of electromagnetic radiation are produced by 'exciting' matter in different ways. For example, radio waves are produced when rapid movements of electric currents are made in certain substances. A high-frequency (10 kHz or more) electric current can radiate, i.e. travel through space. This is used in **radio transmission**. To communicate by radio there has to be both a radio receiver and a radio transmitter for two-way communication. The workings of the transmitter and receiver are shown on the next page.

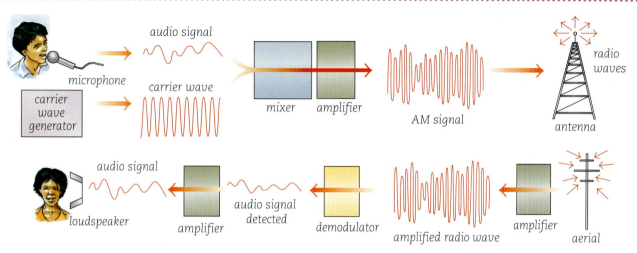

AM radio – above: transmission, below: reception

How does radio work?

The transmitter Sound waves are used to produce alternating electric currents of a certain frequency.

These currents can be transmitted along electric wires but are at frequencies too low to be radiated through space, therefore a much higher frequency current must be used. This is called a **carrier wave**. Low-frequency sounds (audio frequency) are then used to **modulate** the higher frequency sounds (radio frequency). The radio signal follows the 'contours' of the sound (audio) signal (see above). Finally, the signal is amplified and then conducted to the transmitting antenna and sent out (radiated).

The receiver In the radio receiver the reverse happens. A weak signal is picked up by the receiving aerial or antenna. The signal is amplified and the carrier wave separated, leaving the much lower frequency audio signal, which is amplified, enabling it to activate a loudspeaker.

This type of radio (transmission and reception) is called **AM radio**. AM stands for **Amplitude Modulation**. Can you think why? Look at the illustrations again and notice how the height (amplitude) of the wave form changes.

Another type of radio is **FM radio**. FM stands for **Frequency Modulation**. In this system, the frequency of the carrier wave varies at a rate depending on the frequency and amplitude of the applied modulating signal. The amplitude of the carrier wave does not change because of the modulation (see diagram in the next column).

How is the strength or amplitude of the audio signal conveyed to the carrier wave? The answer is, by the amount of frequency variation on the carrier wave. If, for example, the amplitude of the audio signal (b) were less, then the variations in the waveform at (c) would also be less, although the time interval between the maximum and minimum differences would remain the same.

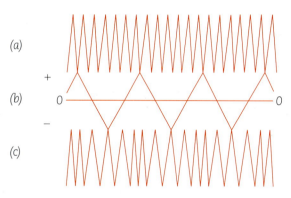

The principle of FM radio – (a) is a radio signal which is modulated by (b) the audio signal to produce (c) the FM signal which is a combination of (a) and (b). The FM signal increases in frequency when (b) is positive and decreases in frequency when (b) is negative

Activity | AM and FM radio

On a radio, compare reception on the AM band with that on the FM band. Make the comparison between local stations. Which type of reception is clearer? Which produces better sounds? Find the stations on a map. How far away are they from you? Explain the differences in the quality of the reception between the two stations.

Questions

1. Explain how a telephone earpiece works.
2. Explain how a carbon microphone works.
3. What are radio waves?
4. What is an amplifier?
5. Discuss the advantages and disadvantages of (a) AM and (b) FM radio.

How do we communicate via radio waves?

Can radio waves bend?

The behaviour of radio waves was predicted by the Scottish scientist James Clerk Maxwell in 1864. Some years later, the first radio waves were artificially produced and they behaved as expected. They travelled in straight lines like other types of electromagnetic radiation. In 1901, Guglielmo Marconi, after having successfully used radio waves to send wireless signals over short distances, surprised the scientists of the day by sending Morse code signals across the Atlantic. This was unexpected because there is not a straight line between England and North America, owing to the curvature of the Earth.

Later it became known that radio waves could travel in a curve like this because they are reflected by ionic layers high up in the Earth's atmosphere which act as a 'radio mirror' to reflect waves back to Earth. These ionised layers are called the ionosphere. These layers are made up of ionised particles caused by solar disturbances. There are at least four distinct zones of ionised layers in the ionosphere. The most important are the E and F layers. The E layer extends from 88 to 144 km above the Earth. The E layer is used during the day to transmit signals about 2 400 km away. The E layer disappears during the night so the higher F layer is used. The F layer is especially useful for transmitting over long distances and, depending on the season and the time of day, frequencies from 2 to 30 MHz can be transmitted around the world. This takes about $\frac{1}{7}$ second. The F layer extends from 144 to 400 km above the Earth.

Look at this picture. You will see that for reflection to occur, the transmitter must be pointed at a **critical angle** to the ionosphere. There are also skip zones (places on Earth where the signals cannot be received).

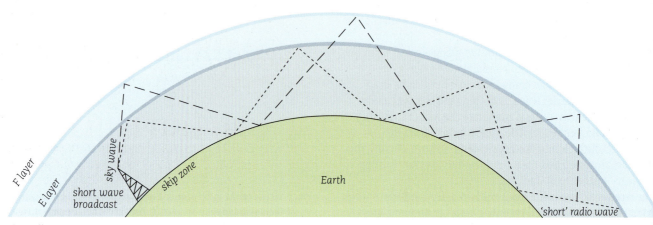

The radio spectrum

Frequency band	Frequency range	Wavelength	Typical uses
Low frequency (long wave)	30–300 kHz	10 000–1 000 m	Marine, navigational aids
Medium frequency (medium wave)	300–3 000 kHz	1 000–100 m	Broadcasting, marine
High frequency (short wave)	3–30 MHz	100–10 m	Communication of all types
Very high frequency	30–300 MHz	10–1 m	TV, FM broadcasting radar, air navigation, short wave broadcasting
Ultra high frequency	300–3 000 MHz	1 m–10 cm	Radar, microwave relays, short distance communication
Super high frequency	3 000–30 000 MHz	10–1 cm	Radar, radio relay, navigation, experimental

Activity | How do radio waves behave?

Look at the table of the radio spectrum on the opposite page. Compare the range and wavelength of the short wave frequency band with other frequency bands.

In the table, typical uses of radio waves are shown. The long waves travel along the ground following the curvature of the Earth. These waves are capable of travelling over a long distance. These long waves are of low frequency and are not the best for broadcasting music, because their reproduction of high frequency sounds is not very good. Medium and short waves are much better for music. Short waves are also suited for sky wave transmission; in these cases, amplitude modulation (AM) is used.

For the best radio reception, where the highest sound quality is faithfully reproduced, FM radio is used. FM is also used in TV broadcasts.

These and higher frequencies are not reflected by the ionosphere because they pass through the sphere. This is why TV cannot be transmitted over long distances in the same way as radio.

Short wave radio transmission has its own problems: it may be seriously affected by bursts of radiation from the sun. These bursts can cause 'fade outs', and sometimes the signals disappear altogether.

Lightning flashes also affect short wave (AM) reception. FM radio is not affected by radiation from the sun, nor lightning flashes.

Activity | What kind of aerials are needed?

Visit a radio and television station and look at the different types of aerial (antennae). Alternatively, look at pictures of the different aerials.
1. Which aerials do you think are for (a) sending and (b) receiving radio signals?
2. Which aerials are for television?
3. Which aerials receive signals from a satellite?

Types of aerial The simplest kind of aerial is a **long wire**, insulated from the ground and supported by trees or poles. This kind of aerial can be used to receive AM radio. This aerial works by intercepting a portion of the passing radio waves. The size of the aerial depends on the wavelength of the radio waves it is designed for. Short wavelengths are adequately received with short and compact aerials.

The most compact aerial of all is a **dish aerial**. This aerial is curved like a saucer. It is like a spherical mirror used for focusing light (see page 277). A dish aerial is used for focusing very short radio waves at super-high frequencies.

A TV aerial

Satellite communication

Super-high frequencies are used for communication by satellites. These are artificial satellites, which are bodies put into orbit around the Earth. (The moon is a natural satellite, see page 304.) Satellites which stay in the same relative position to a point on Earth (called **geostationary** satellites) are used to relay radio or TV signals around the world. Super-high frequencies are picked up by the satellite, amplified, and transmitted back to Earth. The signals are then picked up by local stations and recorded or re-broadcast at more convenient, lower frequencies. An ordinary home radio or television receiver cannot decode the super-high frequencies. However, some homes have acquired special equipment for receiving satellite TV directly.

Questions

1. What is the ionosphere?
2. Why is the ionosphere important in broadcasting?
3. What causes fading in AM short wave radio?
4. Explain how a radio signal is able to travel right around the Earth.
5. How are satellites used in radio and TV communications?

How does light behave?

How do we see objects?

Light has to travel from the source, for example a lamp, to the object and then to our eyes. Objects look different because they reflect or transmit light of different wavelengths (page 132). Objects may look bright or dim according to the amount of light they reflect. Mirrors are bright because they can reflect almost all the light shining on them. There are laws governing the behaviour of light just as there are for friction and movement.

Activity | What are the laws of reflection?

1. Set up a strip of plane mirror (flat mirror) vertically, with its silvered surface on a line MM' drawn on a sheet of white paper on a drawing board (see below). Place a pin O, the object, about 7 or 8 cm from MM'. Look from some convenient position E_1, and place two pins P_1 and P_2 into the paper so that they are in a straight line with the image I of the pin O seen in the mirror. Place these sighting pins P_1 and P_2 as far apart as the paper will allow. Remove the pins P_1 and P_2 and mark their positions by small pencil crosses and letters.
2. Carry out step 1 again with your eye in several other positions on either side of the object, for example E_2. Afterwards, remove the mirror.
3. Join up the points P_1 and P_2 and so on, cutting MM' at B_1, etc. Continue these lines backwards behind the mirror. Do they intersect at I?

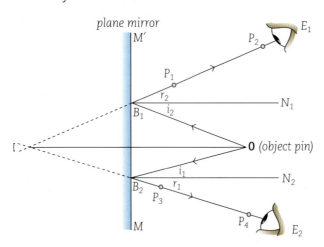

The lines OB_1, OB_2 and so on are called **incident rays** and show the direction in which light falls onto the reflecting surface. B_1P_2, B_2P_4 and so on are corresponding **reflected rays**. Draw in normals (perpendicular lines) B_1N_1, B_2N_2. Measure the angles of incidence and reflection for each pair of rays. Enter them in a table. Is the angle of incidence equal to the angle of reflection each time? Experiments like this tell us the laws of reflection.

The laws of reflection

1. The incident ray, the reflected ray and the normal at the point of incidence all lie in the same plane.
2. The angle of incidence is equal to the angle of reflection.

Activity | Using plane mirrors

Carry out other experiments with plane mirrors, using blocks, small toys and printed materials.
1. Find out if the object is the same size when reflected in the plane mirror.
2. Does it look as if the image is the same distance behind the mirror as the object is in front?
3. Can you easily read the writing in a reflection?
4. Can the image in a plane mirror be reflected onto a screen?

Experiments with plane mirrors show that light travels in a straight line. If light is shining from a point in an otherwise darkened room, even a mirror will appear dark unless the light is being reflected directly from the source via the mirror into the eyes of the observer. How is it then that the inside of houses can receive sunlight during the daytime even when the sun is not visible from inside the house? The answer lies in the nature of reflection.

In a mirror, reflection is regular, because the light is reflected from a completely flat or regular surface. However, most surfaces are not regular. If you look at most surfaces (for example, this paper), with a powerful magnifying glass, you will see that they are uneven; full of bumps and hollows. Light striking these surfaces is scattered in all directions. Therefore, objects may appear well illuminated by a point source of light in a darkened room. Reflection from such surfaces is called **diffuse reflection**.

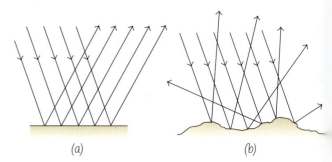

Reflection – (a) regular reflection, (b) diffuse reflection

Can mirrors produce 'real' images?

Real and virtual images You see an object because the light coming from the object passes through the lens in your eye and forms an image on your retina (page 278). This image, called a **real image**, is formed inside your eye.

If you hold a screen in front of an object, an image will not form on it. If you hold a plane mirror near an object at a certain angle, an image will appear in the mirror. But again, this will not form on a screen. We say that the image is not real but **virtual**. However, if a concave mirror is used, at certain positions real images can be made to appear on a screen.

Concave and convex mirrors A concave mirror may be thought of as part of a hollow glass sphere, the *outside* of which has been silvered and covered. The mirror appears on the inside surface. A convex mirror is also part of a hollow glass sphere. In this case, the *inside* of the sphere is silvered and covered.

In a curved mirror, the centre of the sphere of which the mirror is a part is called the **centre of curvature** (see below). Half the distance between the centre of curvature and the mirror's surface is called the **focal length**. The halfway point is called the **focus**. Parallel rays of light from a distant object strike a concave mirror and meet at the focus forming a real image.

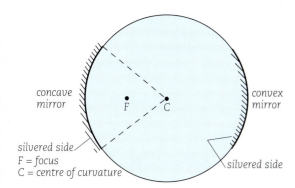

Concave and convex mirrors

Activity | A concave mirror
1. Examine a small concave mirror and find the distance from the focus to the mirror's surface. This is called the **focal length**. What is the centre of curvature?
2. Use a candle and a screen in a darkened room to investigate the images formed by the mirror. Observe the kind of images produced when the candle is placed at different distances from the mirror.
3. Discuss your results in class. Which arrangements produce real images? When are magnified images produced?

Activity | Can lenses form an image?
1. Select a number of lenses of different size and thickness (see next column). Look at some objects through each of these lenses in turn. Look at very close objects as well as distant ones.
2. Try to focus the rays of the sun with each lens onto a piece of dark paper. Do *not* look at the sun. Record your observations. What do you notice? Try using a bright object, for example a candle flame close to a lens. Is an image formed? Where?

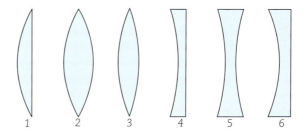

If some of the lenses you used in the Activity were the same shape as 1 to 3 in the picture above, you will have found that only this type of lens will form a **real image**. The other type of lens (4 to 6) can only form a **virtual image** and this is always smaller than the object. **Concave lenses** (4, 5, 6) behave like convex mirrors while **convex lenses** (1, 2, 3) behave like concave mirrors.

How does a projector work?
If a bright object is placed far from a convex lens, an image of the object will be formed close to the lens (near to the focal point). When a bright object is held just outside the focal point of a convex lens a distant image is formed. This principle is applied in the construction of the projector. The object is a brightly illuminated film close to a convex projection lens. The image is **projected** on to a screen.

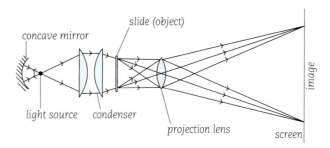

How a projector works

Questions
1. Explain why shadows are not 'pitch black'.
2. What is meant by focal length?
3. Explain why some objects appear very bright and others appear dim.
4. What are the uses of curved mirrors?
5. What are the laws of reflection?

How do our eyes work?

Activity | What is an eye like inside?
Your teacher will show you a mammalian eye such as that from a sheep or a cow.
1. Look at the *front* part of the eye and identify the **iris** (coloured part), the **pupil** (the black hole) and the **sclera** (the white part).
2. Look at the *back* part of the eye and identify the **optic nerve** leaving the eye.
3. Your teacher will cut the outer coat as shown below and fold back the flaps. Identify the parts you can see.

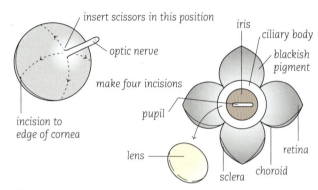

Left: how to cut open the eye   Right: eye cut open from the back

How does our eye focus?
Light rays enter the eye through the pupil. The rays are **converged** (brought together) by the cornea and the convex lens. The rays are focused on the retina to form an inverted, real image like this.

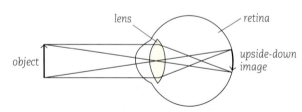

Focusing

Why is our retina important?
The retina contains light-sensitive cells called **rods** and **cones**. The cones respond to bright light and allow us to see colours. The rods respond to dimmer light and allow us to see in black and white. At twilight, when the light is dim we cannot distinguish colours as we are seeing with our rods.

These responses to light occur by **photochemical reactions**. Two other photochemical reactions occur in photosynthesis (page 66) and photography (page 284).

How is the eye like a simple camera?
Look at the diagrams below of the eye and a simple camera. The table underneath compares the structure and functions of the parts.

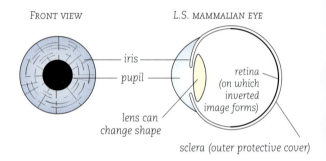

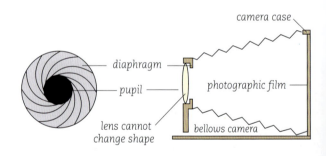

Comparison of the eyes (top) and simple camera (bottom)

Similarities	Eye	Camera
1 Outer protective case	Sclera	Camera case
2 Control of light	Pupil and iris	Pupil and diaphragm
3 Convex lens	Eye lens	Camera lens
4 Formation of image	Inverted image	Inverted image
5 Light-sensitive layer	Retina	Photographic film

Differences	Eye	Camera
1 Structure	Living: blood vessels and optic nerve	Non-living
2 Humours	Aqueous and vitreous humours	No humours
3 Focusing	Lens changes shape	Lens is moved
4 Image	Only 'seen' in the brain	Can be developed on film

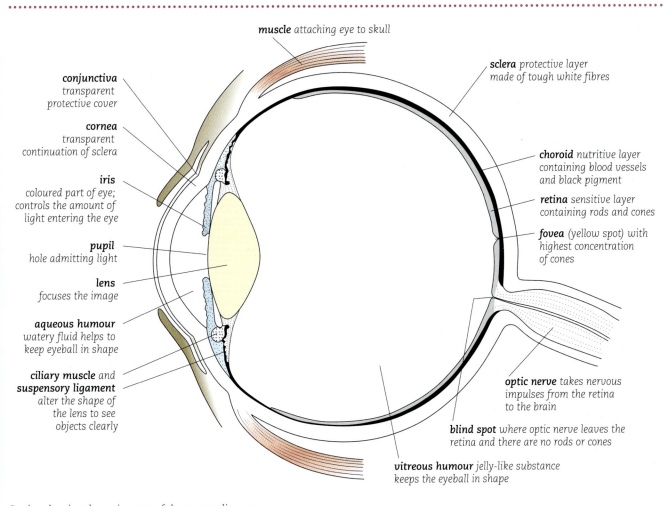

Section showing the main parts of the mammalian eye

How do we see?

The light which falls on the retina causes photochemical reactions in the rods and cones which set up nerve impulses. The impulses pass from the rods and cones along the nerve fibres and into the optic nerve which takes them to the sight centre of the brain (page 125).

The brain receives information about the inverted image from the eye but it interprets the image so that we see it as it is in the real world. The upside-down image is interpreted as a right-side-up object.

There is a place in the retina directly opposite the pupil (the **fovea** or **yellow spot**) where there are a lot of cones, so here we have our clearest vision. You can demonstrate this by using a pin to make a very small hole in a piece of paper and looking through the hole while holding the paper close to the eye. You will see clearly, even if you are short-sighted. At the place where all the nerve fibres leave the retina (the **blind spot**) there are no rods or cones and we cannot see an image which falls here.

Questions

1. Where *exactly* are the following, and why are they important?
   (a) The pupil and iris.
   (b) The suspensory ligaments and ciliary muscles.
   (c) The sclera, choroid and retina.
   (d) The rods and cones.
   (e) The yellow spot and the blind spot.
2. Describe in detail how an inverted image of an object is formed on the retina (refer to page 278 to remind yourself how light rays behave).
3. (a) What is a photochemical reaction?
   (b) Give two examples.
4. Compare the structure and functioning of a simple camera and the eye.
5. (a) Suggest two ways in which diseases in the eye might make people blind. (Also see page 281.)
   (b) Suggest two ways in which diseases in other parts of the body might make people blind.

Problems with our eyes

How our eyes should work

The size of the pupils The pupils become larger in dim light and smaller in bright light (below).

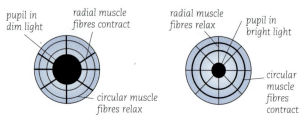

Pupil changes

The focusing of the lens In a normal eye we can see near and distant objects clearly: sharp images are formed. For this to happen the shape of the lens must be changed. This is called **accommodation**.

When we look at a *near object* the lens has to be *fat*. The suspensory ligament relaxes to allow the lens to bulge out and the ciliary muscles contract (see (a) below). The lens must be fatter (more convex) because the light rays are coming from a point close to our eyes and they need to be converged a great deal so as to come to a focus on the retina (see (b) below).

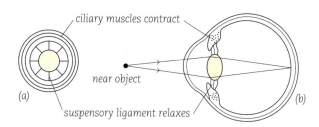

Eye viewing near object

When we look at a *distant object* the lens has to be *thin*. The suspensory ligament contracts and makes the lens thin and the ciliary muscles relax (see (a) below). The lens must be thinner (less convex) because the light rays are coming from a distance and so do not need to be converged very much in order to come to a focus on the retina (see (b) below).

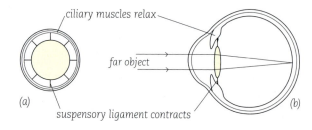

Eye viewing distant object

Short sight

A short-sighted person can see *near* objects clearly, but can only see distant objects as a blur. Short sight usually develops during late childhood. It will cause problems at school as the child will not be able to see the chalkboard clearly.

In a short-sighted person the images of distant objects fall short of the retina (see (a) below). This is either because the eyeball is too long, or because the lens is too fat (too convex) and has converged the rays too much.

What can be done? The image is falling short of the retina. Therefore the light rays need to be spread out before they enter the eye. This can be done with spectacles containing diverging (**concave**) lenses.

The concave lens spreads out the light rays so that they are focused on the retina. The person sees a clear image.

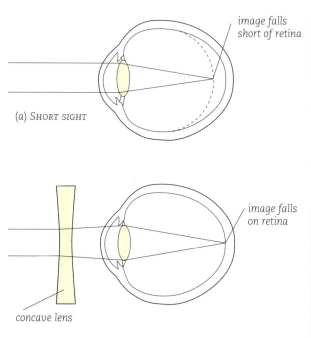

Correction of short sight with a concave lens

Loss of accommodation

In the normal eye the lens is automatically thickened to view near objects and thinned to view distant objects. As people get older the lens of the eye hardens and its shape cannot be changed so easily (loss of accommodation).

This means that a person cannot focus on near objects such as writing in a book and has to hold the book a longer distance away.

What can be done? The person can be prescribed **reading** glasses if his or her vision is otherwise normal. For short or long sight, together with loss of accommodation, **bifocals**, which have two parts with different lenses, may be worn.

Long sight

A long-sighted person can see *distant* objects clearly, but can only see near objects as a blur (out of focus). If a person suffers from long sight he or she has usually had it from birth. A child with long sight may complain of eye fatigue when reading.

In a long-sighted person the images of near objects are focused behind the retina (see (a) below). This is either because the eyeball is too short, or because the lens is too thin and has not converged the rays enough.

What can be done? The image is falling behind the retina. Therefore the light rays need to be converged more before they enter the eye. This can be done with spectacles containing converging (**convex**) lenses.

The convex lens converges the light rays so that now they are focused on the retina and the person sees a clear image.

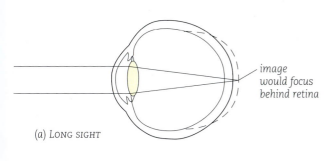

(a) LONG SIGHT

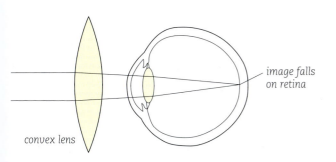

convex lens

Correction of long sight with a convex lens

Astigmatism

Light rays are focused by the lens and the cornea. The cornea may be uneven. The result is that one part of the image may be clearly focused but not another part. This is called **astigmatism**.

What can be done? A person with astigmatism may also be short or long sighted so when the spectacle lenses for the other defects are prepared they will also be shaped so as to counteract the uneven curvature of the cornea. This is why people with the same degree of short-sightedness cannot see clearly with each other's spectacles.

Other eye problems

Glaucoma Glaucoma is pressure in the eye. It is most common in older people. It is caused by a build-up of aqueous humour at the front of the eye. If not treated it can cause the death of blood vessels and nerve fibres in the eye which can lead to blindness.

Symptoms are blurred vision and pain in the eyes. In an operation a new drainage canal is made so aqueous humour drains away.

Cataract Cataract is the gradual clouding up of the lens. This prevents light from passing through the eye and makes it more difficult for the person to see. It is most common in older people.

If a cataract is severe, the lens has to be removed and it may be replaced by a plastic lens inside the eye. Sometimes the lens may not be replaced at all. This makes the eye very long-sighted, as now only the cornea can focus the light rays and cannot converge them enough. The person is therefore given spectacles with very strong converging (convex) lenses or plastic contact lenses.

Care of the eyes

Cleanliness The surface of the eye is washed with tears. If something gets into the eye it should be carefully removed with a clean cloth or washed out in an eye bath. When doing practical work, especially with dangerous chemicals, great care should be taken.

A **stye** may develop if the base of an eyelash becomes infected. This is a painful red swelling like a boil.

'Pink' or 'red' eye is an infection of the conjunctiva. It can be caught by the use of contaminated towels or face cloths. A person with pink eye is given antibiotic drops or ointment.

Light On page 277 you saw how light is focused by a lens. The resulting pinpoint of light can be very hot. This is why you must never look at the sun, especially through lenses, because the rays can burn your retina.

Questions

1. How does the iris help to protect the eye from bright light?
2. What is accommodation? What are the symptoms of loss of accommodation?
3. Describe carefully (using diagrams) the causes and correction of (a) short-sightedness and (b) long-sightedness.
4. What are some ways in which we should take care of our eyes?

What is colour?

How are different colours produced?

We saw on page 132 that white light is made from other colours. These colours are called the **visible spectrum** to distinguish them from the rest of the electromagnetic spectrum (page 320). Each colour is a certain set of wavelengths that activate specific cones in the retina (page 278). White light can be made by mixing coloured lights of **red**, **green** and **blue**. This can be done by projecting the colours onto a white screen. These colours are known as the **primary colours of light** because they cannot be made by mixing any other coloured lights together (see below).

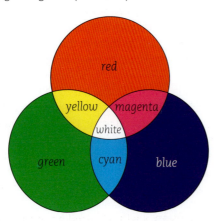

The primary colours of light: red, green and blue

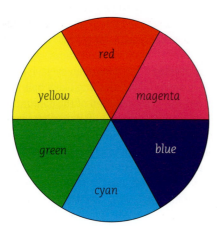

The colour wheel – red, blue and green are the primary colours. Magenta, cyan and yellow are the secondary colours

Activity | The colour wheel

Colours may be arranged in a colour wheel (above). The colours opposite the primary colours are known as the **secondary colours**. These are **yellow**, **cyan** (a blue-green colour) and **magenta** (a reddish-blue colour). These are formed by mixing light of the two colours on either side. For example, yellow is formed from mixing red and green light.

Yellow is called a **complementary colour** to blue because yellow and blue light mixed together will give white light. Similarly, magenta and cyan are complementary colours of green and red respectively.

1. Use the colour wheel to derive all the colour combinations which would produce white light.

Can pigments (paints) produce different colours?

Paints are made from finely divided particles called **pigments**. These pigments absorb (i.e. subtract from white light) some colours and reflect others. The colour mixing described above is *adding* coloured *lights*. Colour mixing with *pigments* is *subtracting*. Coloured objects or paints subtract from white light the colours they absorb (and cannot reflect). For example, yellow and blue *lights* mixed together add to give white light but if yellow and blue *paints* are mixed, they subtract colours from white light and green is produced. Blue pigment in paint absorbs yellow and red light but reflects some green and violet as well as blue. Yellow pigment absorbs violet and blue light but reflects some green and red. Thus, the only colour that is not fully absorbed by the yellow-blue pigment mixture is green. Therefore, the mixture of yellow and blue pigments looks green.

What are the primary pigments?

Just as there are primary colours (lights), there are **primary pigments**. These are **magenta**, **yellow** and **cyan**. The primary pigments cannot be made by mixing other pigment colours. (Try to mix yellow from other pigments!) The primary pigments, yellow, magenta and cyan, are normally referred to as yellow, red and blue respectively. While a mixture of the three primary colours of light will produce white, a mixture of the three primary pigments will produce black.

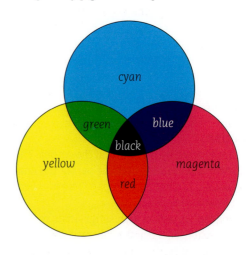

The primary pigment colours: yellow, magenta and cyan

Activity | Does coloured ink contain different pigments?

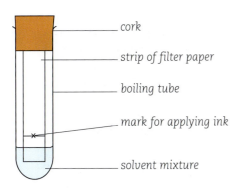

1. Get a boiling tube, a stand for holding the tube, filter paper, a cork with a split in the underside, and a solvent (such as alcohol-water mixture) from your teacher.
2. With clean dry hands and a sharp pair of scissors, cut a rectangular strip of filter paper about 2 cm wide and 12 cm long. Draw two light pencil lines near one end of the paper 1 cm and 2 cm along. Make a light cross in the middle of the line 2 cm from the end (see above).
3. Use a black ink pen or felt tip marker to put a tiny spot on the cross. Carefully build up a dense spot by allowing the ink to dry between applications so that there is very little spread.
4. Insert the other end of the paper into the split cork (wedge it in if necessary). Try the cork and paper in the boiling tube to check the length. Adjust the length and width of the paper so that it touches neither the side nor the bottom of the tube. Make a mark on the outside of the tube corresponding to the position of the ink spot.
5. Remove the cork and paper and place the solvent mixture in the boiling tube, about 1 cm below the ink spot. Replace the cork and paper and keep the tube upright (in a stand).
6. The solvent will rise steadily up the paper. When it nearly touches the cork, remove the paper and make a light pencil mark to show the final position of the solvent. Dry the paper using a fan.
7. How many colours do you see? Do you see different shades of the same colour?
8. Do the colours separate well? Why do you think the colours separated?
9. Do you think the results would be the same if other solvent mixtures were used? (Perhaps you could repeat the Activity using a different solvent mixture, for example a water-acetic acid mixture, to find out.) This process of colour separation is called **chromatography**. It is simple, but very useful in chemistry and biochemistry.

Can chromatography be used to separate colourless substances?

Using coloured inks is one of the best ways of demonstrating chromatography. Different coloured inks are chemically different substances. They are adsorbed with different amounts of vigour by the fibres of the filter paper. As the solvent rises up the filter paper, it will push along each chemical substance at a different rate, thus spreading them along the paper. The separation of the different substances in ink is easily observed because the chemically different substances are all coloured. To separate a mixture containing colourless substances, a chromatogram may be prepared in the same way but it will have to be 'developed' at the end of the run. Certain chemical sprays may be used to convert the colourless substances into coloured compounds so that their positions on the chromatogram can be seen.

Activity | How are coloured pictures printed?

1. Examine a coloured picture postcard or a coloured picture from a magazine. List the different colours you can see: include colours such as light brown, beige, light yellow and all the different shades of blue and green. If you wanted to paint a picture like this, how many shades of paint would you need?
2. Examine the picture with a magnifying glass or a hand lens. What do you see now?

You probably noticed that all the colours and shades in the list you made have been reproduced from **dots** of just three colours: the primary pigments magenta, cyan and yellow plus black, which gives accent and depth.

To print a coloured picture, four **printing plates** must be prepared to produce different sets of dots. The picture is separated into four colours: cyan, magenta, yellow and black. A reproduction house produces four sheets of film, one sheet for each colour. Each film is then burnt onto a plate. These plates are used to print the photo. The paper is sent through the printing machine four times, using a different plate each time, so that the four colours are printed over each other to form the final picture.

Questions

1. Distinguish between (a) primary and secondary colours (lights) and (b) primary and secondary pigments.
2. How are the primary (a) coloured lights and (b) coloured pigments mixed to give new colours?
3. Outline the process of colour printing. What evidence can you give to support your account?

How are images reproduced?

How are photographs made?

If you were living in the USA in 1838 and were fortunate enough to have your photograph taken, you would have had to be prepared to sit perfectly still for half an hour at least. If afterwards you were proud of the result, this would have been as much a tribute to your patience and endurance as to the skill of the photographer. In the few years following 1838, the sitting time was reduced from 30 minutes to a mere 30 seconds and photography became big business. At this time, photography was called **daguerreotyping**, after the Frenchman Daguerre who invented the process.

The progress in photography over the last 160 years has been tremendous. But the process remains basically the same.

Activity | How does photographic paper work?

1. Acquire a piece of photographic paper, about 6 cm square (in its lightproof wrapping) from a photographic shop.
2. Open the paper and quickly arrange a number of small, preferably flat objects on it. Small dried flowers, a transparent object like a button, and even a photographic negative will work well. If possible, cover with a sheet of clear glass to keep the objects in place.
3. Place the whole arrangement in bright sunlight or under a powerful lamp. Observe what happens over 10 minutes or so. Does the film get dark?
4. Remove it from the bright light and continue to observe. Take away the objects. Immediately examine the unexposed areas. Can you explain your results?

The opaque objects tend to give outlines that only show the edges clearly. Transparent objects give a reverse image called a **negative**, while photographic negatives give a positive image.

Photographic paper Photographic paper is sensitive to light. The areas which are exposed to light become dark. Once you had removed the objects obstructing the light, the whole piece of paper would eventually have become dark. That is why it is very important that unused paper or film is kept in lightproof packets or boxes.

Photographic film When a photograph is being taken, a camera is used, with photographic film rather than paper. Film is much more sensitive than paper and an **image** can be formed in less than 1/100 second.

However, if you could examine exposed film immediately after a picture was taken you would see nothing on it because the image formed is 'latent' (this means *hidden*). This image first has to be **developed** and then 'fixed'. The exposed photographic film is treated with special mixtures of different chemical solutions to make this latent image show up.

As in the previous Activity, this image will eventually disappear, but it can be fixed by another set of solutions.

Photographic negatives. Black and white and coloured areas are the reverse of how they will appear in the final photograph

Photographic chemicals

What are the light-sensitive substances used?

Photography is based on the light-sensitive properties of silver salts, particularly of silver chloride and silver bromide. When these crystals are exposed to certain electromagnetic radiation, they absorb the radiant energy and their crystal structure changes. The silver salts used (called **halides**) are particularly sensitive to high-energy radiations, for example ultraviolet and X-rays, but are made sensitive to visible light by adding special chemicals called **sensitisers**.

How are colours photographed?

To produce a colour photograph the principles of colour mixing using the three primary colours are applied (page 282). A colour film has three light-sensitive layers called **emulsions**: a blue-sensitive layer, a green-sensitive layer and a third red-sensitive layer. Between the blue and green layers is a yellow filter, which absorbs excess blue light rays and prevents them from reaching other layers.

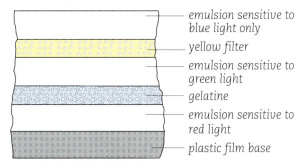

The emulsion layers and filters in a colour-sensitive film

So, the three layers record the different amounts of light of different colours given off by an object. The three emulsions are then developed to **negatives** and then reversed to **positives**. In this process, the three layers are changed to yellow, magenta and cyan. The result is a colour transparency, and from this colour prints can be made.

How are television pictures produced?

You have seen how a picture can be reproduced by altering the proportion of different coloured dots from which it is composed (page 283). Also, the eye has **persistence of vision**, which means it can see movements from a fast-moving sequence of still pictures. These principles are used in the production of television pictures.

A television picture is produced from electronic signals which come first from a television camera. In the camera, the image is broken up into small picture 'elements' of different brightness. The picture is then **scanned** with an **electron gun** and the variation in the elements is converted into different electrical or video signals. Camera synchronising signals are added to the video information to help with the reassembly of the picture at the receiver. These signals from the camera are amplified to a high level and are used to modulate the carrier wave from the transmitter (compare page 273).

The picture is scanned horizontally in 525 lines in the American system (used in the Caribbean) or 625 lines in the British and European system. An entire picture is transmitted line by line in $\frac{1}{25}$ second. Every second 525×25 (= 13 125) horizontal lines are scanned and each line may consist of as many as 700 points of different brightness. To reproduce a picture, the television receiver picks up the signal via its antenna and amplifies it. Two signals are separated: the audio signal, which activates the speaker, and the video signal, which guides the action of the **video tube** inside the TV set. Inside the video tube, the electron gun (see picture opposite) directs a beam of electrons to a special coating on the inner surface of the tube. The coating emits light in response to an electrical field. The deflection coil around the tube produces magnetic fields, which guide the electron beam across the screen. The information in the video signal enables the original scanning action of the camera to be reproduced in the television receiver.

In **colour television** the principles are the same but of course much more complex. The colour television camera has three tubes and the picture is scanned in the three primary colours, red, green and blue. Three picture signals are therefore transmitted over the air to the receiver. The picture is then reconstructed using a TV tube with three electron guns.

The screen of a colour picture tube contains hundreds of sets (triads) of three different spots – red, green and blue. The brightness of the components of each triad relative to each other will affect the colour perceived by the eye and brain and thus the range of colour.

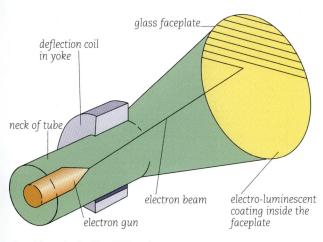

The video tube inside a TV receiver

Questions

1. What are the principles photography is based on?
2. How is a range of colour produced in photography?
3. Do photographic and television cameras have anything in common?
4. Why is picture quality better on a small television set?
5. Explain why a colour television receiver is more complex than a monochrome (black and white) receiver.
6. How are many different colours produced in colour TV?

Why do we need transport systems?

On page 68 we looked at how particles can move around by diffusion. Particles move or diffuse from an area where they are in high concentration to an area where they are in lower concentration. We have seen many examples of diffusion in plants and animals – such as the movement of oxygen in leaves and lungs, the loss of water vapour in transpiration, and the movement of food particles into the villi in the gut.

If diffusion is so important, why do we need transport systems such as the bloodstream in animals, and the xylem and phloem tubes in plants? The answer is that diffusion is useful only over short distances.

Activity | Looking at diffusion
1. Collect some different shaped glass containers.
2. Pour a measured volume of 100 cm³ of water into each container like this.

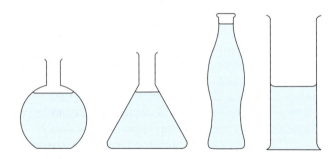

Diffusion in water

3. Using forceps, and being very careful, drop *one* crystal of potassium permanganate into each container. Or you can use a grain of instant coffee.
4. Observe the containers. In which one does the colour diffuse most quickly? Why do you think this is so?

Diffusion with agar blocks
Agar blocks of two sizes were made and placed in a beaker containing dye.

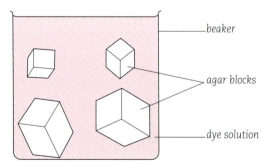

Agar blocks

After 2 minutes, one cube of each size was removed with forceps, and cut lengthways like this.

Cut agar blocks

The colour had diffused all the way through the smaller cube, but only into the outermost part of the larger cube. After 8 minutes the remaining cubes were removed and cut. The smaller cube was fully dyed, but the centre of the larger cube was still clear of dye.

Questions
1. Why do you think the dye travelled more quickly into the small cube than the large one?
2. Which do you think would cook more quickly: a whole potato, or potato slices? Why? Set up a fair test to find out.

The importance of surface area and volume
Substances diffuse more quickly into the centre of a small cube than a larger one. The smaller cube has a larger **surface area** in relation to its bulk or **volume**: we say it has a larger surface area to volume ratio.

The distance to the centre of a small cube is also shorter than in a large cube and so diffusion can occur more quickly.

The cube below has sides of 2 cm. Its volume is length × breadth × width ($2 \times 2 \times 2$) = 8 cm³. Each side has an area of $2 \times 2 = 4$ cm² and there are six sides, so the total surface area is 4 cm² × 6 = 24 cm².

The surface area to volume **ratio** (S.A. : Vol ratio) is therefore 24 : 8 or 3 : 1.

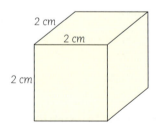

Vol = $2 \times 2 \times 2$ = 8 cm³
S.A. = 4 cm² × 6 = 24 cm²
S.A. : Vol ratio = 24 : 8 = **3 : 1**

Large cube

The picture below shows how the surface area to volume ratio *increases* as the blocks become *smaller*, and the distance from the outside to the centre becomes shorter.

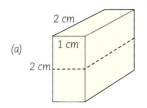

(a)
Vol = $2 \times 1 \times 2 = 4$ cm^3
S.A. = $(2 \times 4$ cm$^2) + (4 \times 2$ cm$^2)$
= 16 cm^2
S.A. : Vol ratio = 16 : 4 = **4 : 1**

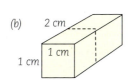

(b)
Vol = $1 \times 1 \times 2 = 2$ cm^3
S.A. = $(2 \times 1$ cm$^2) + (4 \times 2$ cm$^2)$
= 10 cm^2
S.A. : Vol ratio = 10 : 2 = **5 : 1**

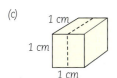

(c)
Vol = $1 \times 1 \times 1 = 1$ cm^3
S.A. = 1 cm$^2 \times 6 = 6$ cm^2
S.A. : Vol ratio = **6 : 1**

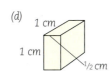

(d)
Vol = $1 \times \frac{1}{2} \times 1 = \frac{1}{2}$ cm^3
S.A. = $(4 \times \frac{1}{2}$ cm$^2) + (2 \times 1$ cm$^2)$
= 4 cm^2
S.A. : Vol ratio = 4 : ½ = **8 : 1**

Diffusion

Diffusion is effective for small organisms, such as *Amoeba*, or across thin surfaces such as the lungs in larger organisms (page 100).

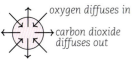

oxygen diffuses in
carbon dioxide diffuses out

Small organism with high surface area to volume ratio, can exchange gases by diffusion across its surface

Large organism has low surface area to volume ratio so it cannot exchange gases quickly enough through its surface

The surface area is increased, so there is more surface through which diffusion can occur

Diffusion in small and large organisms

Besides the lungs, there are other places in the body of a mammal where the surface area to volume ratio is large so that diffusion can occur easily. For example, the villi in the small intestine (page 84) and the long tubules in the kidney (page 112). Chewing food and emulsifying fats also increases the surface area for the enzymes to work on (pages 84 and 87).

Diffusion is also important in flowering plants, so there are places where the surface area to volume ratio is high. For example, leaves are thin and have many loose-packed cells (page 71) which give a large surface area in contact with the spaces inside the leaf. This allows gaseous exchange to take place more easily. The large surface area of the root hairs in contact with the soil (page 90) also helps intake of water and mineral salts.

But even the increase in surface area to volume ratio is not sufficient for large organisms.

Transport systems

In large organisms substances have to be transported around the body. The distances involved are large, and diffusion would not occur quickly enough for the needs of the organism.

In flowering plants the transport system is very simple, and consists of a series of tubes, the xylem and phloem (pages 90–1). These tubes transport water, mineral salts and food substances around the plant.

Flowering plants are rooted in the ground. Animals are much more active than plants and so they need more oxygen and food for the release of energy. They must get oxygen quickly and get rid of carbon dioxide. The animal's respiratory surface (lungs) which allows for the diffusion of these gases must be supplied with blood which picks up the oxygen and brings back the carbon dioxide (page 101).

Substances such as gases, food, and waste products must be transported around the body in the blood (page 95). For the blood to circulate quickly enough there is a powerful pump: the heart.

Animals have to search for their food, and escape from their predators. They therefore have to respond quickly to what they see, hear and feel in their surroundings. They need a nervous system (pages 122–5). Nerves carry messages chemically and electrically.

Questions

3 (a) What is the surface area to volume ratio? (b) Why is it important?
4 What is the surface area to volume ratio of a cube which has sides of 3 cm?
5 Give two examples where diffusion occurs in (a) plants and (b) animals.
6 How are transport systems (a) similar and (b) different in flowering plants and mammals?

How are transport systems protected?

The blood system (pages 92–5) and nervous system (pages 122–5) are our transport systems. Blood vessels and nerves are soft and easily damaged. Many parts of the blood and nervous systems are enclosed by the hard bone of the skeleton or given at least some protection by the ribs or vertebrae. As well as protection, the skeleton is important for movement, using the principle of levers (page 227).

3 **Neck vertebrae** Find the space through the centre of the vertebrae. This is where the spinal cord runs. Look for two small holes, one on each side of this large space. These holes protect the small blood vessels in the neck.

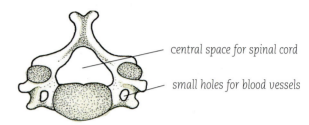

Front view of neck vertebra

4 **Back vertebrae** Fit together two back vertebrae to see how the central space can protect the spinal cord, and how the spinal nerves have space to come out from between the vertebrae.

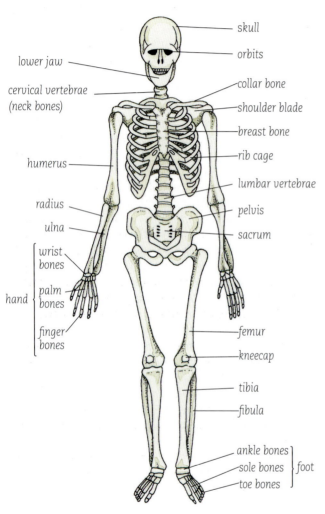

The human skeleton

Activity | Protection by the skeleton

You will need a model skeleton and some cervical (neck) vertebrae and lumbar (back) vertebrae.

1 **Skull** See how the skull is made from flat bones joined together. Look inside to find the hollow space. Find the **orbits** at the front of the skull.

2 **Rib cage** See how the ribs are joined to the thoracic (chest) vertebrae at the back, and to the **sternum** (breast bone) at the front. These structures make up the rib cage.

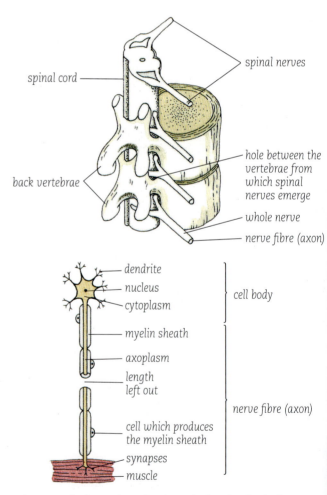

Above: two back vertebrae showing spinal cord and spinal nerves
Below: one motor nerve cell; in bundles these form nerves

Protection from mechanical damage

Protection of the nervous system
1. The brain is inside the cavity of the skull.
2. The eyes are protected by the bony orbits of the skull.
3. The spinal cord runs down inside the central space in the vertebrae which make up the backbone.

Protection of the blood system
1. The blood vessels in the neck run through the holes in the neck vertebrae.
2. The heart, and the blood vessels to and from the lungs are protected inside the rib cage.

Protection from micro-organisms

Our body is continually under attack from micro-organisms (germs) such as viruses and bacteria. Here is a photograph of the surface of human skin magnified to show bacteria on the surface.

If the skin is cut then these bacteria can infect us

Clotting If we cut our skin we are likely to break a blood vessel. We will bleed, but the blood soon begins to thicken up and forms a **clot** (page 92). This forms a plug over the cut which then begins to heal. This is our first defence against germs.

White cells If germs enter our bloodstream either through our skin, or through the respiratory or digestive system, they are attacked by our white blood cells as shown below. (See also page 161.)

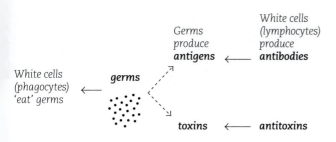

Protection against germs

Results of damage or infection

The blood system Germs that enter the bloodstream are attacked by the white blood cells. Some of these, the lymphocytes, are produced in the **lymph gland**s (page 128). These glands, especially in the neck and groin may become swollen when they produce more lymphocytes to deal with the infection.

The time during which the germs increase in numbers is called the **incubation period**. The germs may then settle in a particular part of the body and cause the **symptoms** of the disease. As the white cells fight the germs the person may have a fever, and the poisonous substances produced may make the person feel ill and weak. Some medicine or other treatment may be needed (page 160).

The blood system can be damaged if arteries become narrow due to plaque (page 96). As a result parts of the brain may not get enough oxygen and the person may have a **stroke**. If the part of the brain is concerned with speech then the person can no longer talk properly. If the part is concerned with movement of the legs, the person could becomes paralysed.

The nervous system Paralysis can also be caused by the **polio** (poliomyelitis) virus which attacks motor nerve fibres which control the muscles. **Parkinson's disease** damages control of movement because nerve centres in the brain are destroyed. In **multiple sclerosis**, the myelin sheath covering the nerve fibres becomes swollen and damages the nerves.

Unlike the repair and growth of blood vessels, nerves cannot easily repair themselves. However, although new nerve cells are rarely formed, there is regrowth of connections, **synapses**, between damaged nerve cells, for example in the brain.

Alcohol affects the functioning of the nervous system. It reduces inhibitions. In larger amounts it causes slurred speech, blurred vision and loss of muscle coordination (pages 80 and 253).

Other drugs also affect the nervous system, causing either the speeding up, slowing down, or distortion of mental processes (pages 80–1).

Questions

1. What are the protective functions of the skull, rib cage and vertebrae?
2. How does the body respond to:
   (a) a cut blood vessel?
   (b) micro-organisms entering the body?
3. Name five diseases which are associated with damage or infection of the blood and/or nervous systems.

How do collisions transfer energy?

Can particles transfer energy?

A stack of cans on a supermarket shelf or at a fair will obey Newton's first law of motion. They will remain where they are unless they are disturbed by a force (which could dislodge them). A moving ball (thrown to hit them) will easily dislodge and scatter them. However, a stationary ball will not disturb them. The difference between a ball at rest (stationary) and a moving ball is that the moving ball has **kinetic** (movement) **energy**. Some of this kinetic energy may be transferred to the other objects with which it may collide.

Activity | A peashooter

Make a peashooter using a hollow bamboo tube about 10 to 15 cm long with a small bore (inner hole) about 5 mm in diameter.

1. Put a small pea or pellet into the tube. Blow down the tube to expel the pea. Aim the peashooter at a suitable target (for example, an empty match box). Can you knock it over?

Energy is transferred from the moving pea to the box just like the ball and cans described earlier.

This transfer of energy from moving to stationary objects may be demonstrated like this. A loudspeaker is connected to a sound generator. The sound makes the loudspeaker move in and out. At a low sound level, when the loudspeaker moves with little energy, there is very little movement of the spheres and the disc of polystyrene stays at the bottom of the transparent tube. As the strength of the sound signal is increased, more energy is transferred and the spheres acquire more kinetic energy and so push the disc further up the tube.

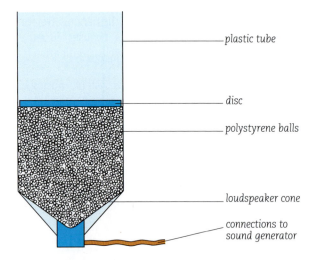

A moving body carries kinetic energy. It will pass on some or all of its energy to an object with which it collides. This is called the **conservation of momentum**.

How is energy transferred?

On pages 16–17, you saw the evidence for believing that matter is made of particles. Solid matter transmits heat energy by particle-to-particle contact, which is known as **conduction**. Conduction also occurs in liquids and in gases, but to a lesser extent, because the particles of matter are further apart from each other. Heat energy also causes matter to expand and phase changes to take place, for example solid to liquid, or liquid to gas. If heat makes matter expand too rapidly, an explosion occurs. Fast-moving particles can provide the thrust for a rocket or jet engine (page 299) or push upon a piston in the internal combustion engine (page 192).

Activity | What is momentum?

Ask three or four of your friends to help you push a garden roller, the kind of roller that is used for preparing a cricket pitch. How quickly can you get it to stop? Alternatively, try to get a heavy-laden trolley or box cart moving. Now try to make it stop suddenly.

Both tasks are very difficult even with a lot of help. This is because once it is moving the roller has a lot of **momentum**. This is a combination of mass and velocity.

Momentum = mass × velocity

A large momentum requires a large force to overcome it.

Have you noticed how easily cars can be wrecked even when travelling at low speeds? The large mass of the car will produce a large momentum even at low speeds, and will therefore require a large force to overcome it. When a car crashes into a wall, its high kinetic energy is expended either in demolishing the wall or in bending and twisting the metal of the car (or both).

The greater the mass of a vehicle, the greater the damage it can cause in an accident. Smaller vehicles, because of their smaller mass, can accelerate and decelerate more quickly. They are thus more manoeuvrable and easier to control.

Activity | How can we calculate momentum?

1. Place one of two trolleys in the middle of a slope, slightly tilted to compensate for friction (friction-compensated slope). When a trolley is gently pushed on such a slope, it will continue moving at the same speed.
2. Attach one end of the other trolley to ticker tape and pass the other end through a ticker-tape timer. Fit a pin to one trolley and a cork to the other. The apparatus is shown below.
3. Give the first trolley a sharp push so that it runs down the track at constant speed and collides with the second trolley and they both move off together.
4. Use the ticker-tape to find the velocity before and after the collision.

$$\text{Velocity} = \frac{\text{distance}}{\text{time}} = \frac{\text{length of tape}}{\text{time measured on tape}}$$

A certain time interval is given by successive dots on the tape. In this case it is 0.02 seconds.

5. Repeat the experiment using one, two and three additional trolleys stacked on top of the stationary one to give two, three and four units of mass respectively after collision. The mass of each trolley is known (in kg), the distance travelled can be measured (in m) and the time in seconds (s) is known. The momentum can be calculated in kg m/s.

Taking into account the limitations of the experiment (for example, not being able to compensate properly for friction), the total linear momentum of the system before collision is equal to the total linear momentum after collision. This gives the important result that the total linear momentum of the colliding bodies is unaltered by the collision. This fact is called the **principle of conservation of linear momentum**. This is one of the most important principles in the branch of science concerned with motion and force, called mechanics.

This type of collision, where two objects collide and stick together, is called a **completely inelastic collision**. The other extreme is a **completely elastic collision**, for example the collision of two billiard balls, or a rubber ball that bounces back up when dropped from a height onto the floor.

Questions

1. Are all forms of energy transferred by particles?
2. How does an object travelling through the air, for example a bullet from a gun, lose energy and come to rest?
3. How could you measure the speed of a trolley other than by ticker-tape? (Hint: think of a tape recorder with an external microphone.)
4. Do perfectly elastic collisions really occur?
5. Why is the law of conservation of linear momentum so important?

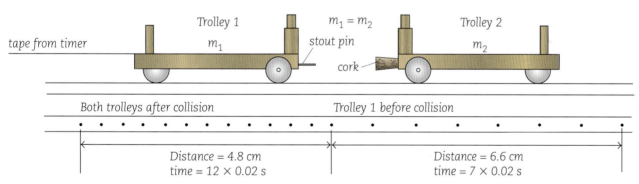

Inelastic collision experiment

Enter your results in a table like this:

	Distance d (metres)	Time t (secs)	Velocity = $\frac{\text{distance}}{\text{time}}$ (m/s)	Momentum (kg m/s)
Trolley before collision				
Trolley after collision				
Two trolleys together				

Does gravity affect balance?

What is gravity?

A ball dropped from a height will fall to the ground. An object attached to a suspended spring causes the spring to stretch. The falling of the ball and the stretching of the spring are both due to a force called **gravity**. The Earth exerts a gravitational force on all objects near it, such as the falling ball. We say that the objects are in the Earth's **gravitational field**. It is the strength of this gravitational field that gives **weight** to matter. Weight is a force. When a 1 kg mass is attracted by the Earth's gravitational field it does so with 10 N of force (9.8 N, to be exact).

The Earth and the planets in our solar system are subject to gravitational forces. They are in the sun's gravitational field. They move around the sun rather like objects being whirled at the end of a string.

Newton developed the law of **universal gravitation**, which states that: *Every particle in the universe attracts every other particle and the force of gravity between two objects is related to their masses (multiplied together) and the distance between them.*

Thus, every particle on Earth acts on every other particle of every object on or above the surface of the Earth. The result of these numerous small forces is a single large force directed towards the centre of the Earth.

All objects have a **centre of gravity**. This is the point through which gravity is directed. In the case of regular geometric objects, the centre of gravity corresponds with the **geometric centre** of the object, like this.

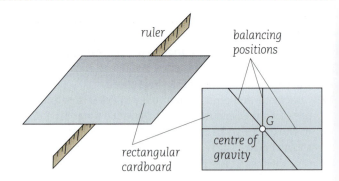

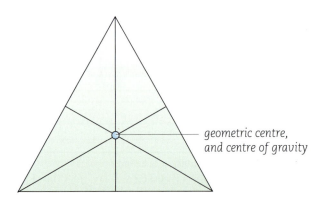

geometric centre, and centre of gravity

Activity | How can you find the centre of gravity of an object?

1. Obtain a rectangular piece of cardboard. Carefully balance it horizontally on the edge of a ruler and mark a line along the edge of the rule where the object is balanced.
2. Rotate the object slightly and find a new position where it is again balanced on the ruler's edge. Mark the second line.
3. The point where the two lines intersect is the centre of gravity. Any other lines along which the object is balanced will pass through this point. If you make a hole through this point, the object will spin freely.

Activity | The centre of gravity of an irregular flat object

The previous method can also be used for irregular flat objects. However, it is not always easy to get the flat object (**lamina**) to balance on the edge of a ruler. There could therefore be a problem in drawing the line correctly through the centre of gravity.

1. Make a hole in the lamina at three different places along the edge as shown below. Suspend it from one hole from a pin held by a clamp. The lamina will come to rest with its centre of gravity along the vertical line.
2. Repeat the procedure with the lamina hanging from a different hole. The centre of gravity is where the two lines intersect. Any other vertical line will also pass through the centre of gravity.

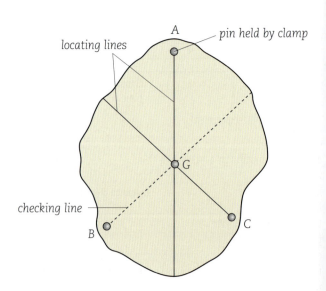

Must the centre of gravity always be within the object?

Find the centre of gravity of an L-shaped lamina (flat object). Where is it located?

Three-dimensional objects, laboratory stools for example, may also have a centre of gravity outside the material of the object.

How does the centre of gravity affect balance?

Stability and instability An object is stable when it returns to its original position after being displaced. An object is **unstable** and likely to fall over when its centre of gravity falls outside the base area on which it was standing.

The stick below is in a state of **unstable equilibrium**. As soon as it is tipped, the centre of gravity is lowered and it falls. The ball has a centre of gravity, which is neither raised nor lowered when it is pushed along. The ball is said to be in neutral equilibrium.

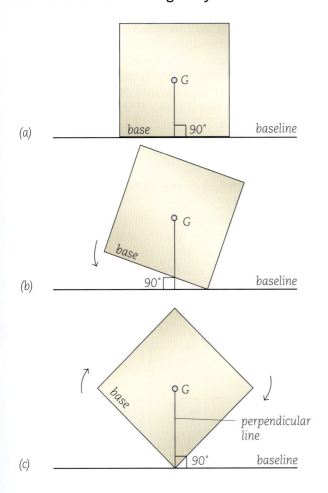

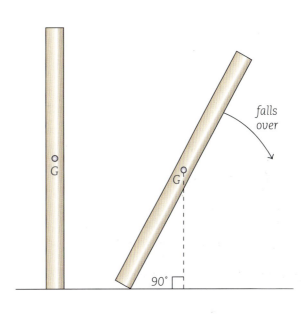

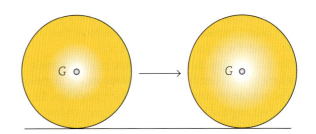

Another word for balance is **equilibrium**. Look at this diagram of wooden blocks. The position of the centre of gravity is shown by the black dot on the block. Part (a) shows the block in a state of stable equilibrium. If the block is tilted slightly as in (b), the centre of gravity (G) is raised. The perpendicular line from the centre of gravity falls within the base area of the block. If the block is let go, it will fall back to its former position at (a). However, if the block is tipped so that the vertical line from its centre of gravity comes outside the base (of the block) it will fall over as in (c). In (c) it is in a state of unstable equilibrium.

Questions

1. Is gravity a property of all matter?
2. Explain why gravity on the moon is weaker than gravity on Earth.
3. How can you tell whether an object will balance? Hint: consider the centre of gravity.
4. 'Objects identical in shape and size but made of different materials have their centres of gravity in the same place.' Discuss.

movement | *movement on land, in water and in air* | *does gravity affect balance?*

Which objects are most stable?

Activity | Is stability related to the centre of gravity?

1. Make an adjustable inclined plane. A temporary one can be made by using tape and two pieces of cardboard, while a more permanent one can be made from plywood using a hinge between the two boards.

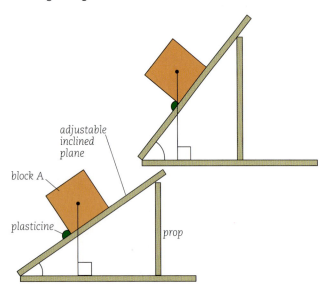

Adjustable inclined plane – left: stable block   right: unstable block

2. Use the inclined plane to find the angle of tilt of a block, i.e. the angle to which you can tilt a block before it falls over. Place a wooden block, block A, on the plane. Put a small piece of plasticine at the front edge of the block to prevent it from sliding.
3. Raise the upper portion of the inclined plane gradually and hold it in position with a suitable prop. This increases the angle of the inclined plane.
4. Continue raising the slope until you reach the angle where block A just topples over. Measure the angle using a protractor. Try another block. Which block has the greatest angle of tilt? Which block is most stable?

Activity | Does mass affect stability?

1. Obtain some blocks with the same dimensions but made from different materials. Try blocks of wood, aluminium and iron. Do the masses differ? Measure the angle of tilt of each block. If blocks are not available, make your own blocks using matchboxes filled with sand, marl, clay, iron filings, sawdust, lead shot, etc. Find the angle of tilt of each block using the adjustable inclined plane. Enter the results in a table. (Note that if the blocks are not cubes, they must all be positioned in the same way on the inclined plane in order to compare the angle of tilt.)

Activity | Does height affect stability?

1. Make up blocks of different heights by taping blocks of the same base area together. Use blocks of the same material. Find the angle of tilt as before. Enter your results in a table.
2. Which block is the least stable? Which block is the most stable? Is height related to stability?

Activity | Does the distribution of mass in a column (stack) affect stability?

1. Take two pairs of blocks of two different densities but with the same measurements. Arrange them in different sequences as shown below. X and Y have different densities (X is denser than Y). Tape blocks X and Y together. Draw diagrams to show the arrangement of blocks X and Y in your stack. Find the angle of tilt as before. Enter your results in a table.
2. Which stack of blocks is the most stable? Which stack of blocks is least stable?

Does the position of the centre of gravity affect stability?

Look at the picture below. Stack 3 is 'top heavy' with two heavier blocks on the top. This stack is the least stable. The centre of gravity will quickly fall outside the base when the column is tilted. The most stable stack is stack 2, which has an arrangement of blocks the opposite of stack 3. The heavier blocks are at the bottom and the centre of gravity is therefore nearer to the base.

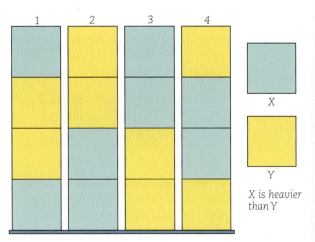

Imagine trying to balance a cone on its tip. Even if you could do this temporarily, the slightest vibration or draught would make it fall over. We say a cone on its tip is in **unstable equilibrium**. However, a cone on its base would not fall over easily. We say a cone on its base is in **stable equilibrium**.

Is stability important in loading vehicles?

When loads are being transported, it is important to think about the distribution of the load so that the vehicle does not topple over. The centre of gravity is kept as low as possible. In order to do this the loader thinks about (a) the maximum height of loading, (b) the density of the goods being loaded, and (c) the way the goods of different densities are distributed. The denser materials are best placed at the bottom near the base and the less dense materials are best placed at the top. The type of terrain over which the vehicle is travelling is important because unevenness and slopes in the road can upset the equilibrium. Just as steep slopes and rough terrain affect land vehicles, so rough seas affect boats, and attention must also be given to the loading of boats. Heavy loads on deck tend to shift the centre of gravity upwards, making it more likely that the boat will capsize.

What is the maximum loading weight allowed?

In most Caribbean countries, vehicles that transport loads are allowed to do so only after being issued with a special permit. They have to carry a sign specifying the mass of the vehicle and the carrying capacity in kilograms. For passenger vehicles, the number of people is limited.

Tare refers to the mass of the vehicle, **net** (or **nett**) to the cargo carrying capacity and **gross** to the total mass of the vehicle and cargo. There are limits to the quantity of material that can be carried to ensure that the centre of gravity is not raised beyond safe limits. Too many passengers in a bus will also have the same effect: that of raising the centre of gravity and causing instability over rough roads and around corners. However, there are other reasons why load may be restricted: too much weight over a small area of the road could damage the surface, and the power of the engine and strength of the brakes must be matched to the load transported.

Goods vehicle permit on the side of a lorry

Questions

1. How does the centre of gravity affect the stability of an object?
2. Do standing passengers affect the stability of a bus?
3. 'In proportion to their numbers, tractors cause as many deaths as cars.' Discuss this statement.
4. Does speeding affect stability?
5. How are vehicles constructed to enhance their stability?
6. How are pieces of furniture constructed to enhance their stability?

Loading ships. How should the load be distributed so the ship does not capsize?

How is movement opposed?

Does friction oppose movement?

Newton devised the law which states that: *A moving object will continue to move unless an opposing force causes it to stop.* He also said: *An object at rest will remain at rest unless an applied force causes it to move.* One such force, which opposes the motion in moving objects and prevents motion in stationary objects, is **friction**.

If one object is on top of another, for example a block of wood on top of a wooden bench, the object will remain stationary unless it is given a push or pull by some force. It can be shown by experiment that the object will only move and the two surfaces will only begin to slide over each other when a precise, measurable amount of force is applied.

In the following Activity, you will find the force which is needed just to overcome the frictional force and allow movement to begin.

Activity | How can we measure frictional force?

1. Place a small rectangular block on a horizontal board near the end of the bench. Tie a length of string to a hook or a screw eye on the block. Let the string pass over a free-running pulley (as in the diagram).

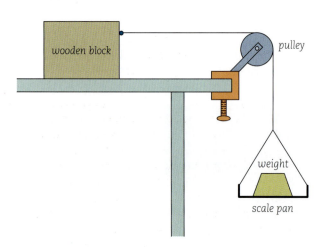

2. Add a few grams at a time to the scale pan until the block first moves. When this happens, the force of friction F (in newtons) has reached its maximum or limiting value and is equal to the total force exerted by the pan and weight. This force is called **static friction**. The total force applied is equal to the weight (in newtons) of the scale pan plus the added weight.
3. Find the mass of the wooden block.
4. Repeat the experiment and find the frictional force with other masses on top of the wooden block. Enter your results in a table.

R (newton)	F (newton)	F/R = μ_s
1		
2		
3		

Take the average of three readings.

$$R \text{ (in newtons)} = \frac{\text{mass of block in grams (+ mass on top, if any)}}{100}$$

$$F \text{ (in newtons)} = \frac{\text{mass of pan in grams + contents}}{100}$$

$$\frac{F}{R} = \mu_s$$

μ_s is called the **coefficient of static friction**.

Using the method above, it is possible to find the coefficient of static friction (μ_s) for many different substances. Repeat the Activity using blocks of other materials, for example rubber and metal, sliding them on different surfaces.

Kinetic friction

It is possible to measure the frictional force when an object is already moving, rather than at rest. The same apparatus as before is used, but the procedure is to give the block a slight push every time weight is added to the pan. In this way, you find the smallest value of the force that can make the block continue to slide steadily at constant speed. The results can be used to calculate the **coefficient of kinetic friction** (μ_k).

$$(\mu_k) = \frac{F}{R}$$

In most cases, μ_s is greater than μ_k, which means that a greater force is required to make an object move than to sustain the movement once it has started.

Can frictional forces be altered?

If you put drops of water between a wooden block and the surface beneath it, the nature of the surfaces in contact is changed. Instead of wood being in contact with wood, water is in contact with water. Similarly if oil is used the interacting surfaces are changed.

Question

1. Can you design an experiment to test whether or not oil, water or sand reduces friction?

How can friction help us?

Many accidents are caused because of over-smooth road surfaces or because of loose gravel chips on the road. Vehicles skid very easily on such roads. Can you explain why friction is low?

Properly made and maintained roads have few loose chips and good drainage to remove water. Roads may also be roughened to promote good contact between the tyre and road surface. Most modern tyres are made with specially designed treads to enable a vehicle to maintain good frictional contact in adverse conditions, for example on wet roads.

Friction enables you to walk without continually slipping and falling over. Friction enables drive belts in motors to work to transmit power and also enables brakes to work.

When is friction undesirable?

Frictional forces are undesirable when they cause breakdown through excessive rubbing of mechanical parts in machinery. Fluids (gases and liquids) also offer resistance to objects moving through them due to friction with the molecules. This resistance is known as **drag**.

Activity | *How is friction reduced?*

1. See if you can find pictures of objects that move swiftly through fluids, including water, and air. Draw or if possible paste pictures of these objects in your notebook. Also look at pictures of slow-moving vehicles, for example lorries, for comparison. Try to find out the speeds of all these objects. Do they have similar shapes?

Supersonic and subsonic flight Supersonic means travelling *faster than (the speed of) sound.* **Subsonic** means travelling *slower than sound.* Look at pictures of different types of aircraft. Notice the difference in shape between the subsonic and supersonic aircraft. There are special problems associated with supersonic aircraft. They create a cone of disturbed air called a **shock wave**, which trails from the nose, leading edges of the wings, the tail fin and other points. Aircraft designers find that by sweeping the wing back, the onset of shock waves is delayed. This is why supersonic aircraft, for example the **Concorde**, have thin triangular wings called **delta wings**. These wings also create less drag, or resisting force, at supersonic speeds.

Streamlining Streamlining makes it easier for objects to move through fluids (liquids or gases). It reduces the resistance or drag of the object as it flows through the medium by encouraging flow over the surface.

Activity | *The effects of streamlining*

1. Make this balloon rocket.
2. Take the shape of aircraft or submarines into consideration and then stick various paper shapes onto the rocket. Go outside and string up the rocket between two poles so that it can run freely on the wire.
3. Sketch the overall shape of the balloon rocket with its attachment.
4. Plan an experiment to record the distance travelled. Work out how to compare rockets with different shapes attached to each. Which shape makes the rocket move furthest?

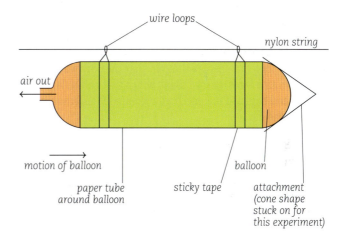

The greater the resistance to airflow, the greater the effort (force) needed to push the object along and therefore the greater the demand for energy. For advanced aircraft, streamlining is of paramount importance as high resistance to movement at these speeds will produce considerable stresses on parts of the aircraft and will also reduce fuel economy.

Are fish streamlined? Some fish, for example those adapted to life on the reef, have very strange shapes. They are not well streamlined as they may be camouflaged and this protects them from pursuit by predators. For objects moving through air, only the very fastest objects have a lot to gain by being streamlined. But in water, which is a much denser medium than air, even the relatively slow-moving fish have much to gain from being streamlined (see page 142–3). Thus migratory, ocean-going fish and sea mammals are very well streamlined.

Questions

2. What is (a) static friction and (b) kinetic friction?
3. When is it desirable to (a) increase friction and (b) decrease friction?
4. Do spaceships need to be streamlined?
5. What is the most streamlined shape?
6. Why are supersonic aircraft shaped differently from subsonic aircraft?
7. Some reef fish, for example the puffer fish, are not streamlined. Why do you think this is so?

How do objects and animals fly?

What are the problems of flight?
There are two basic problems to overcome: (1) the effect of gravity and (2) the effect of frictional forces.

People have always admired animals that fly. For thousands of years human flight was the ultimate challenge. It was only in the last century that the principles of flight were sufficiently understood so that the problems could be solved. In the early part of the 20th century, scientists firmly believed that flight by machines heavier than air was impossible. But in 1903 the Wright brothers proved that this was not so.

How do birds fly?
For a very long time people thought that the only way to fly was to imitate birds, i.e. by having wings that could flap. Birds, insects and bats use this method to create lift and movement through the air.

The **downward stroke** of the wing has the maximum resistance to the air. It creates a forward movement by a backward thrust. Then the **upward stroke** happens and air passes between the wing feathers. The wings of birds are shaped to be convex at the top and concave at the bottom and airflow over these surfaces creates a **lifting force**.

The bird's wing movement is a complex one. People tried to imitate birds by building wings that could be attached to their arms, but these flights always ended in failure.

A pigeon has powerful flight muscles in its wings

These birds are usually grounded in the early hours of the morning but as the land heats up, it heats the air above it and soon rising air currents called **thermals** are produced. Large birds such as hawks seek out these thermals and use them to gain height. Birds can be carried up into the sky for several hundreds of metres on these thermals, and they can then easily glide towards their destination.

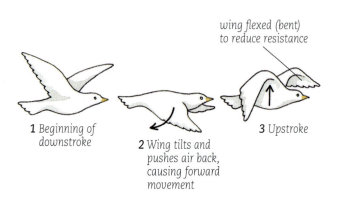

Downstroke and upstroke of flapping flight

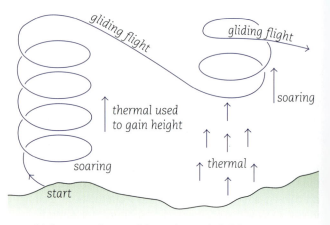

Large birds may make use of thermals to gain height

How is a bird's body adapted for flying?
Think of a typical flying bird, for example a pigeon. The body is **streamlined** for movement through air, properly **balanced** and has a **strong**, **light skeleton**. The most powerful muscles in the body are the flight muscles, which are attached to the wings. Larger birds need a longer wingspan. Long wings are very difficult to flap, and as a result larger birds, like hawks and vultures, rely less on **flapping** flight and more on **soaring** and **gliding**.

Lighter-than-air flying machines
Leonardo da Vinci, a 15th century Italian artist, engineer and scientist, had many good ideas about flight. During his lifetime, lighter-than-air machines were developed, first **hot air balloons** and later **hydrogen balloons**. Buoyancy provided the necessary upthrust to overcome the force of gravity. In the early part of the 20th century **airships** were developed, until a series of disasters lead to their almost total abandonment. Fortunately, at this time heavier-than-air machines were being developed, namely the **aeroplane**, which makes use of airflow to create lift.

Activity | How does airflow create lift?

1. Tape a piece of string to each of two table tennis (ping pong) balls and suspend them so that they are about 4 cm apart.
2. Blow between the two balls.
3. Do they move apart? Do they come together? Can you explain your results?

As air is pushed quickly through the gap, there is less air than usual between the balls. This means that air pressure between the balls is lower than outside, so the outside air pushes the balls together. You probably expected them to move further apart. If there were rapid airflows at the top of each ball, wouldn't you expect a lifting force to be produced?

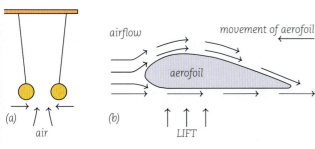

How airflow moves the ping pong balls

An aerofoil – how airflow can create lift

Aeroplane wings The shape above (b) is a cross-section of a typical aircraft wing. This is called an **aerofoil**. The small arrows indicate the airflow over the wing as it moves through the atmosphere. Note how the airstream divides at the front of the wing. The air around the top of the wing has furthest to travel, following the contours. It reaches the back of the wing at the same time as the air flowing along the bottom of the wing. Air going over the top of the wing must therefore move much more rapidly. This difference in speed creates a difference in air pressure and this in turn creates the lift – just as in the case of the table tennis balls in the Activity. Air pressure is greater below the wing than above, so it is lifted.

For sufficient lift, a high forward speed must first be reached. This explains why aircraft must thunder quickly along the runway before they can take off. The propellers screw their way through the air, producing forward motion.

Racing cars The body of the modern racing car has the basic aerofoil shape. So, when these cars travel at high speed, lift is generated by rapid airflow over their tops. The result is that the tyres grip the road less firmly, therefore the control and steering of the vehicle is affected. This is why a small upside-down aerofoil is usually fitted to the top of most racing cars, to counteract the lift generated by the shape and movement of the car.

How do jet aircraft fly?

The way jet aircraft propel themselves forward is similar to the balloon rocket (page 297). The backward movement of air from the balloon produces a forward thrust to push the balloon forward. This follows Newton's third law of motion: *For every action there is an equal and opposite reaction.* We say that the balloon moves by the **action-reaction principle**.

The simplest type of jet engine used in some aircraft, is the ram jet type. Air is taken in, mixed with fuel, and ignited. The rapidly expanding gases leave the engine, creating a forward thrust. In some types of propeller plane a jet engine is used, but the expanding gases drive a turbine, which has the shaft attached to a propeller, which provides the thrust.

In jet planes, lift is generated by the airflow over the wings as previously described. In this type of aircraft, very high speeds are attained by paying special attention to streamlining to reduce air resistance.

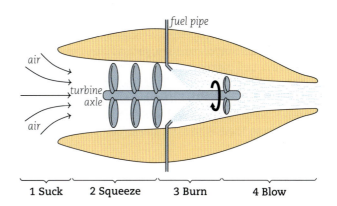

How a jet engine works:
1. **Suck** Air sucked in.
2. **Squeeze** Air squeezed by compressor fans.
3. **Burn** Fuel burns continuously.
4. **Blow** Burnt gases blow out of the back and push the engine forward; they also turn the turbine which keeps the compressor valves going.

Questions

1. How is lift created in lighter-than-air machines?
2. How is lift created in (a) an aeroplane and (b) a helicopter?
3. How are birds adapted for flight?
4. How does streamlining assist flight?
5. How is forward motion achieved by different types of aircraft?
6. Why are larger birds dependent on soaring and gliding flight?

How can we travel in space?

Activity | How could we escape from Earth?
1. Tie a rubber bung *securely* onto a 2 m length of string. Find an open field well away from glass windows.
2. Whirl the string above your head. Steadily increase the speed of whirling. Does the pull on the string increase? How can you tell?
3. Whirl it faster. Notice how the pull increases.
4. If it is safe, let go of the string and observe the path of the bung as it moves away. How does the bung move?

The string can be compared with the gravitational force holding a rocket in space 320 km above the Earth, orbiting Earth and travelling at a speed of 28 000 km/h. At this speed, the motion of the rocket just balances the pull of gravity. If the rocket increased its speed, it would overcome the pull of gravity and speed out into space.

What is it like in space?
In space, objects are weightless and do not fall. You cannot drop anything and there is no right way up. Instead of air there is a perfect or near-perfect vacuum. There is also a considerable amount of radiation, such as ultraviolet, X-rays, gamma rays, and cosmic rays. Most of this radiation comes from the sun or other stars. Most of these rays do not reach Earth because they are filtered by the atmosphere. In space, temperatures can rise very high in direct sunlight or fall very low in shadow. There is no night and no day; only light and darkness.

How can people survive in space?
The conditions in space are hostile and so unlike what people are accustomed to on Earth that humans cannot survive without assistance. (In the same way people cannot survive the strange and hostile conditions at the bottom of the sea, see pages 256–7). So, space travellers need a specially created suitable environment in a spaceship or spacesuit, with Earth-like conditions.

Lack of gravity Most astronauts are fascinated by weightlessness in space. They seem to enjoy performing all kinds of acrobatics that are not possible on Earth. Unfortunately, prolonged weightlessness affects the body. Astronauts returning from long periods in space are usually physically weak, and have to be helped out of their spacecraft. The human body needs to have the stresses caused by gravity in order to remain adapted to normal Earth conditions. Once these stresses are removed, adverse changes in the body begin. Calcium is lost and the bones become weaker. The muscles all over the body, including the heart, degenerate because they have less stress on them in space and therefore have to work less hard. However, this loss of calcium and muscle tissue can be reduced by exercise. Exercise in space is now part of the regular routine for astronauts. The wasting of muscles and the weakening of the skeleton is a common condition in older people here on Earth. From studies on astronauts, scientists now think that the lack of adequate exercise in elderly people may be one of the main causes of this degeneration.

Protection from space Astronauts need to leave their spacecraft to carry out certain tasks such as performing experiments and making repairs to the spacecraft. Spacesuits with a life-support system and protection from radiation enable them to spend hours outside the spacecraft.

How must a spacesuit be designed to protect an astronaut?

How are spacecraft powered?
Spacecraft make use of a **rocket engine** for propulsion. The principle is the same as for the jet engine (see page 299). Hot gases are produced by burning fuel in a combustion chamber. The sudden outrush of these gases pushes the spacecraft forward. In jet aircraft, oxygen is obtained from the air. Spacecraft have to fly beyond the atmosphere in space and therefore have to carry their own supply of oxygen for combustion.

In practice, rockets are multi-stage: several rockets are mounted one on top of the other. Each stage rocket has its own fuel tanks and engine. When the fuel from the first stage is used up, it is discarded and the second stage fires. The whole rocket becomes lighter as it travels upward and can reach a far greater speed and height than single-stage rockets.

In July 1969, after only eight years, the United States Apollo programme successfully achieved its objective of putting a man on the moon. The Saturn Apollo rocket they used weighed over 3 000 tonnes at launch but only a tiny part, the **Command Module**, completed the journey and returned to Earth.

In 1981 a new development in space travel came into being with the launching of the **space shuttle**. This has a major advantage over rockets in being extremely versatile and also reusable. It was able to launch satellites and retrieve them for repair. In January 1986 the programme was suspended following a disaster. The programme was later resumed.

The space shuttle lifting off

How do spacecraft find their way?

Navigation in space involves (a) **maps**: working out suitable routes for the spacecraft, (b) **location**: continually checking the actual flight path during the course of the journey and (c) **communication**: making corrections to the flight path.

Maps Maps for space flight are complicated, they are constantly changing because the planets are constantly moving relative to each other (see page 302) and so they have no fixed position. Maps using the sun and the stars are employed.

The compilation of an accurate map has been done with the help of very advanced technologies. These include large radio telescopes, ultra-sensitive radar receivers and sophisticated signal detectors able to detect radiation as weak as 2×10^{-21} watt (equivalent to the radiant power from a lighted match on Mars).

Location Knowing the precise speed of the spacecraft and the time and direction of travel, the exact location in space can be calculated. Radar is vital to navigation in space but this in turn is extremely dependent on accurate time measurements. Clocks capable of measuring nanoseconds (10^{-9} seconds) over several hours are used. Radio pulses are sent out to the spacecraft, received, amplified and retransmitted. As the rocket moves relative to Earth, there is a change in signal and a difference develops between the transmitted frequency and the received frequency. This shift is like the shift in frequency of a car horn as it goes past. The shift is known as the **Doppler shift** and the timing of the signal when it gets back to Earth can supply vital information about the speed and position of the spacecraft.

Communication As a spacecraft travels higher and further, not only does it become more and more difficult to send and receive signals, but communication also becomes delayed. The exact location of the spacecraft becomes less well known back on Earth. For these reasons, spacecraft have navigation computers and control systems, which make immediate adjustments to their flight paths. They also employ hybrid (i.e. combination) guidance systems, which enable them to use the sun and recognise star patterns for navigation.

Questions

1. How could astronauts exercise in space?
2. Why is the heart affected by weightless conditions?
3. What is the advantage of the space shuttle over other rockets?
4. Is space navigation different from terrestrial navigation?
5. What is the Doppler shift?

Where are we in space?

To describe where we are in space we can make a diagram like the one on the opposite page. We start from the country in which we live, and then work up.
(a) We live in one of the countries of the Caribbean region.
(b) The Caribbean region lies in the northern hemisphere of the Earth between 22° N and 1° N, and stretches from 89° W (Belize) to 56° W (Guyana).
(c) The Earth is our **planet** and has a natural **satellite**, the moon, which travels around it in an anti-clockwise direction.
(d) The Earth is one of the planets that travels around the sun, which is a **star**. Each planet has a particular path which is called its **orbit**. Between the orbits of Mars and Jupiter are many small rocky planets called **asteroids**.
(e) The nine planets and their satellites, the asteroids and the sun make up our **solar system**.
(f) This is only one of the solar systems that forms a part of our **galaxy**. Our solar system is actually on one side of our galaxy in one of the 'arms'. We can see part of our galaxy in the sky, we call it the Milky Way.

Our galaxy is only one of the very many that are in the **universe** which extends far further than we are able to observe with the most powerful light telescopes and radio telescopes. In the universe there are many millions of galaxies, solar systems and planets.

The planets of our solar system
The planets can be divided into two main groups (see below). The inner four terrestrial planets are small, rocky and have a high density. The next four gaseous planets are much bigger, consist largely of liquid and solid hydrogen and have a low density. Pluto is in a group of its own.

Our sun is an average-sized star. It appears large to us because it is much closer than other stars. Stars produce heat and light: **solar energy** (pages 194–5).

Most of the planets have satellites (moons) which orbit around them. These are **natural** satellites. The Earth also has **artificial** satellites such as weather and communication satellites that humans have put into orbit (page 275).

Activity | Making a model of the solar system
1. Eleven students should make labels: the Sun, Mercury, Venus, Earth, Moon, Mars, Jupiter, Saturn, Uranus, Neptune and Pluto. The whole class can go outside. The 'sun' should stand in the middle. The planets arrange themselves in order and walk clockwise around in their orbits – this is easier if you make chalk lines first to show the path of each planet. The moon should walk anti-clockwise around the Earth. The rest of the class act as 'asteroids' and orbit between Mars and Jupiter.
2. You can also make circular discs, or plasticene balls to represent the planets. Use the table below and let 1 mm represent 1 000 km. The diameters will be approximately: Mercury 5 mm, Venus 12 mm, Earth 13 mm, Mars 7 mm, Jupiter 13.8 cm, Saturn 11.4 cm, Uranus 5.2 cm, Neptune 5.0 cm and Pluto 3 mm.
3. Mark a position in the classroom to represent the surface of the sun. If we use a scale where Pluto is 1 m away from the sun then the other planets are approximately these distances away: Mercury 1 cm, Venus 2 cm, Earth 3 cm, Mars 4 cm, asteroids 7 cm, Jupiter 13 cm, Saturn 24 cm, Uranus 50 cm and Neptune 80 cm.
4. Attach cotton threads to the planets and suspend them the correct distances from the sun. (Note that the distances given are the *average* distances from the sun.)

Some characteristics of the planets

Planets	Diameter at equator (km)	Average distance from the sun at equator (millions of km)	Approximate time to orbit the sun (in earth years)	Number of satellites (moons)	Mean temperature (°C)	Atmosphere
Mercury	4 876	58	$\frac{1}{4}$	none	350 (day) −170 (night)	None
Venus	12 102	108	$\frac{2}{3}$	none	480	Mainly carbon dioxide
Earth	12 756	150	1	1	22	Mainly nitrogen and oxygen
Mars	6 794	228	nearly 2	2	−23	Carbon dioxide
Jupiter	142 016	778	12	16 and rings	−150	Mainly hydrogen and helium
Saturn	120 000	1 427	$29\frac{1}{2}$	30 and rings	−180	Mainly hydrogen and helium
Uranus	52 400	2 870	84	15 and rings	−210	Mainly hydrogen and helium
Neptune	48 600	4 504	169	8 and rings	−220	Mainly hydrogen and helium
Pluto	3 276	5 900	248	1	−230	None detected

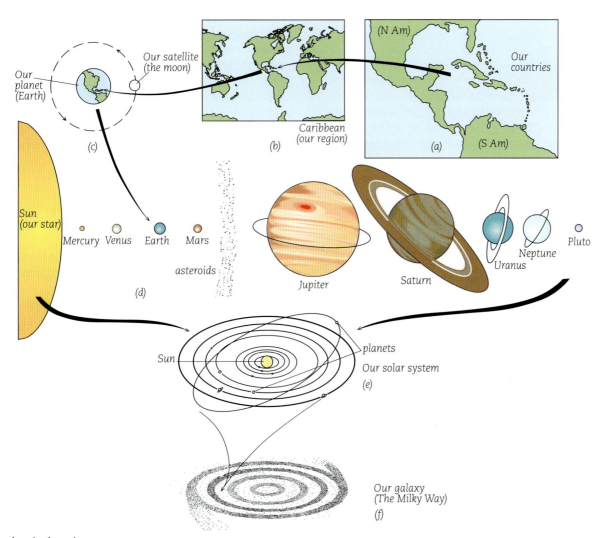

Our place in the universe

How do planets keep in orbit?
Planets are attracted by the force of gravity of the sun, so they do not fly off into space. The planets move in orbits around the sun (see below; also see page 300).

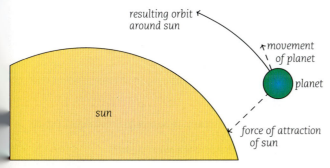

A planet in orbit

Questions

1. Define and give examples of the following:
   (a) natural satellite (b) planet (c) star (d) galaxy.
2. Which planet or planets:
   (a) is/are similar to the Earth in size?
   (b) is/are closer to the sun than Earth?
   (c) take more than one earth year to orbit the sun? Why?
   (d) are on either side of the asteroids?
   (e) are called the gaseous planets? Why?
   (f) are hotter than the Earth? Why?
3. (a) Why is life like ours possible on Earth?
   (b) Do you think life like ours exists on other planets in our solar system? Why? How could we find out?
   (c) What problems do humans face in trying to find out about life on other planets?

How is Earth affected by other bodies?

The Earth is a planet in the solar system (page 302) and it has its own natural satellite, the moon. The movements of the Earth and the moon cause changes which we can observe every day, every month and every year.

The Earth is tilted on its axis. It travels around (**orbits**) the sun. Because of the tilt, the southern hemisphere is closer to the sun than the northern hemisphere at certain times of the year, and vice versa at other times. This gives us the **seasons**.

The Earth also spins (**rotates**) on its axis. This means that at different times during 24 hours the sun shines on different parts of the Earth, and gives us **day** and **night**.

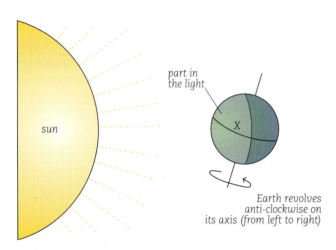

Day in the Caribbean

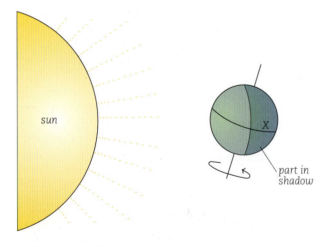

Night in the Caribbean

The Caribbean is marked with an **X** on the Earth. In the top picture **X** is on the side of the Earth which is receiving light from the sun, and it will be day in the Caribbean.

In the bottom picture the Earth has rotated halfway round. **X** is now in shadow and it is night in the Caribbean.

Phases of the moon

The moon travels around the Earth in an anti-clockwise direction. It makes one complete orbit in about 28 days (a **lunar month**). The moon does not produce light of its own. We can only see the moon because it reflects light from the sun. At different times of the month different parts of the moon are lit up by the sun. Then, on Earth, we see the sun's light reflected from these parts of the moon. So the moon appears to be different shapes during the month and we call these the **phases of the moon**.

Activity | Observing the moon

1 On a page in your notebook make a drawing of the shape of the moon you see every evening for 30 days. Write the date next to each drawing.
2 If the evening is very cloudy, or if you do not see the moon, then leave a blank, and make the next drawing for the next evening.
3 When you have completed your drawings compare them to those shown below and try to identify the different phases of the moon.

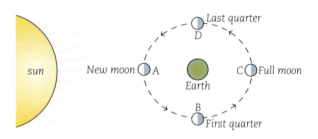

In position **A** the sun is shining on the side of the moon which is facing away from the Earth. So we cannot see the moon from the Earth: this is **new moon**.

In position **B** the sun is shining on the right side of the moon as seen from Earth. When we look at the moon we see ☽ this is the **first quarter**.

In position **C** the sun is shining on the side of the moon which is facing towards Earth. When we look at the moon it is all lit up ○ this is **full moon**.

In position **D** the sun is shining on the left side of the moon. When we look at the moon we see the left side lit up ☾ this is the **last quarter**.

Questions

1 Your younger brother says he sees the 'moonlight'. Explain to him what he is seeing.
2 What does 'phases of the moon' mean?

Tides

Tides refer to the distance that the sea comes up the shore. At high tide the sea comes far up the shore, whereas at low tide it is the farthest out. Tides have effects on shore organisms (pages 306–7).

Tides are due to the pull or attraction (**force of gravity**) of the moon and the sun. Because the moon is much closer to the Earth than the sun, the moon's effect is more than twice as important as the sun's.

As the Earth spins on its axis the moon will be overhead at a certain time in the day. It will pull on the water in the oceans and cause high tide on that side of the Earth. There will also be high tide on the opposite side of the Earth at the same time.

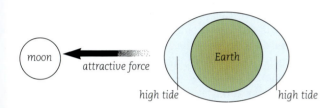

Formation of tides

Approximately 12 hours later, the moon will be overhead on the opposite side of the Earth, and again causes high tide. There are two high tides and two low tides every 24 hours and 50 minutes.

When the pull of the moon and the sun are more or less in the same direction there will be the greatest difference in tides. The attractive force of the sun is added to the attractive force of the moon. This happens every month during new moon and full moon. These are called the **spring tides**: the high tide is highest and the low tide is lowest.

When the sun and moon are roughly at right angles, their attractive forces oppose each other and there will be the smallest difference in tides. The attractive force of the sun is subtracted from that of the moon. This happens every month during the first and last quarters. These are called **neap tides**: the high tide is lowest and low tide is highest.

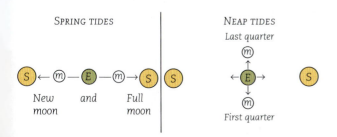

Waves

The surface of the Earth is approximately 70% water and most of this is in large ocean basins. The water is in continual motion to form **crests** (the tops of the ocean waves) and **troughs** (the hollows in between).

On windy days the waves are highest. The heights of the waves and how fast they travel are related to the speed of the wind and the distance of sea over which the wind has been blowing. Thus bigger waves are formed on the Atlantic-coast side of the islands than on the coasts facing the Caribbean Sea (page 264). Also, bigger waves will form during a hurricane (page 312) than during calm weather.

As the waves reach closer to the shore, where the water is shallow, they become higher and hollow in front. The waves collapse or **break** onto the shore with a lot of force. (See also page 264.)

Effects of moon phases and tides on organisms

Plants Calendars have been produced for gardeners and farmers which link the phases of the moon with different activities, but these ideas are not necessarily scientific.

The new moon is said to be a time when growth is greatest in the top of the plant and that this is a good time for taking cuttings.

Once the crescent of the new moon is visible seeds can be sown. In the first quarter grain and seed crops can be sown. The full moon is a time when root systems store food. It is a good time to harvest above-ground crops and to plant root crops. The last quarter is the resting period which is best for pruning and spraying.

Animals Many shore animals are affected by the tides which vary with the phases of the moon. Female grunion fish produce eggs just after the high point of a spring tide. The eggs are buried in the sand for two weeks until they are washed out by the next spring tide in time for hatching. Some corals also have a monthly reproductive cycle. There are also daily cycles, for example, limpets actively feed every $12\frac{1}{2}$ hours at high tide when water covers their part of the shore. They avoid day-time low tide when they might dry out.

Questions

3. What effects are associated with (a) the rotation of the Earth (b) the moon orbiting the Earth and (c) the Earth orbiting the sun?
4. Describe a scientific experiment which could be done to test whether the first quarter is a good time to plant grain and seed crops.

How are organisms affected by the sea?

The coastal regions at the edge of continents and islands are continually washed by the sea. What are the characteristics of the sea and what effects might it have?

1. The sea is **saline**, it contains about 3.5 g of dissolved salt in each 100 cm^3 of water.
2. The sea is in constant motion and waves break upon the land with a lot of force.
3. The sea comes in and out (ebbs and flows) with the tides. Organisms must survive total immersion in water, as well as exposure to the sun.

Rivers At the mouth of a river, the **estuary**, the full effects of the sea can be felt. The sea water that comes into the estuary at high tide is very salty: has a high salinity. But when the tide goes out, freshwater from higher up the river flows down, and the salinity drops. Organisms such as crabs and oysters living in the estuary have to be able to live in water with a wide range of salinity.

The movement of the water produces **currents**. Organisms must avoid being washed away. They either live in sheltered pools on the river bank, are attached to rocks, are able to burrow into the soil, or are strong swimmers.

Mangrove swamps Swamps form in the tropics where mud has been brought down to the coastal region. There are currents and changes in salinity.

The mud level is always shifting and changing, and the plants that grow there best are red mangroves (*Rhizophora*), which have **prop roots** (see (a) below). Animals such as mudskippers and oysters live attached to the roots. Also, the mud has a low oxygen concentration compared to normal soil, and the white mangroves (*Avicennia*) are adapted with 'breathing roots' (see (b) below) which can take oxygen from the air.

(a) Prop roots   (b) Breathing roots

Seashore

The seashore can be sandy or rocky. On sandy shores, animals such as crabs and clams burrow into the sand as the tide comes in, and come up to the surface as the water retreats. So they are protected from the force and current of the waves.

Different organisms are found at different places up the shore depending on how much they are covered or uncovered by the high and low tides. We can see the effects of high and low tides best on steep rocky shores where the height of the tides vary a lot (see opposite page).

Splash zone Up near the top of the shore is a place where even the high tides do not reach. The only water that reaches this part is the spray from the waves. This is called the **splash zone**. The organisms that live here are lichens and small sea snails that can close up their shells to protect themselves from the drying effects of the sun.

Intertidal zone Between the area where the high tides reach and where the low tides come to, is a zone called the **intertidal zone**.

Organisms at the *top* of this zone are only covered by the sea at high tide. For most of the day they are left uncovered and are affected by the heating and drying effects of the sun. In this zone there are many sea snails and also limpets which can pull down tightly onto the rock so that they do not dry out.

Towards the *bottom* of the intertidal zone the organisms are covered by water for a longer period of the day. Here there are seaweeds (also called sea moss) which are less resistant to drying out than shelled animals. Crabs and sea urchins will also be found here.

Sub-littoral zone This zone is below the intertidal zone. It is always covered by the sea, even at low tide. Here there are many different kinds of seaweeds, and soft-skinned animals such as sea anemones and sponges.

Rock pools Some of the same organisms as those found in the sub-littoral zone may be found in rock pools at different levels of the shore.

Rock pools are hollowed-out areas in the rocks filled with sea water. The conditions are not exactly the same as in the sub-littoral zone because the water in the rock pools may not be replenished by the sea very often. It will therefore tend to get hot and very salty as some water evaporates.

If you live near a rocky shore you can visit it to collect organisms from the different **zones** and see how they are adapted. The different zones will be most easily identified on steep rocky shores with a large variation between high and low tide levels.

Take plastic buckets and bags, and a sharp knife. Plan to arrive at the shore at high tide, then you can follow the water as it goes down towards low tide.

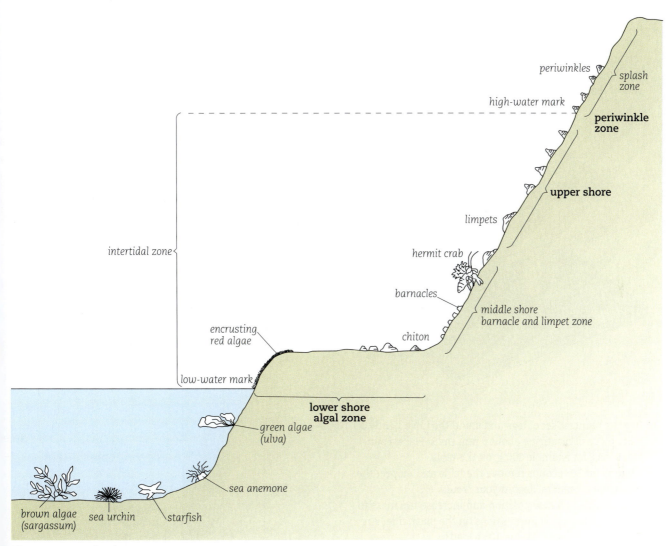

The zones on a rocky shore

High tide and low tide

Approximately every 24 hours and 50 minutes there are two high tides and two low tides. So on each succeeding day the corresponding tides will be a little later than those of the previous day.

For just over 6 hours the water comes in (**flows**) further up the shore until it reaches high tide. Then for just over 6 hours the water goes out (**ebbs**) further and further until it reaches low tide. This is then repeated.

The difference between high and low tides varies greatly around the world, and is comparatively little in the Caribbean. For this reason it is often difficult to see the zones shown above. This is because splash from the waves may produce as much difference in water heights as the tides. (See also page 305.)

Questions

1. Describe how a named organism from (a) an estuary and (b) a mangrove swamp is adapted to the conditions found there.
2. What are the conditions in these areas of a rocky shore? (a) The splash zone (b) The rock pools (c) The intertidal zone How is a named organism from each zone adapted to the conditions found there?
3. What organisms that live on the shore do humans eat? How does the behaviour of the organisms help us to catch them?
4. If it is high tide at 6 a.m. and 6.25 p.m. on a particular day, approximately when will the high tides be on the next day?

How does pressure vary?

Activity | What is pressure?

1. Build a miniature wooden three-legged table as in the picture. Use a 4 cm square piece of thin plywood or polystyrene. Use wooden pencil-sized legs which fit snugly into holes bored for them to fit in the top. Make other sets of legs with ends ranging from blunt to sharp points.

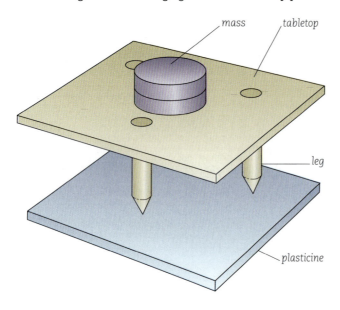

2. Use the bluntest set of legs and stand the table lightly on a flat piece of plasticine about 5 mm thick. Place a suitable mass, for example 100 g, on the table.
3. After about 30 seconds remove the table and inspect the plasticine for marks. Record your results.
4. Put the table in a new position and increase the mass to 150 g. Are there now impressions on the plasticine? Are these greater than with the 100 g load?
5. Repeat the procedure using other loads. Devise a simple way to record your results. Perform a similar experiment, but instead of varying the mass on the top of the table, interchange the legs so that the points in contact with the plasticine vary. Discuss your results in class.

In this Activity, you used masses to exert force. How do you calculate the force exerted by a known mass? The force exerted by a mass of 100 g is about 1 N (see page 292). The forces used were acting on specific areas of plasticine. How can you measure the area of plasticine acted on? Draw up a table for force and area, and force divided by area for each trial. You will find that when force/area is greatest the holes made in the soft plasticine are deepest. This force/area ratio is known as the **pressure**. You will have noticed that the greatest masses used in the Activity do not necessarily lead to the greatest pressure.

Does the atmosphere exert pressure?

If the air is removed from a can or plastic bottle, it will collapse. If a little water is placed in a can and boiled vigorously and the can is then corked and cooled, the can will collapse. This is because the air is first pushed out by the steam, which then condenses to water when the can is cooled, leaving a partial vacuum.

Both of these experiments show that air exerts pressure. You are so accustomed to the air pressure on your body that you hardly notice it. Yet, your ears can detect even very slight changes in air pressure (see also pages 257 and 268).

How is atmospheric pressure measured?

The mercury barometer Look at the diagram below. The space in the capillary tube above the mercury is a vacuum. You might think that all the mercury in the tube would run out into the reservoir, but it is kept in by atmospheric pressure. Atmospheric pressure at sea level is equivalent to 760 mm of mercury. Careful measurement of this column of mercury will show that the column length does in fact vary by very small amounts, as the pressure of the atmosphere varies. This is how the mercury barometer works.

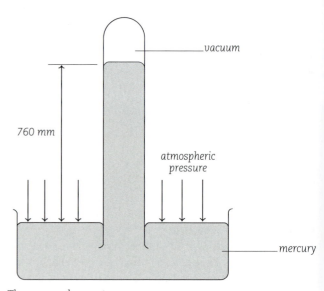

The mercury barometer

Mercury has a density of 13.6 g/cm^3. Other liquids may be used in a barometer but the smaller the density of the liquid the longer the column will be. For example, if a water column was used its length would be 10.3 m!

The aneroid barometer This is a more common barometer. It has no liquid. Look at the picture at the top of the next column. How do you think it works?

Barometers are designed to measure very small changes in pressure.

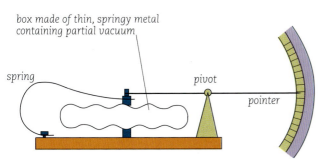

The aneroid barometer

What are the units of pressure?
In the Activity you performed earlier, you could express the pressure exerted, force/area, in units of newtons per square centimetre (N/cm^2). Another more acceptable unit of pressure can be calculated like this:

$$1\ N/cm^2 = 10^4\ N/m^2 \text{ (because } 1\ cm^2 = \tfrac{1}{10^4}\ m^2\text{)}$$

The unit N/m^2 is usually known as the **Pascal (Pa)**.

The average value for atmospheric pressure at sea level is 760 mm of mercury. This is called the **standard atmospheric pressure** or one atmosphere. However, for meteorological purposes the unit of pressure is called the bar and is defined as a pressure of $10^5\ N/m^2$. The **millibar**, which is one thousandth of a bar, is more commonly used:

$$1\ \text{millibar} = \frac{10^5\ N}{10^3\ m^2} = 100\ N/m^2$$

How can air pressure vary?
Land and sea breezes In the atmosphere, changes in the density of air occur due to the heating effect of the sun during the day on the land and the sea. The relative difference in pressures between these warm and cool air masses results in airflows. These airflows are land and sea breezes (see page 115).

Air currents Local heating of air masses, for example by factories, can set up small but important air currents which can disperse pollutants over great distances. These pollutants include toxic gases like sulphur dioxide and microscopic particles of soot as well as relatively large particles of ash dispersed by sugar cane factories in the Caribbean.

Water vapour in the atmosphere is unevenly distributed. This creates pressure differences in the atmosphere because water vapour has a density less than that of dry air. A barometer will therefore show a low-pressure reading in moist air (for example, when rain is expected) and a high-pressure reading in dry air (when fair weather is expected).

How are winds useful? The force exerted by moving air masses (wind) has been successfully exploited by humans and other living things. Wind is used by plants for the dispersal of seeds and pollen, and by birds for soaring and gliding flight. We have used wind to propel sailing ships and to power windmills and wind turbines from which other types of energy may be obtained.

Fluid pressure
Liquid columns The higher the column of a liquid, the greater will be the pressure it can exert. For columns of different liquids of the same height, the denser the liquid, the greater will be the pressure exerted.

Hydraulic brake One of the most important applications of fluid pressure is found in the **hydraulic brake** (see below). Notice the use of levers in the foot pedal and brake shoe, to multiply force (page 226). To apply the brake, the foot pedal is pressed. The hydraulic (liquid) pressure in the system is increased and the wheel pistons are pushed outwards against the brake shoe. The brake shoe makes contact with the revolving brake drum (not shown), which is attached to the wheel, impeding its movement, thus slowing it down or bringing it to a stop.

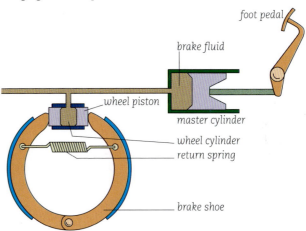

The hydraulic brake

Questions
1. Use the idea of pressure to explain why sharp knives and needles are more effective than blunt ones.
2. A pressure gauge commonly used is the pocket pressure gauge for measuring the pressure of car tyres. Examine one. How do you think it works?
3. Why is mercury a good choice of liquid for a barometer?
4. Explain how barometric readings (of pressure) can indicate changes in the weather.
5. Explain how changes in atmospheric air density can occur and how these changes can bring about air movement.

What causes changes in the weather?

Weather describes the day-to-day changes in the temperature, humidity and movement of air in a particular place. What the weather is like influences what we wear, what we do and how we feel. A study of the weather includes examination of land and sea breezes along the coast (page 115), the cold and wet conditions on the windward sides of the islands compared to the warmer and drier conditions on the leeward sides, and the extreme power of hurricanes which can cause death and destruction.

Air pressure

Underlying many of the changes in the weather, there are differences of air pressure (page 308) of different masses of air. Average air pressure at sea level is 760 mm of mercury (one atmosphere), but for meteorological purposes we use the **millibar** (mb) which is a pressure of 100 N/m^2 (page 309).

Standard atmospheric pressure at sea level is about 1 013 mb. The lowest low pressure (as in the eye of a hurricane) can be below 960 mb.

On a weather map, places which have the same atmospheric pressure are joined together by lines called **isobars**. Isobars are drawn at 2 mb or 4 mb intervals.

Winds

Direction Winds blow from areas of high pressure (more dense) towards areas of lower pressure (less dense). The winds are deflected a little to the right (east) in the northern hemisphere by the spinning effect of the Earth (below).

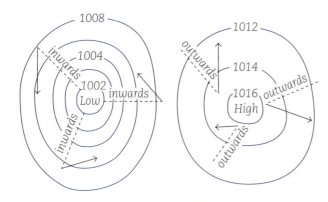

Isobars and winds – Left: low pressure system with a steep pressure gradient. Strong winds tend to blow inwards but are deflected to the right. Right: high pressure system with a gentle pressure gradient. Light winds tend to blow out but are deflected to the right

Speed The closer together the isobar lines, the stronger the wind. The **Beaufort scale** describes wind speed all the way from light breezes to storms and hurricanes.

Clouds

Clouds are floating collections of very small water droplets. They are formed when water evaporates from land, water surfaces and living things, and later condenses. (See also the water cycle on page 139.) There are different types of cloud.
Cirrus Narrow ribbons of thin white clouds high in the sky.
Cumulus Fluffy clouds low in the sky. If they look like separate balls of cotton wool they indicate fine weather. If they clump together from a grey base they may bring rain.
Stratus Sheets of clouds at a medium height in the sky.

A common sight in the Caribbean – cirrus clouds above and cumulus clouds below

If the water droplets inside a cloud join together, the cloud changes from white to grey. These are called rain clouds and they are common, especially in the afternoon, during the rainy season. The droplets become too heavy and fall as rain.

The word *nimbo* or *nimbus* can be used to indicate rain-bearing clouds, thus **nimbostratus** are grey sheets of cloud and **cumulonimbus** (below) are grey balls of cloud.

Rain-bearing cumulonimbus clouds

Fronts

Over the Caribbean we usually have warm, moist air, which has been heated by the sun and has picked up water vapour from the sea (page 309). But colder, drier air may come in, for example from North America. When two kinds of air meet this is called a front. There are four kinds of fronts, and all of them bring rain.

Cold front Where cold air comes into a region of warm air and begins to displace it, it is called a cold front. This is the most common front in the Caribbean.

A cold front is preceded by a rapid fall in pressure, deep dense clouds (often cumulonimbus) and rain. As the front passes there will be heavy thunderstorms and intense rain for 5 to 15 hours, and there will also be a drop in temperature.

The windward sides of hills and mountains always get the worst rain. The leeward sides are protected and get less rain. In the Caribbean the Windward Islands get more rain than the Leeward Islands for the same reason.

Warm front This is when warm air comes into a region of cold air and displaces it. It is not common in the Caribbean. It is preceded by a slow fall in air pressure, by cirrus clouds which change to become thicker and darker, and by drizzle. As the front passes there will be showers for 10 to 25 hours, the wind may change direction and the temperature will rise.

Occluded front Cold fronts travel more quickly than warm fronts and a cold front may overtake a warm front and lift it off the ground. It is said to be occluded (cut off). These fronts are not common in the Caribbean.

Stationary front If a cold front, passing, for example, from North America over the Caribbean Sea, warms up so that the temperature difference between the cold and warm air disappears, then it is called a stationary front.

The symbols used for the different fronts are shown below.

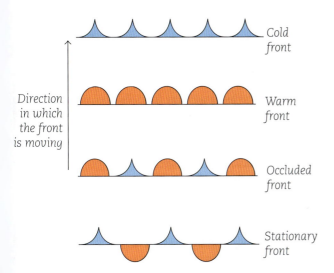

Predicting the weather (weather forecasting)

We can get a good idea of what the weather will be like for the next few hours by looking at the sky. If it is blue with no clouds or only a few cotton wool clouds (cumulus) it is likely to stay fine. If the sky is covered over with balls (cumulus) or sheets (stratus) of clouds which are grey (nimbus) then we are in for some heavy rain or even thunderstorms.

We can predict the weather a little further ahead by looking at a barometer. When the barometer is *falling* (that is when air pressure is decreasing or low pressure air is arriving) there is likely to be rain. A falling barometer is associated with the different fronts which all bring rain. When the barometer is *rising* (that is when air pressure is increasing or high pressure air is arriving) there is likely to be fine weather.

To predict the weather farther ahead we make use of aerial photographs taken by weather satellites. The white areas on the photograph below show rain clouds, associated with an area of low pressure. By examining photographs taken at different times, meteorologists are able to observe the direction and speed at which the low pressure area is travelling and so make weather forecasts.

Aerial photograph of a hurricane. You can see the eye on the centre right

Even longer term effects on the weather can be caused by El Niño: changes in the direction of flow of the currents in the South Pacific (page 319). These can upset weather patterns.

Questions

1. What kind of weather would you predict if it was hot and sticky (33 °C), with a strong wind from the NE and a sky covered with cumulonimbus cloud?
2. Explain how the barometer can help us to predict the weather.
3. Why are weather satellites important?

What are hurricanes?

Hurricanes are an extreme form of cyclonic storm. A cyclone is a region of low atmospheric pressure (page 308) surrounded by rotating winds. The word hurricane comes from Spanish *huracán* meaning 'big wind'. In eastern Asia they are called typhoons from the Chinese *tai fung* meaning 'big wind'. That is the major characteristic of a hurricane: the wind speed is over 120 km/h.

A cyclonic storm, and a hurricane in particular, can form over tropical seas heated to 28 °C. They begin when moist air, heated by the sun, rises from the water (below, left). As the air rises, it cools and condenses into rain. This releases energy which powers the hurricane.

As the air rises upwards, more hot moist air rushes in from all sides to replace it. These winds are deflected to the right so that they come to spiral around a central calm 'eye' (below, right).

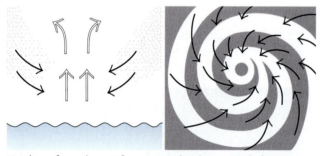

Hurricane formation – Left: warm, moist air over tropical ocean rises. Surrounding air flows in, becomes warm and moist, and rises. Right: later, top view. As seen from above, winds whirl around a calm 'eye' in the centre. Winds are strongest near the eye but calm in the very centre of the eye

Hurricanes in the Caribbean

On average five hurricanes affect the Caribbean each year during the hurricane season from July to November. The damage done by a hurricane is caused by high winds which can blow down buildings and destroy crops, and by flooding which is caused by the torrential rains and the high waves produced in coastal areas.

For example, when hurricane David struck Dominica in August 1979, over 40 people died and 3 000 were injured. The majority of houses and many public buildings were damaged and most of the banana crop was destroyed. Roads were impassable, and electricity and water works were put out of action. Hurricane Allen in 1980 caused widespread damage, especially in St. Lucia, Barbados, Haiti and Jamaica.

Coastal areas are usually the worst hit as the effects of winds, floods and waves cause the most damage in these areas. During a storm, surges of water are the main cause of destruction. Long-term effects are due to loss of revenue because of flood damage, destruction of crops and the expense of rebuilding and of repairing services.

Hurricanes form to the east of the Caribbean and then sweep across in a north-westerly direction (see solid lines on map below). Other unusual tracks are also seen, for example hurricane Flora in 1963 (dotted line).

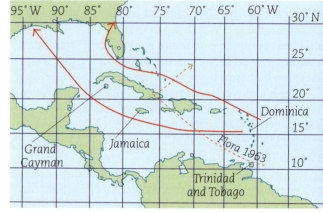

Hurricane tracks in the Caribbean

Hurricane safety measures

Always have a battery-operated radio available (as well as spare batteries) on which you can listen to weather forecasts. A **hurricane watch** indicates your area may be in danger, and a **hurricane warning** means there is likely to be direct damage. Usually the safest place is at home, if it is well built and above flood level.

If you hear a hurricane warning, bring things inside from outside, such as furniture, plants and garbage cans. Stick wide paper tape over the windows or board them up with strong wood.

When a hurricane occurs there will be widespread damage, especially from flooding. Electricity, gas and water may be cut off and you may not be able to travel on the roads. You will therefore have to prepare for a possible emergency. For example, keep a supply of tinned food that does not need cooking or refrigerating. Make sure you have flashlights, lamps, kerosene and candles. You also need a supply of water for drinking or possible fire-fighting. Keep a gas-operated camper burner for boiling water.

After a hurricane stay away from disaster areas and if you drive, take care to avoid fallen electricity cables and other debris.

Follow any instructions from disaster preparation groups. Keep alert for a few days in case of further problems. Boil drinking water until advised otherwise.

Activity | Tracking hurricanes

1. Use the map opposite to track any hurricanes using weather reports on the radio or in the newspapers.
2. Find out the names and dates of and the damage done by two hurricanes in the Caribbean.

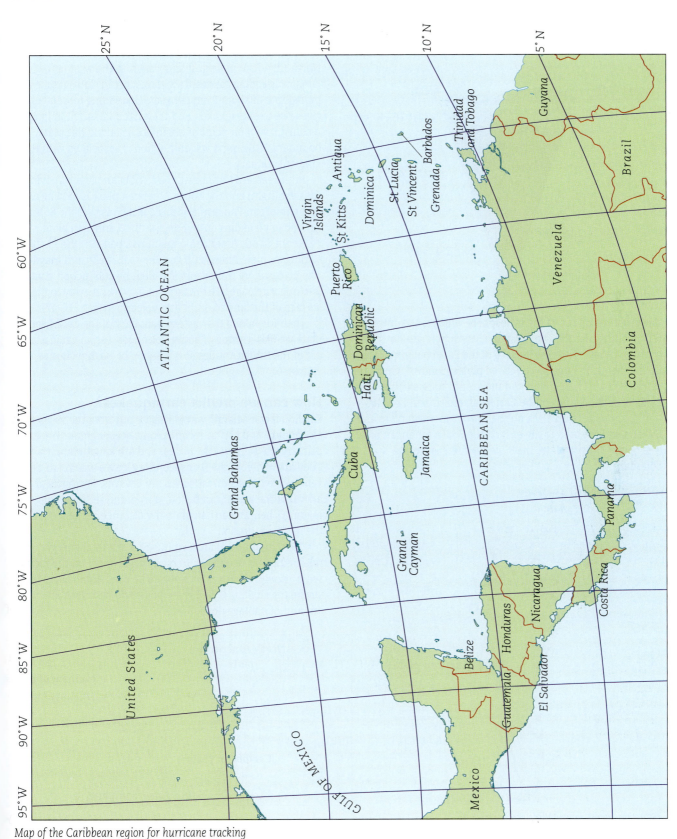

Map of the Caribbean region for hurricane tracking

movement | land and air **movements** | *what are hurricanes?*

What are earthquakes?

Earthquakes are movements of the Earth's surface. If we cut an imaginary section through the Earth we would see three parts, shown below.

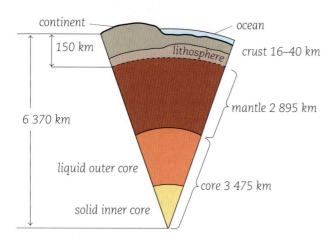

The **crust** lies on top of the **lithosphere** which is the outer region of the **mantle**. The crust and lithosphere are not a continuous layer over the surface of the Earth but are broken up into about 15 large pieces called **plates** (some of which are shown below). These plates can move and cause earthquakes.

Most of the islands of the Caribbean are situated on the Caribbean plate, with Cuba on the North American plate and Guyana and the area to the east on the South American plate (see below). Where plates meet we say there is a **plate boundary**.

What is the relationship between earthquakes and volcanoes?

The Caribbean plate is moving to the east (right), and the North American plate is sliding past it to the west (left). Jamaica is sliding eastwards with respect to Cuba at an estimated rate of about 2 cm per year. This sliding movement gives rise to earthquakes, and the earthquakes felt in Jamaica usually have their point of origin (focus) along this plate boundary. Little or no volcanic action is associated with this kind of boundary.

In the eastern Caribbean, the Caribbean plate is moving to the east (right) and the South American plate is moving west, straight into it. The result of this collision is that the South American plate dips down beneath the Caribbean plate.

As the South American plate moves beneath the Caribbean plate it heats up and the rock melts into **magma**. The place where this occurs is called the **subduction zone** (see opposite page). The magma then works its way to the surface of the land to erupt as a **volcano** (pages 316–17).

Therefore along the eastern boundary of the Caribbean plate we find a string of active volcanoes, the volcanic island arc of the Lesser Antilles. It is a region of earthquakes and volcanoes.

How can we predict earthquakes?

Earthquake or **seismic waves** radiate out from the focus of an earthquake and can be recorded on a seismograph. The **P** or **primary** waves travel the fastest and are longitudinal waves. They arrive before the **S** or **secondary** waves which are transverse waves. The slowest waves are **L** waves. The times of arrival of the waves can be found from the seismograph reading. Then we can calculate the distance of the focus from the recording station.

The **magnitude** of an earthquake is the total energy released and is usually expressed by a number on the **Richter scale**, though other scales can also be used. **Intensity** is how severe the shaking is at a given point and describes the amount of damage done.

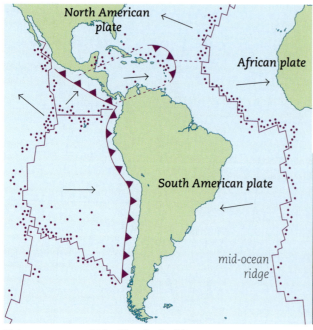

Plate movements which affect the Caribbean

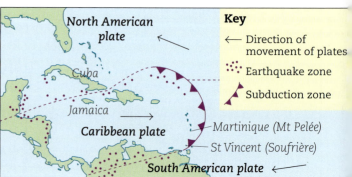

movement | land and air *movements* | *what are earthquakes?*

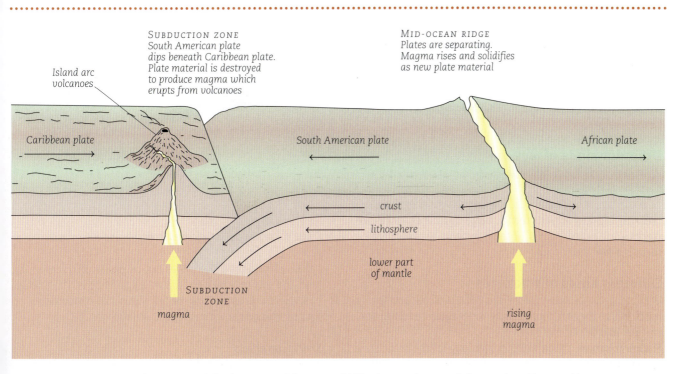

Strato-volcanoes (page 316) form at the subduction zone and fissure or shield volcanoes (page 316) form at the mid-ocean ridge

What effects do earthquakes have?

Most of the problems of an earthquake are caused by actual movements of the land. But an earthquake under water can also cause high tidal waves called **tsunamis**.

Tsunamis are caused by the rapid rise or fall of the sea floor during an earthquake. They travel at around 750 km/h. When they come to the shore they rapidly rise to heights of 12 m or more, causing massive destruction.

Tsunamis can sweep people and buildings into the sea and so it is very dangerous to be near the coast during an earthquake.

Short-term effects of earthquakes
1 The earth cracks open and so destroys the foundations of buildings. Buildings collapse and debris is thrown around causing injury or loss of life.
2 Fires can start very easily because of damage to electricity cables and appliances, or when escaped gases catch fire.
3 Tsunamis may cause flooding which can lead to destruction of crops and loss of life.

Long-term effects of earthquakes
1 Earthquakes cause rivers to change course, which can spoil farmlands and ruin crops.
2 People may move away from earthquake areas.
3 Buildings have to be designed either to withstand earthquakes or to be very light and easy to rebuild.

Earthquake safety measures

We know places where earthquakes are likely to occur, but not exactly when they will occur. We therefore have to be prepared.

If you are inside, stay there
1 Stand in a strong doorway or get under a desk or table.
2 Watch out for falling plaster, bricks, lights and other objects. Protect your head and face.
3 Do not light a match or switch on a light. Use a torch.
4 Check your house for serious damage – leave it if you are in danger.
5 Clean containers and collect water.

If you are outside, stay there
1 Stay in the open to avoid flying objects.
2 Keep away from buildings that may have been weakened by the earthquake.
3 Do not go to the beach to watch for tsunamis as you may get swept away.

In general
1 Stay calm.
2 Check for injuries and give first aid where necessary.
3 Use a battery-operated radio to listen to earthquake reports.
4 Use the telephone only in an extreme emergency.
5 Be prepared for additional earthquakes.

What are volcanoes?

Volcanoes are often found where plates are moving apart, or one is moving under another. At these places hot material from inside the Earth, **magma**, comes to the surface. The magma contains different amounts of dissolved gases and when magma comes to the surface as **lava**, the gases escape to cause an explosive **eruption**. The build-up of lava over time forms a volcano.

Different kinds of lava

If the lava is thin and runny, gases escape easily and the eruption is mild.

If the lava is thick and stiff, the trapped gases explode, breaking the lava into fragments. Large fragments – rock-like pieces – are called 'bombs'. If the fragments are very small there is a fine dust or ash.

Different kinds of volcanoes

We distinguish between the different kinds of volcano according to where they are formed, how the magma comes to the surface, and the kind of lava.

Fissure volcanoes These build up over long cracks in the crust of the Earth. The magma comes up along the length of the crack. The lava is usually thin and runny and spreads out to form a flat plateaux. These volcanoes are associated with mid-ocean ridges where two plates are moving apart from each other (see page 315).

Central volcanoes The magma comes up through a single channel to build a cone around a central crater. Central volcanoes are of two kinds:

Shield volcanoes form a flat cone (top of next column). The lava is usually thin and runny and spreads out. These volcanoes are associated with mid-ocean ridges where two plates are moving apart from each other (page 315). They are therefore usually under water and may go unnoticed. Because the lava is thin, the eruption is fairly mild.

Strato-volcanoes form a pointed cone (opposite page). The lava is usually thick and stiff and piles up around the central crater to form steep slopes. These volcanoes occur along a subduction zone (page 315) where one plate is pushed down beneath another. The island arc volcanoes at the edge of the Caribbean plate (page 314) are of this type, for example Soufrière in Montserrat (photo on opposite page).

In older volcanoes a lot of pressure is needed to push the magma up to the top crater. In this case it may break out along subsidiary channels to form **parasitic cones** on the slopes.

What effects do volcanoes have?

It is hard to predict exactly when a volcano will erupt. Besides the molten lava a volcano may also produce hot volcanic

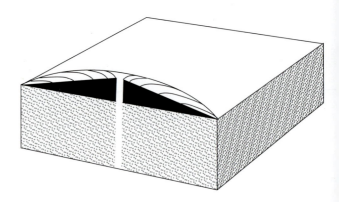

Shield volcano: a flat cone formed from thin and runny lava

gases, ash and mud. Volcanoes can cause disturbances of the land and this may lead to tidal waves and floods. The effects of a volcanic eruption may be felt thousands of kilometres away, and dust can be carried around the world.

Short-term effects

1. Molten lava is ejected from the volcano and will kill living things in its path.
2. Super-heated gases which come out under pressure can cause damage many kilometres away.
3. Volcanic ash may blow into cars, machinery, sewers, etc. and cause them to be blocked up. The ash also gets into people's lungs and bronchial tubes and can cause death.
4. Mud, formed from ash mixed with water, flows into rivers and clogs them up. This can then cause floods, with loss of crops, homes and life.
5. The associated winds can uproot mature trees and blow them away.
6. A huge tidal wave (tsunami, page 315) may form, which can cause great damage to coastal regions.

Long-term effects

1. Volcanic eruptions can produce new land masses, as in the eastern Caribbean. As the lava cools down it becomes part of the land.
2. Some of the ejected materials remain rock-like and sterile and are unsuitable for farming.
3. Other materials decompose. The minerals, such as sulphur and phosphorus which came from the lava, then become available for plant growth. The soils which are formed may be very fertile.
4. Where the ash falls there may also be long-term benefits in soil fertility.
5. Where rivers flood onto the land they may cause damage, but will also increase the soil fertility.

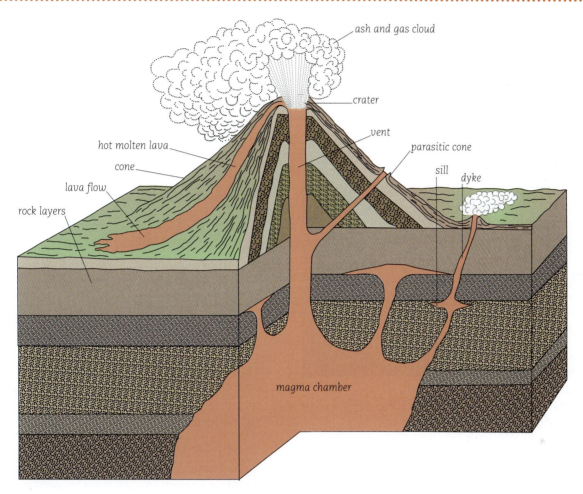

A strato-volcano with a pointed cone. Magma rises up the vent and explodes as fine ash and lava. Some lava may also emerge from side vents to form parasitic cones. Other magma is forced into surrounding rock layers as sills (horizontal sheets) and dykes (vertical sheets) which can also reach the surface

Devastating eruption of Soufrière Volcano on Montserrat in 1995. Like many islands of the Lesser Antilles, Montserrat is volcanic in origin. Most Windward Islands have an active volcano called soufrière from the French word for sulphur

Questions

1. Describe the differences in origin and structure of shield and strato-volcanoes. Include diagrams.
2. In what different ways could floods arise as a result of earthquakes and volcanoes?
3. For each of the following islands say whether you would expect them to have earthquakes and/or volcanoes and explain why. (a) Jamaica (b) St Vincent (c) Trinidad
4. Draw a large-scale map of the eastern Caribbean and label the islands. Which volcanoes are found in or near each island? Add them and label them.
5. Would a large earthquake *always* cause more damage than a small earthquake? What other factors might be involved?
6. In your country, what organisations help to deal with disasters such as hurricanes, earthquakes and volcanoes?

Living in the Caribbean

The Caribbean is a region of islands, coastlines, and coral reefs. Many of the islands are volcanic in origin, and some still have active volcanoes. Parts of the Caribbean experience earthquakes and hurricanes. How is our life affected by land, sea and air movements? And how do changes in other parts of the world affect us in the Caribbean?

Erosion of the land Our land is under threat from soil erosion (pages 239–41) which is made worse by the flood waters from hurricanes, and by the effects of the sea, as shown in this photo of the coastline of Barbados.

Aerial photograph showing coastline erosion

Volcanic activity We have volcanic activity in St Vincent (Mount Soufrière), Montserrat (Soufrière Hills), Guadeloupe (La Soufrière), Martinique (Mount Pelèe) and the Grenadines (Kick'em Jenny). Lava and mud slides affect crops, buildings and people. In Montserrat and Dominica, orioles and mountain chickens (edible frogs) declined in numbers due to volcanic activity and were taken into breeding programmes.

Hurricanes Hurricanes devastate crops and buildings, and cause economic loss which can take several years to overcome. In Belize, a hurricane caused so much damage to Belize City that the capital was moved to Belmopan. Regional tracking stations can help to predict when hurricanes will occur. Hurricanes are likely to become more frequent and of greater severity due to global warming (see below).

Greenhouse effect and global warming

You know that if you close the windows of a car or greenhouse, the temperature inside becomes greater than that outside. This is called the **greenhouse effect** and occurs because the energy from sunlight can come in through the glass, but the heat reflected from objects inside cannot escape. So the temperature inside rises.

A similar thing happens with the Earth. The Earth's envelope of clouds lets in some of the sun's energy, while reflecting some back into space. The layer also reflects some energy back to the Earth.

But another layer of gases is building up: carbon dioxide and, to a lesser extent, methane. Carbon dioxide comes from burning fossil fuels, respiration and the destruction of rain forests. Methane comes from the decay of organic material. These 'greenhouse gases' do not let so much heat escape, but rather reflect it back to the Earth. This causes a rise in the temperature of the atmosphere, land and water. This rise in temperature is called **global warming**. Dealing with greenhouse gases is a problem for the whole world, and makes it necessary to examine the way in which countries generate electricity. For example, it is better, where possible, to use wind turbines and solar energy rather than burning fossil fuels.

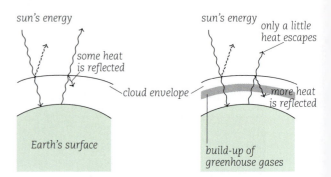

The greenhouse effect. More heat is reflected to Earth, causing global warming

Global warming has several effects.

Upsetting the weather Changes in sea temperatures upset the direction of currents and this causes changes in the direction and speed of winds. As a result we have unpredictable weather. (See also El Niño, page 319.) Because of these changes in the currents some countries can actually have colder weather, even though the world is being affected by global warming.

Hurricanes As you saw on pages 310 and 312, the formation of winds and hurricanes depends on differences in temperature. Hurricanes can form when sea temperatures are above 28 °C. Because of global warming we are more likely to see a longer, more severe hurricane season.

Rising sea levels The increase in the global temperature is beginning to melt the ice at the polar caps. If more of this ice melts, it will raise the sea level.

This is of importance in the Caribbean where many islands could experience severe flooding as a result. Deeper water is also a threat to coral reefs which may not be able to grow quickly enough to maintain their position in fairly shallow water. Mangroves are also under threat from rising sea levels, and the protection they give to wildlife, and to the mainland behind them, could be threatened.

Effect on plants and animals If temperatures increase in some areas, it means we could grow different crops, but it also means that different pests could spread. Mosquitoes could become more widespread in the Caribbean, and other pests could become a threat to livestock.

El Niño

El Niño is the name given to occasional changes in the direction of flow of warm currents in the Pacific Ocean. These in turn have been linked to climatic changes in many parts of the world.

Normally, surface currents move westwards across the South Pacific (see below). This causes moist air over South-East Asia. It also allows cooler water to come up the South American coast carrying nutrients for the fish populations.

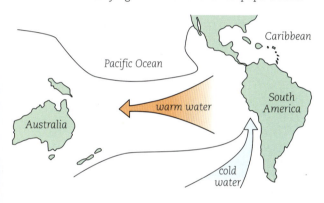

Normal surface currents in the South Pacific

During El Niño, the warm current flows eastwards (see below). This causes floods in the Americas and the Caribbean and droughts in South-East Asia. El Niño can sometimes help prevent the formation of hurricanes, but mostly it upsets weather patterns over a large area.

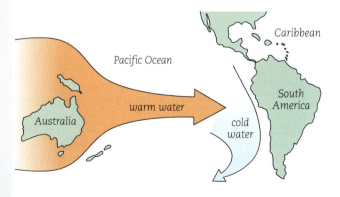

Reverse of warm current during El Niño

Ozone hole

Ozone is a gas similar to oxygen, but each molecule of ozone is made up of *three* atoms of oxygen instead of two. There is a layer of ozone high in the stratosphere which gives the Earth protection against solar ultraviolet radiation. Increasingly, there is concern that certain chemicals, in particular CFCs (chlorofluorocarbons), are thinning the ozone layer and causing holes in the ozone, especially over the Antarctic, and now over the Arctic region. CFCs were used in aerosol sprays, refrigerators, cooling systems and plastic packaging. In many cases other chemicals have now been used to replace them, though some of these replacements may still cause problems. Other gases, such as methane and carbon monoxide, also affect the ozone layer.

What can humans do?

1 Avoid pollution Burning fossil fuels contributes to the greenhouse effect and global warming, as well as producing acid rain (page 193). Using alternative sources of energy can help the environment (see also pages 194–5). Cutting down on the use of aerosols may also help stop the ozone hole from spreading.

2 Practise conservation Suggestions to reduce soil erosion have been given on pages 239–41. Any activity which might upset our coastlines – such as mining peat, building near the coastline or making new beaches for hotels – should be carefully considered before action is taken. (See also page 249.)

3 Take precautions Because the ozone layer is becoming thinner, damaging ultraviolet rays can come through more easily. People with light skins should be especially careful because of this. Wear hats and keep covered in light clothing. Sunscreens can be used, but these have to be chosen carefully to be both effective and also not cause damage to the skin – recently some ingredients in sunscreens were found to be harmful.

4 Make predictions Scientists can predict where earthquakes and volcanic eruptions will occur by studying the movements of plates, and by using instruments (pages 314–16).

Scientists can predict when hurricanes might occur by using weather satellites and ocean temperature readings. For example, early summer hurricanes usually begin in the Gulf of Mexico and western Caribbean where the shallower waters warm up quickly. As the Atlantic waters warm to 28 °C or more in late July and August more hurricanes form much further east, gathering energy as they travel towards the Caribbean.

5 Be prepared In the Caribbean we cannot be sure exactly when earthquakes, volcanoes and hurricanes will occur, but we can be prepared by making sure we have a battery-operated radio and emergency rations available (see also the hints on pages 312 and 315).

The electromagnetic spectrum

increasing wavelength ↓

Radiation		Typical wavelength	How produced	How detected
10^{-14} m		10^{-12} m (a million-millionth of a metre)	Radioactive substances, nuclear reactions	Geiger-Müller tube
10^{-12} m	Gamma rays			
10^{-10} m	X-rays	10^{-10} m (a ten-thousand-millionth of a metre)	X-ray tube	Photographic film
10^{-8} m	Ultraviolet	10^{-7} m (a ten-millionth of a metre)	Sun, sparks, mercury lamp	Photographic film, sun tan, fluorescent substances glow
10^{-6} m	Visible light	5×10^{-7} m (a two-millionth of a metre)	Sun, electric lamps, hot objects	Eyes, photographic film
10^{-4} m	Infrared	10^{-5} m (a hundred-thousandth of a metre)	Hot objects, sun	Skin, photo-transistor
10^{-2} m	Microwaves	10^{-2} m (a hundredth of a metre)	Microwave ovens	
1 m	TV	1 m	TV transmitter	TV aerial
10^{2} m	Radio	300 m	Radio transmitter	Radio aerial
10^{4} m				

Sample questions in Integrated Science

CXC-style structured questions
Write answers to these questions on your own.

1 The diagram below shows the foetus, umbilical cord and placenta.

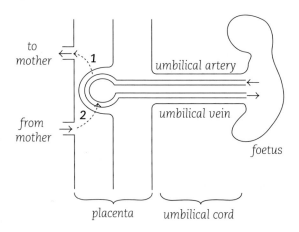

(a) Name two substances found in the umbilical artery which would be passed to the mother at (1).
(b) Name two substances needed by the foetus that would pass from the mother at (2).
(c) What may be the effect on the foetus if:
   (i) in the second month of pregnancy the mother comes in contact with a child suffering from German measles?
   (ii) the mother suffers from syphilis?
   (iii) the mother's diet has insufficient calcium?
   (iv) the mother is a heavy smoker?

2 Susan got home late. She had made a casserole the evening before, but she still had to cook the potatoes and spinach.
(a) (i) Where should she have stored the casserole overnight? Give reasons for your answer.
   (ii) What precautions should she take in warming the casserole?
   (iii) What dangers are there in eating warmed-up food?
   (iv) If there were problems resulting from (iii), what symptoms might the family show?
(b) (i) Susan washed the potatoes and then left the skins on. Why do you think she did each of these things?
   (ii) Because she was in a hurry, instead of leaving the potatoes whole, Susan cut them into slices. How and why would this affect the cooking time?
   (iii) What scientific principle was she using in (ii)?

(c) (i) She washed the spinach and cooked it with a small amount of water for just long enough to make it soft. What is the advantage of cooking it in this way?
   (ii) Name two constituents of spinach. Describe how each is useful in the body.

3 Urea is called a nitrogenous waste.
(a) (i) Why do mammals produce urea?
   (ii) Why don't plants produce urea?
(b) What is the role of each of the following mammalian structures in water balance (osmoregulation)?
   (i) lungs
   (ii) skin
   (iii) kidneys
(c) (i) What is a kidney machine?
   (ii) Give a simple description of how it works.

4 The diagram below shows half a flower.

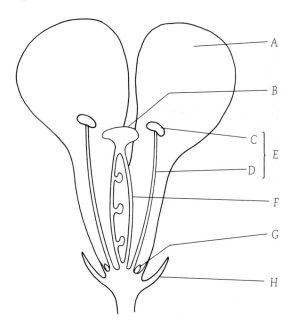

(a) Label the parts numbered A to H.
(b) Three similar flowers have had different parts removed. State the effect on pollination and fruit production of the removal of parts:
   (i) A
   (ii) C before they were mature
   (iii) F before it was mature.
(c) Describe the steps by which food manufactured in the leaves is made available to a germinating seed.

5 (a) What would be the effects of:
  (i) a baby having too little protein in its diet?
  (ii) adults eating too much fat in their diets?
(b) The table below lists the vitamins present in an egg and the percentage of daily needs which these vitamins provide. Use graph paper (see below) to prepare a bar chart of the information in this table.

Vitamins	
A	10%
B$_1$	8%
B$_2$	10%
Niacin	4%
D	12%

(c) State a use for each of two vitamins and the results of their deficiency.

6 In a heating experiment small quantities of boiling water were placed in each of two containers, A and B. The temperature of the cooling water was measured at intervals in each container and recorded as shown.

Temperature °C	A	77	74	70	65	58	52	42
Temperature °C	B	87	81	75	68	59	52	42
Minutes		1	2	4	6	10	15	30

(a) Using the data in the table, plot a graph of temperatures for A and B (vertical or y-axis) versus time in minutes (horizontal or x-axis).
(b) Predict the temperatures of A and B after 40 minutes.
(c) Explain the difference in temperature between A and B after 1 minute.
(d) Suggest how the heating experiment could be modified to produce identical curves.

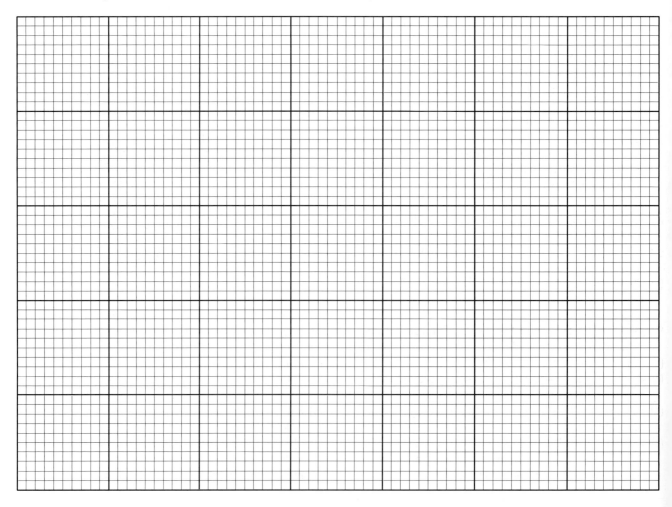

7 (a) List (in order) the structures in the mammalian respiratory system through which air passes when it is breathed in through the nose.

(b) Below is a diagram of an alveolus with its blood supply. Complete the labels.

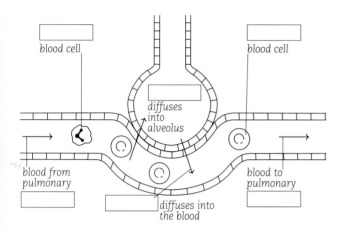

(c) How does the blood leaving the lungs differ from the blood entering the lungs?

(d) How does smoking cigarettes affect a person's health?

8 Electricity is usually sold in units.

(a) Explain what a unit is.

(b) A thermostatically controlled refrigerator uses 8 units per day at a rate of 16 cents per unit. What would be the running cost for one month of 30 days if there is also a fuel-adjustment charge of 10 cents per unit?

(c) Why would a fuel-adjustment charge vary from month to month?

(d) Explain why voltage is not an important consideration in your calculation in (b).

(e) Outline the principle of the operation of a thermostat used in a refrigerator such as in (b).

9 (a) What are the differences between the fluorescent lamp and the incandescent lamp?

(b) Which kind of lamp is more efficient?

(c) Draw a labelled diagram to show the structure of an incandescent lamp.

(d) State the function and the importance of the parts you have labelled.

(e) Given the formulae: $I = \frac{V}{R}$ and $W = VR$, calculate the resistance in ohms of a 100 W spot lamp operating at 12 V through a transformer connected to the mains.

10 (a) Write a word equation summarising photosynthesis.

(b) Write a chemical equation summarising photosynthesis.

(c) The table below shows the results of an experiment done with leaves. Leaves of equal surface area were chosen. Some surfaces were coated with Vaseline as shown in the table. The leaves were weighed before and after several days.

Surfaces covered	Both	Upper only	Lower only	Neither
Loss in mass (g)	0.01	0.30	0.08	0.40

(i) What mass of water is lost when both surfaces of the leaf are coated with Vaseline?
(ii) What mass of water is normally lost from the leaf?
(iii) Which of the surfaces has the most stomata?
(iv) Justify your answer in (iii).

11 The diagram below shows the cut surface of a mammalian heart as seen from the front.

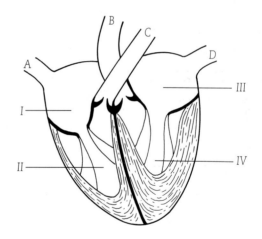

(a) What are the names of the chambers: I, II, III and IV?

(b) What kinds of blood would be found in the vessels A, B, C and D?

(c) Why is it important that:
  (i) the heart is divided into atria (auricles) and ventricles?
  (ii) the blood on the two sides of the heart do not mix?

(d) (i) Where does the heart muscle get its blood supply from?
  (ii) What would happen if this blood supply was insufficient?
  (iii) How is the heart important for overall health?

12 (a) Steel is widely used in the construction industry. One use is as reinforcement (as bars or rods) for concrete.
  (i) If the steel to be used is partly corroded, would you use it? Give a reason for your answer.
  (ii) If you decide that you will use the steel only if the corrosion has been removed, suggest a method for doing so.
  (iii) Assuming that the corrosion-free steel is now available, would you protect it from corrosion before using it in the construction of a building?
(b) (i) The metal hulls of ships are likely to corrode. Suggest two ways of protecting these metal hulls.
  (ii) In a number of islands of the Caribbean you will find a careenage. A careenage is a place were wooden ships are removed from the sea and then turned on their sides so that their hulls can be cleaned and treated. Give one reason why this treatment might be necessary.
(c) Each substance in the list below should be stored in a sealed container. Suggest an effective storage material for each substance, and give a brief reason for your answer.
  (i) sea water
  (ii) kerosene
  (iii) dried and salted fish
  (iv) liquid bleach
  (v) fresh yam

13 (a) (i) Make a list of the following metals in order of their reactivity. Start with the most reactive metal: silver, lead, sodium, copper, zinc, iron.
  (ii) Gold is one of the least reactive metals. Would you expect to find gold as a compound in the Earth? Give a reason for your answer.
  (iii) Platinum is a metal which is used in fine jewellery and also in a range of high-temperature chemical reactions. How reactive would you expect platinum to be?
  (iv) If you were given a solution of a copper salt, explain briefly how you would displace the copper from the solution.
(b) (i) What is an alloy?
  (ii) Copper and glass (for example, Pyrex) are commonly used in cooking utensils. Which properties of these materials make them suitable for this use?
(c) (i) The density of steel is quite high whilst that of aluminium is quite low. How would you decide whether to use steel or aluminium when making the body of a car? Give at least two reasons.
  (ii) Fibreglass and other composite materials are now used in many car bodies. Suggest two reasons for using these rather than metals or alloys.

14 (a) In the water cycle explain what happens as the sun heats the water surfaces on Earth.
(b) (i) With which kinds of fronts is rain associated?
  (ii) Which is the commonest kind of front in the Caribbean? Describe what happens when this kind of front passes.
(c) (i) In which direction do winds blow?
  (ii) What are the main differences between a wind and a hurricane?
  (iii) How could you predict that rain is expected?
(d) List four safety precautions you should take if a hurricane is predicted for your area.

15 (a) (i) What is a parasite?
  (ii) Describe the control measures for one named parasite.
  (iii) How are the control measures effective?
(b) (i) Name a household pest.
  (ii) Describe the damage that it does.
(c) (i) List three instructions for the safe handling of pesticides.
  (ii) Describe why each instruction is important for the health of the person using the pesticide.
(d) Research into genetically modified crops continues to produce crops that are resistant to disease. What is your opinion on the usefulness and/or danger of this kind of work?

16 The diagrams below illustrate machines that use different lever systems. Using a pencil, indicate on each of the diagrams:
(a) the position of the pivot (fulcrum) (use the letter P)
(b) where force is applied (use the letter F)
(c) where the load may be found (use the letter L).

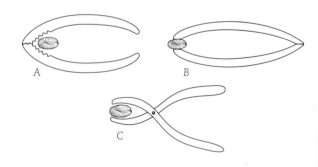

(d) Which of the three machines has the least mechanical advantage? Give reasons for your choice.

17 (a) Complete this diagram showing the aerofoil shape.

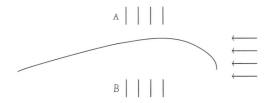

(b) Use a pencil to draw arrows indicating the forward movement of the aerofoil in an air stream; without turbulence.
(c) Use a pencil to draw arrowheads on the groups of lines A and B, in order to show how lift is generated.
(d) How is the aerofoil used on racing cars to prevent lift?

18 (a) A and B are incomplete diagrams of *the same eye* showing the path of light onto the retina.
  (i) Complete diagram A to show defective vision.

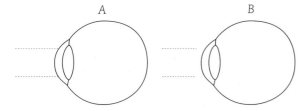

  (ii) Complete diagram B by inserting an appropriate lens and show corrected vision.
(b) Explain why some people need bifocal lenses.

19 (a) (i) You have been asked to heat an organic solvent (flammable) with a boiling point of less than 100 °C. State two safety precautions you would take.
  (ii) People who work in the local fire service need to be able to work near, and sometimes in, burning materials. They have to wear protective clothing to do so. Suggest two properties for the materials of the protective clothing.
(b) The store for a science laboratory was built next to the laboratory, does not have any windows, has a dirt floor and a flat wooden roof. What advice would you give about improving the design of the store?
(c) One label on a can of hairspray reads, 'no residue or stickiness when used', but there is also a hazard label showing that the hairspray is 'extremely flammable'.
  (i) What precautions would you take when using the spray?
  (ii) How would you try to check the claim about no residue or stickiness?

20 (a) (i) What is meant by pH?
  (ii) Describe how to find the pH of a soil sample.
(b) (i) Compare and contrast a sandy and a clay soil.
  (ii) Give two ways in which a sandy soil could be improved and explain the reason in each case.
(c) Plants growing on sand above the high-tide level are very close to the sea. Many of these plants show modifications to restrict water loss. Explain why this is necessary.
(d) Describe how conditions in a rock pool in the middle zone on a beach would change during a 24-hour period.

21 (a) Suggest four factors to consider when selecting recreational activities. Give a reason for each factor.
(b) Describe four effects of exercise.
(c) Explain the reasons behind each of these pieces of advice:
  (i) Do not carry out strenuous exercise such as swimming straight after having a heavy meal.
  (ii) People, especially those with fair skin, should use sunscreen or wear hats if they spend long periods in the sun.
  (iii) In hot weather you should drink more water.
  (iv) People with diarrhoea, especially children, should drink cooled, boiled water containing a little sugar and a little salt.

22 (a) How are the ionised layers in the Earth's atmosphere caused?
(b) How are the ionised layers useful in radio broadcasts?
(c) Explain why FM radio is unsuitable for long-range broadcasts.
(d) What gives FM radio superior quality over AM radio?

23 (a) What causes motor vehicles to skid?
(b) Why does a car travelling at a high speed take longer to stop than one travelling at a low speed?
(c) Why do high-speed collisions cause much more damage than low-speed collisions?
(d) Explain how seat-belts can help to reduce injuries.

24 (a) (i) What is gravity?
  (ii) How does a rocket escape Earth's gravity?
(b) A tin of red peas has a mass of 200 g. How will its mass and weight on the moon compare to those on Earth? Explain your answers.
(c) Distinguish between the following pairs of terms and give an example of each:
  (i) star and planet
  (ii) natural and artificial satellites
  (iii) rotate and orbit.

Free-response questions

Use these questions for class discussion or for further research.

1. Would you call viruses living things? Support your answer.
2. Cloning of humans should not be allowed. Discuss.
3. Do you agree with genetic engineering of food crops? Give reasons for your answer.
4. What are the advantages and disadvantages of the use of inorganic fertiliser?
5. How do biological enzymes in washing powder work? What precautions are needed for their use?
6. Every adult should be a blood donor. Discuss.
7. Why is your blood pressure reading a useful indicator of your over-all health?
8. How can high blood pressure be reduced?
9. Smoking is bad for your health, so why do people smoke?
10. Why are most leaves green and not red, and flat like a disc and not round like a ball?
11. 'Food is a fuel.' What do we mean by this statement?
12. Why is the elimination of faeces not an example of excretion, whereas sweating is?
13. How is the liver important to the whole body?
14. Explain why and how a dialysis machine is used.
15. Why do lizards bask in the early-morning sun, but hide away in the middle of the day? Why do birds not have to behave in a similar way?
16. 'We see with our eyes and hear with our ears.' Why is this statement only partly true?
17. Why is the pituitary called the 'master gland'?
18. How is an Amoeba similar to and different from a single cell in the human body?
19. How do pests and parasites cause economic loss?
20. What is the importance of preservatives in food?
21. How could you manage during a power outage without the use of a refrigerator?
22. Why is it important for babies to be breast-fed?
23. Why is it important for babies and children to be immunised?
24. What are the short-term and long-term effects of clearing large areas of rain forest?
25. Prepare a poster of hints for safety in the home.
26. When visiting the beach, how should you minimise the effects of the sun?
27. How could you reduce your electricity bill?
28. How does water drainage affect plant growth?
29. What different problems do fresh-water and sea-water fish have?
30. It is important to use nets with mesh of a certain size to catch fish. Discuss.
31. Factories should clean up the water they return to streams and rivers. Discuss.
32. Without bacteria, life on Earth would be extremely difficult. Discuss.
33. Without plants, life on Earth would be impossible. Discuss.
34. What is 'the balance of nature'? Why is it important?
35. Do you think there is life elsewhere in the universe? Give reasons for your answer. Why is it difficult to find a definite answer to this question?
36. The moon affects the Earth more than the Earth affects the moon. Discuss.
37. Write a 24-hour account of life in a rock pool.
38. How do clouds help us to predict the weather?
39. Prepare a list of hurricane safety measures. Why do hurricanes in the Caribbean tend to occur only in certain months?
40. Why do volcanoes form?
41. Why do some people object to the transport of nuclear waste through the Caribbean?
42. Why are radioactive materials harmful to life?
43. Can one form of energy be converted into any other?
44. Explain what is meant by **renewable** and **non-renewable** forms of energy.
45. What safety features are necessary in a microwave oven? Explain why.
46. 'Really, there is no such thing as heat.' Discuss.
47. Show that the smaller the particle, the greater the surface area to volume ratio.
48. 'Of all the regular shapes, the sphere has the smallest surface area per unit volume.' What are the implications of this for living things?
49. What are the properties of the liquids most likely to be used in the liquid-in-glass thermometer?
50. Explain the reason for the irregular isotherm patterns on the global radiation map (page 194).
51. How would you calculate the cost effectiveness of a solar water heater?
52. To what use would you put a solar furnace?
53. Is there an ideal temperature for comfortable living? Discuss.
54. Why are air conditioners needed?
55. Is there a difference between a body builder and a weightlifter? Explain.
56. Should athletes ever use drugs? Discuss.
57. Why can a needle or a razor blade float, but not a nail?
58. What precautions must a scuba-diver take for safe diving?

59. Explain what makes a ball, for example a cricket ball or tennis ball, swing or curve during play.
60. What factors may cause variable bounce on a playing surface?
61. Why do players in games such as cricket and tennis have a problem with variable bounce?
62. Are X-rays harmful? Discuss.
63. What gives musical instruments their distinctive sound?
64. The digital sound recording on a compact disc is far superior to that of the vinyl record. Explain.
65. On FM radio the sound quality is superior to AM radio. Explain.
66. Explain why AM radio-'fade out' sometimes occurs. Why is FM radio not so affected?
67. Explain the difference between a real and a virtual image.
68. How is chromatography useful?
69. Is colour real or is it all in the mind?
70. What precautions must be taken with a truckload when the centre of gravity is high?
71. Can we save energy by streamlining motor vehicles?
72. How does the rotor of a helicopter create lift?
73. 'A stiletto heel can cause more damage than an elephant's foot.' Discuss.
74. Why does atmospheric pressure vary?
75. What information does a cell pass to its 'offspring'?
76. How did the fossil fuels originate?
77. What are the factors thought to be contributing to global warming?
78. Can temperature affect buoyancy? Explain.
79. Size seems to limit flight in birds, but not in aircraft. Discuss.
80. 'Magnetism is indeed a strange phenomenon.' Discuss.
81. Many Caribbean countries have floods in one season and drought in another. What advice would you give to (a) town planners and (b) households to help reduce these problems?
82. Polysaccharides are examples of naturally occurring polymers. Find out what you can about some important properties of polysaccharides.
83. Silicon is a very common element in the Earth's crust. How has it been used in computers?
84. The technique known as 'controlled burning' can be used to try to control bush fires. How does this technique work? Would it be useful for fires taking place very far from fire services?
85. Cars and trucks powered by fuel cells are relatively rare at the moment. Would you expect to see more such vehicles in the next 20 to 30 years? Why?
86. You have to clean an old oven which has been used for cooking food for many years. The walls of the oven are covered in grease and food remains. What would be the critical properties of a material to be used as an oven cleaner?
87. Used car tyres and old ships have been used in various parts of the world to create artificial reefs. How could these items be prepared before they are placed in the sea?
88. There is some evidence that water type (hard or soft) is linked to the incidence of certain types of disease. How would you try to test this?
89. In many parts of the world increasing use is being made of very long bridges or tunnels under the sea. In the Caribbean, what would be the advantages and disadvantages of continuing to use ships as ferries for short journeys of 15 km or less?
90. You have been invited out to eat in a restaurant in a country you are visiting. A speciality of the restaurant is a fish which you do not know, cooked by a method you are not familiar with. How would you decide whether to order the fish or not?
91. Many supermarkets and other shops provide plastic bags for carrying goods, food items, etc. What impact might this have on the environment? How could this be changed?
92. Take one room in your home and identify all the fittings and other objects made of plastic. Could these be replaced by objects made from other materials?
93. You are given a substance which you are told contains an insoluble carbonate and a soluble sulphate. Describe and explain the steps you would take to confirm the presence of the carbonate and sulphate. Also explain how you would identify the metals present.
94. Assume that you are living in a house that gets its water supply from a stream in the hills. The water is being distributed by gravity feed. What precautions would you take before using this water for drinking?
95. Imagine you are a carbon atom. Describe your structure, how you can be made into carbon dioxide, and then what happens to you when you enter the leaves of a plant.
96. What are the advantages and disadvantages of using fertilisers compared to compost?
97. 'The quality of the air we breathe can have a serious effect on our health.' Discuss.
98. What suggestions would you make for reducing the negative effects of using cars?
99. Why do some metals corrode more easily than others?
100. Some people think of chemicals as causing pollution. Explain three important and useful uses of chemicals.

Index

Where several references are given the main one is in bold type.

ABO blood groups 93
abortion 59
absorption (villi) 84
accidents 162–3, 170–1
accommodation, of eye 280
acid rain 147, 193
acids 220–1; and corrosion 213, 214; handling of 164, 165; reactivity with metals 198–9
active immunity 161
active uptake of salts 90
additives, in food 76, **77**, 224
adolescence 46, **48–9**
adrenaline 127; in stress 97
Aedes mosquito 151
aerials 265, 275
aerobic respiration 104–5, **106–7**
aerofoils 299
aeroplanes 298, 299
afterbirth 64 *see also* placenta
agglutination 93
AIDS (acquired immune deficiency syndrome) 52, 53, 81, 161; and age expectancy 54
air: composition 99; conditioning 136–8; in diving 256–7; pollution 102, 135, 191, **193**, 248; pressure **308–9**, 310, 311, 312; resistance 258; in soil 235
airships 298
alcohol: and athletes 253; effects of 80, **253**, 289; and pregnancy 64; production of 106–7
algae **13**, 144, 232
alimentary canal 84–7
alkalis: and corrosion 213, 214; handling of 165; properties of 220–1
allergies 103
alloys **202–3**, 215
alternating current 186–7
aluminium 210
alveoli 99, 100
amino acids 74; absorption of 84; breakdown of 108; essential 76; in blood 129
ammeter 175
ammonia: and cleaning (solution) 211; diffusion of (gas) 16, 17
amniotic fluid **63**, 64
amperes (current) 175
amphetamines, and athletes 80–1, 253
amphibian 15, 118; gaseous exchange in 100
amplitude: modulation (AM) 273, 275; of sound waves 266, 267
amylase 87, 88–9
anaemia 79
anaerobic respiration 106–7

analysis and interpretation 31
analysis of chemicals 225
aneroid barometer 308–9
angina 96–7
angle of throw 259
animals: cells 10; classification of 12, **14–15**; life cycle of 46; and plants 10, 12 *see also* soil organisms
Anopheles mosquito 151
anorexia 78
antibiotics 80, 152, **160**, 208; and STDs 52, 53
antibodies 161; in blood 93; Rhesus factor 61
antigens 161; in allergies 103; in blood 93; and Rhesus factor 61
antihistamines 103
anti-knock agents 189, 193
antiperspirants 208
antiseptics **208**, 209
antitoxins 161
aqualung 256
aqueous solutions **22**, 216
aquarium 246
aquatic habitats: freshwater and sea water 142, 144–5; seashore 306–7
arteries 94–5, 96, 97
arthropods 15
artificial asexual reproduction 39
artificial respiration (mouth-to-mouth resuscitation) 170
artificial kidney 113
Ascaris (a nematode) 150
asexual reproduction 7, 38, **39**, 73
aspirin use 80
asthma 103
astigmatism 281
astronauts 300
astronomical unit 9
atheroma **96**, 97, 250
athletes: and drugs 253; and throwing 259; and training 252–3
atmospheric pressure 308–9 *see also* humidity; weather
atoms **18–19**, 23; structure of 19
auditory nerve 122, 269
autonomic nervous system 125
autotrophic feeding 6, **66**
axes on graphs 36, 37

babies **64–5**, 110, 138, 160
bacteria 13; and antibiotics 160; and disease 52, 151, 152; and food spoilage 153, **158–9**; in soil 243; and STDs 52
balance, sense of 122, 263; and inner ear 268, **269**
balance of nature 245, 249
balanced diet 78–9
balloons, flight 298

bandages 171
bar (pressure unit) 309
barbiturates 81, 253
bar charts 37; of soil components 232
barometer **308–9**, 310, 311
base (chemical) 220
batteries *see* electrochemical cells
bauxite/alumina plant 213; pollution from 147
BCG (tuberculosis vaccine) 161
Beaufort scale 310
bedbug 149
'bends' (in diving) 256, **257**
Benedict's solution (reducing sugar test) 74, **75**
beri–beri 78
bicarbonate indicator *see* sodium hydrogen carbonate solution
bile 86, 87
bilharzia (schistosomiasis) 148, **152**
Billings method (contraception) 56
bimetallic strip 131
biodegradable wastes **157**, 205, 207
biogas (methane) 153
biological control measures 154–5
biological wastes **152–3**, 156
birds 15; temperature control in 118; flight in 298
birth (parturition) 53, **64**; rate 54 *see also* babies
birth control: methods 56–9; and populations 54, 55
bites 163, **170**
Biüret test (for proteins) 75
bladder 110
blindness *see* eyes
block and tackle 230
blood: boosting 253; circulation **94–7**, 128; clotting 289; constant composition of 128–9; donors 53, **93**; groups 93; pressure 96, 97; Rhesus factor 61; structure and function of **92**, 93; white cells 53, **92**, 161, 289 *see also* blood system
blood fluke (*Schistosoma*) 14, 148, 150, **152**
blood system 92–7; damage to **96–7**, 289; and exercise 250; gaseous exchange in 100, 287; protection of 288, 289
body shape and temperature control 121
bones, broken 163, **170** *see also* skeleton
bottle-feeding 65 *see also* babies
brain 28, **125**, 262; damage 64
brakes 309
brass 202
bread-making 107
bread mould **38**, 39; and food spoilage 158
breaking strain 260
breast cancer 51

breast-feeding 65; and antibodies 161 see also babies
breathing: in fish 100, 143; in mammals 98–9; problems 102–3 see also gaseous exchange; respiratory systems
bromine, diffusion of 16, 17
bromochlorodifluoromethane (BCF) fire extinguisher 169
bronchitis 102
bronze 202
bulb, of plant 39, **73**
bulb, electric light 132, 183
Bunsen burner 35, 191
buoyancy 254–5
burette 221
burning: compared to respiration 6, **105**; of foods 104, 105; of fuels **190–1**, 192, 193; of plant cover 239 see also Bunsen burner
burns 163, **170**

Caesarean section 53
caffeine 80, 253
calcium 79; salts in hard water 207, **218–19**; and teeth 83
calories see joules
calorimeter 105
camera, compared to eye 278
cancer 47, 51; of breast, cervix and testes 47, **51**; of lungs 102; of skin 53, 251
cannabis 81; and lungs 102
canning 158, **203**, 213
capillaries **94**, 128; in body temperature regulation 119
capillary action (capillarity): demonstration of 130; in plants 91
carbohydrates 25, 74; digestion of 86–7; importance of 76, **105**; structure of 74; tests for 75
carbon compounds **24–5**, 192–3 see also foods; fuels; plastics
carbon cycle 246–7
carbon dioxide: concentration in blood 129; effects on climate 191, 193, **318**; as fire extinguisher 169; and gaseous exchange **98–101**, 108; and photosynthesis 67; and respiration 98–100, 104, **106–7**
carbon monoxide poisoning 102, **193**
carbon tetrachloride 217
carcinogens 47; and food 77; and lungs 102 see also cancer
carnivores 244; in food chains 144; teeth 82
cars: effects of 192–3; racing 299; and internal combustion engine 192–3
cast iron 215
catalysts 89, 189; chlorophyll as 67, 72; digestive enzymes 89
Cartesian diver 256

cataract 281
cell, in living organisms: division 60; membrane **10**, 69, 90; structure 10 see also white blood cells
cell, electrochemical see electrochemical cells
cellulose: production of 72; as roughage 76, 85; uses of 74
central nervous system (CNS) **122**, 125
centre of gravity 292–3; and stability 294–5
cervical (Pap) smear 51
change of state 17 see also evaporation
characteristics (genetic) 10, 60–1
charcoal 190
chemical control measures 155 see also pesticides
chemical digestion 84–7
chemical energy 20 see also burning; energy
chemical equations: photosynthesis 72 see also word equations
chemical pollution **193**, 233, 248
chemicals: analysis of 225; in food preservation 158, 159; safe handling of 34, **164–7**, 222–4; uses of 220–5
chemotherapy 47
chlorides, testing for 225
chlorine and water purification 140
chlorophyll 6, 67; chloroplasts containing 70, 71; importance of 66, **67**
choking 170
cholera 152; antibodies 161
cholesterol 97
chromatography 283
chromosomes, human 10, 60–1
cigarettes see smoking
circulatory system see blood system
classification 11, 28; of living things 12–15; of matter 17, 23
clay soil 233–6; improvement of 236
cleanliness: of eyes 281; of household appliances 210–11; personal 149, 150, 153, **208**
clinical thermometer 120
clotting, of blood **92**, 289
clothing: on fire 170; and temperature control 120–1
clouds **310**, 311
coal 190 see also fossil fuels
cocaine 81
cochlea 122, **268**, 269
cockroaches 154
cold-blooded animals see poikilothermic animals
collecting live organisms: on land 244; on seashore 306–7; in soil 242
colour: mechanism of seeing 278, 285; primary 282; printing 283; television 285
collisions, elastic/inelastic 291
colloidal solutions 216

combustion see burning
community 245; hygiene 153
compact discs 271
complete flowers 41
compost: heap 157, 239; improve soil **236**, 239
compounds 18, 23
computers 28–9, 263
concave lens 277; and eye 280
concave mirror 265, 277
condensation, as part of water cycle **139**, 246, 247, 310
condom (rubber) 56; and STDs 52, 53
conduction, of heat **114**, 116, 194, 203, 290
conductors (electrical) **172**, 173
cones (retina cells) 122, **278**, 279
conjunctivitis 281
conservation: of momentum 290–1
conservation, general measures of 239, 249, 319; of soil 240–1, 249; of water 138, **141**, 249
constipation 76, **85**
consumers, in food chains 144, 242, 244
contraceptive pills 57; problems with use 57 see also birth control
control: of food spoilage 159; of rusting 214–15
controls: and experiments 32–3; of germination 44; and photosynthesis 66, **67**; and rusting 212–14
contour ploughing 241
convection **114–15**, 116, 194, 195
convex lens 277; and eye **278**, 280, 281
convex mirror 265, 277
cooking: and energy saving 117; with gas 189; utensils **202–3**, 210–11
cooling curves 36, **37**, 114
copper 180, 202; testing for ions 225 see also metals
corms 39
cornea 278, 279; and astigmatism 281
coronary: arteries 96, **97**; thrombosis 97
corrosion 210, **212–13**, 261
corrosive chemicals **164–5**, 213, 214
cramp 251
cretin 127
crop rotation 239, **240**
cross-pollination 43
crude oil: **188–9**, 190; formation of 190, 195; refining of 188
crystals 7, **16–17**
Culex mosquito 151
current (amperes) 175–7 see also electricity
cuts 163, **170**
cutting sections 34
cuttings, from plants 39
CXC Integrated Science, examination 30–1
cyclonic storm 312
cytoplasm 10, 138

Index 329

dams 138
D and C (dilatation and curettage) 51
data processing 29
DDT (dichlorodiphenyltrichloroethane): in food 77; non–biodegradable 155
deafness 268; and noise pollution 183, 267
death 59; rate 54
decay: by bacteria 243; in compost heaps 157; in humus formation 233
decibel 267
decomposers 144; in soil 243
decompression 257
defecation *see* egestion; faeces
deficiency diseases 78–9
dehydration: in diseases 138, 152, 153; of food 76, **158–9**
denitrifying bacteria 243
density: measurement of 254; and temperature 114, **254**
deodorants 208
depressants 81
desalination 141
de-starched plants 66, 67
detergents **206–7**, 218; and plants 138 *see also* soaps
developed/developing countries 54–5
diabetes 113, **127**
dialysis (kidney machine) 113
diaphragm (contraceptive) 57
diarrhoea: in babies 65; causes of 85; in disease 152–3
dicotyledons 13
diet: balanced 78, 79; and tooth decay 209
dietary fibre *see* roughage
diffusion 16, 17, **68–9**, 70; and surface area 286–7
diffusion gradients 68–9; in animals **100–1**, 108–9, 128; in plants 70, 90, 101, 108
digestion 83, **84–7**, 138
digestive enzymes 84, 85–9
direct current 174, **186** *see also* electrochemical cells
diseases 78–9, **160–1**, 289; and pregnancy 64
dish aerial 265, **275**
disinfectants 208; care in using 223
dispersal, of seeds and fruits 44
distillation 107; fractional 188; and solar still 196
diving 256–7
dodder (love vine) 148
domestic waste (refuse) 156–7
dominant genes 61
Doppler shift 301
DPT (diphtheria, pertussis and tetanus vaccine) 161
drag (air resistance) 258, 297
drawings, technique for 30, 34, 41

drugs: and diseases 80, 160; misuse of 80–1, 253; overdose 81, 163; and pregnancy 64; types 80–1
dry cell *see* electrochemical cell
dry cleaning 217
ductless glands *see* endocrine system
duralumin 203
dust inhalation 103
dynamo 185, 186–7
dysentery amoeba (*Entamoeba*) 13, 39, **152**

ears: 122, 257, 262; damage to 267; structure of **268–9** *see also* deafness
Earth: as a planet 302–3; rotation of 304; structure of 314
earthing (electricity) **182**, 183
earthquakes 314–15
earthworms 14; and soil 243
ecology 245 *see also* habitats
ecosystem **245**, 249, 318–19
Ecstasy drug 81
ectoparasites 148–9
effectors 122, **123**, 124
efficiency 230–1
egestion 84, 108
egg: human 51, 61; plant 40, 43
elasticity 205, **260** *see also* collisions
elastic limit 260
electric: current 173; fires 169; shock 163, 164, 165, **170–1**, 173, 182
electrical: appliances 34, 131, **178–9**, 181; circuits 172, 173–7, 180–1, 202; conductivity 172, 173, 202, 205; energy 20, **178–9**; field 263
electricity 172–5; cost of 178–9; generators 21, **186–7**, 190; measurement of 178–9; safe use of 180–3; static 173
electrochemical cells 173, **174–5**; alkaline 186; lead-acid 186; zinc-carbon 186
electrolysis 22
electromagnetic spectrum 115, 194, 272, 282, 320
electromagnets 184, **185**
electromotive force 174
electrons **19**, 173, 186
electron microscope, magnification 12
electroplating 214
elements **18–19**, 23
elephantiasis 150, 151
El Niño 319
embryo: human 62–3; plant (in seed) 44
embryonic membranes 63
emotional changes: in adolescents 49; in new mothers 65; and stress 97, 103, 127
emphysema 102
emulsification **87**, 206
emulsion 217; test (for fats) 75

endangered species 249
endocrine system 126–7, 129 *see also* hormones
endoparasites 148, **150–1**
energy 11, 17, **20–1**; chemical 20; electrical 20, 21, **178–9**; kinetic 20, 290; light 133, 195, 262; measurement of 20, 105; nuclear 21; sea 264; solar 115, 173, **194–7**; sound 272, 290 *see also* heat energy
energy released: from foods 79, 105; from fuels 190–1; in respiration 105–6
energy requirements 78; and exercise 251
engine knock 189, 193
engines: internal combustion 192–3; jet 299; rocket 301
environment 245, 318–19
enzymes 86–9
epidermis: in plants 70, 71; of skin 123
equilibrium (balance) 292–3
equipment, using 34–5 *see also* electrical appliances; chemicals, safe handling of
erosion: of coastline 264, 318; of soil 190, 239, **240–1**
eruptions 316–17
estuaries 306
ethene 204
ethanol 24 *see also* alcohol
Eustachian tube 257, **268**, 269
evaporation: and heat loss **117**, 118–19, 134; and water cycle 134, **139**, 246, 247
excretion 6; in mammals 108–13; in plants 108; products of 108; system 110–12
exercise 250–1
exhalation: in fish 100, **143**; in mammals 98–9
exhaust gas pollution 193
experimental techniques 32–3
explosions 290
eyes: 122, 262, **278–81**; blindness 133, 263, 279; cleanliness of 281; defects of 280–1; injuries to 163, **171**; persistence of vision 285; response to light 133, 280

faeces: and disease 148, 150, **152–3**, 156; and egestion 84, **108**
fainting 171
fair tests **33**, 44, 66, 67, 158, 212, 213, 214
fans (air conditioning) 136
farming: pests 148, 154; practice 236, **239**, 240–1
father's role 49, **65**
fats 74; digestion of 86–7; and heart disease 97; importance of **76**, 97, 105, 118; tests for 75
Fehling's solution (reducing sugar test) 74, **75**
fermentation 106, 107
fertilisation 38; in humans 46, **51**, 60, 62; in plants 43 *see also* birth control

330

Index

fertilisers: pollution by 77, **147**, 248; from sewage **153**, 157, 248; for soil 236
fibre in diet *see* roughage
fibres, natural and synthetic in clothing **121**, 204
film: photographic 284–5; sound tracks for 271
fins *see* fish
fire: and accidents 163; blanket 169; damage 168; extinguishers 169; and plastic foam in furniture 205; types of 168–9
first aid 170–1
fish 15, **118**; catching of 144–5; farming of **145**, 153; fins 143; gaseous exchange **100**, 101, 143; life of **142–3**, 144, 145; osmoregulation of 142; streamlining in 142, 143, 297 *see also* gills
fishing industry 145
flaccid cells 70
flammable chemicals 164, 222
flashlight 174
flash point 168
flatworms 14; parasitic 150, **152**
fleas 149
flies 154; on wastes 152, 153, 157
flight: and aerofoils 299; in birds 298; sub- and supersonic 297
flotation, law of 255
flower structure **40–1**, 42–3
flowering plants, classification of 13
flu 160, **161**
fluorescent lamps 133, 183
fluorides **83**, 209
focal length 277
focusing, of eye 278–81
foetus 62–3
food 25, **74**, 78–9; additives in 76, **77**, **224**; allergies 103; chains **144**, 244, 245; contaminants 77; and developing countries 54; labels **76**, 77; made by plants 66; poisoning **153**, 158; preservation **158–9**, **224**; storage 72, 73; substances 74–6, 78–9; supplements 225; tests 75; web **144**, 245
footwear *see* clothing
force of gravity 230, **292**; and orbits 300, 303; and tides 305
fossil fuels 190–1; in carbon cycle 247–8 *see also* burning
fractionating column 188
frequency (sound) 266, 267; and hearing 268; modulation (FM) 273, 275
freshwater fish 142, 145; in food chains 142
friction 261, 291, **296–7**; coefficients of kinetic and static 296; drag 297
fronts (weather) 311
fruit 40, **43**

fuels 6, **25**, 189, 224; fire material 168 *see also* burning; digestion; fossil fuels; heat energy
fungi 13; parasitic 148; and STDs 52
fuses **180**, 181, 182

galvanised iron sheets 215
gametes 38; human **51**, 60; plant 40, 43
gaseous exchange 98–9, **100–1**, 108, 287
gases 16, **17**, 23; diffusion of 16, 17; and heat transfer 115; solubility of 256 *see also* liquefied petroleum gas (LPG)
gas: explosions 163; leaks 169; and volcanic eruptions 316
gasoline **188–9**, 192–3; cracking of 189
gastro-enteritis *see* diarrhoea
gear wheels 229
generators 21, 187
genes 10, 60–1
German measles: and pregnancy 64; vaccine 161
germination 44–5 *see also* seeds
germs *see* micro-organisms
gills: in fish 100, 142, **143**; in tadpole 100
glands: endocrine **126–7**, 129; digestive **87**, 123
glassware: handling of 34, 165–6; and oven use 203
glaucoma 281
global warming 192, 318; and effects 318–19
glucose: absorption of 84; blood concentration of 127, **129**; in respiration 106; structure of 24
goitre 79
gonorrhoea **52**, 53
government chemist 225
graphs: constructing 36–7; of human growth 46; of seedling growth 45 *see also* cooling curves
gravitational field 292 *see also* force of gravity; centre of gravity
greenhouse effect 318
growth: in animals **46**, 62–3; of crystals 7, **16–17**; hormone 127; of living things 7; in plants 44–5, **238–9**
guard cells, in leaves 70

habitats 245 *see also* aquatic habitats; terrestrial habitats
haemoglobin 92; and carbon monoxide 193; and sickle cell disease 61
hallucinogens 81
hand lens 35
handling equipment 34–5, 165–7 *see also* chemicals; electrical appliances
hard water 207, **218–19**
hardening, of arteries 96

harmonics 266–7
hashish 81; and lungs 102
hay fever 103
head louse 149
hearing 262; centre in brain **122**, 125; and ear 268–9 *see also* deafness; sound
heart 94–7; attack **97**, 103, 170, 251; disease 219; and exercise 250, 251–2; structure 96
heat energy 17, 22, 99, 109, **114**; from liquefied petroleum gas **189**, 190
heat exhaustion **171**, 251
heat-sensitive centre (in brain) 118, **129**
heat transfer **114–15**, 116, 121
heat system (solar) 195–7
helium (in divers' air) 257
herbicides: as contaminants 77, 147; handling of 155; usefulness of 239
herbivores 244; in food chain 144; teeth of 82
heroin 80–1
herpes, genital 52, **53**
heterosexuals 49; and AIDS 53
heterotrophic feeding 6, 66
histamines 103
histograms: construction of 37
HIV (virus causing AIDS) 53
homeostasis 128–9
homoiothermic animals 118–19
homosexuals 49; and AIDS 53
hookworm 150
hormones 48, 77, **126–7**, 129; and contraception 57, 59
host (for parasites) 148
household chemicals 162, 163, 171, **222–4**
house wiring 180–1
Human Genome Project 61
humidity 134–5
humus: importance of 233; measurement of in soil 232, **234**
hurricanes **312**, 313, 318
hydraulic brake 309
hydrocarbons 24–5, 188–9
hydroelectric power 195
hydrogen: atoms as fuel 194; test for 198
hydrogen chloride, diffusion of gas 16, 17
hydrogen sulphide, reaction with silver 210
hydroxides 200
hygiene 152–3; 208–9, and STDs 52–3
hygrometer 135
hypersensitivity (allergies) 103
hypertension 96–7
hypotheses 32, 33, 121
hysterectomy 58

ignition temperature 168
image, mirror 276–7 *see also* inverted image
immunisation *see* vaccination

Index 331

immunity 161; and AIDS 53
impurities in water 140–1
inclined plane **228**, 231, 294
incomplete flowers 41
incubation period 289
indicators 220; litmus 200, **220**, 221; universal 200, **220–1**, 234
industrial wastes (pollution) 147, **156**, 248 *see also* pollution
infectious diseases 160–1 *see also* sexually transmitted diseases
information, storage and retrieval of 28–9; data processing 29; word processing 29
infrared radiation **115**, 132, 320
infrasound 267
inhalation: in fish **100**, 143; in mammals 98–9
inheritance 10, 60–1
inherited diseases 61
inorganic compounds, used by plants 12, **66**, 90
insecticides: handling of 155, 224; and mosquitoes 151
insect pollination 42
instinctive reactions 125
insulators: for electricity 172; for heat 114, 116
insulin 80, **127**, 129
internal combustion engine 192–3
interpretation of results 36–7
interrelation of life processes 128–9
intrauterine device (IUD) 58
invar steel 131
invertebrates, classification of 14–15
inverted image: in camera 278; in eye 278, 279
involuntary (reflex) reactions 124, 125
iodine solution, and starch test 66, 67, 75
ion-exchange resins 219
ionosphere 275
iris, of eye 133, **278**
iron: in diet 79; galvanised 215; rusting of 212–15; and ships 255; testing for 225 *see also* metals; steel
irrigation 138, **239**
isobars 310
IUD *see* intrauterine device

jet engine 299
joints: in exercise 250 *see also* levers
joules: as energy unit 20; in exercise 251; in food 78, **105**; as work unit 227

kidney: disease 59; structure and function 108, **110–13**; tubules **112**, 287
kidney machine (artificial kidney) 59, 113
kilojoules *see* joules
kilowatt hour 178–9
kinetic energy 20, **290**

knee-jerk reflex 124
kwashiorkor 78

labels: on cleaning materials 208; on food 76, 77; on household chemicals 162; on laboratory chemicals 164–5
laboratory safety **164–6**, 180
lacteals (in villi) **84**, 128
lactic acid 106
lamp (light bulb) **132–3**, 175, 183
land and sea breezes 115
latent heat of vaporisation 117; and living things 118
lava 316
leaf litter, organisms in 242
leaf of life 39
leaf structure **70–1**, 118
length, measurement of 8–9, 12
lenses 277; of camera 278; of eye 278, 279, **280–1**; of spectacles 280–1
leptospirosis 151
levers: in body 123, **227**; law of 226–7
life processes 6–7; interrelations of 128–9
light: bulbs **132–3**, 183; colours 132, 282–3; intensity **132**, 133, 183; reflection 276–7; sources 132; street 132 *see also* energy; eyes
lightning 173
lime, added to clay soil 236
lime water (carbon dioxide test) 98, 104, 106, 107
lines of force 184–5
liquids 17; and heat transfer 114–15; and electrical conduction 172–3
liquefied petroleum gas (LPG) 189
litmus 200, **220**, 221
liver: and formation of urea 108; and homeostasis 128–9
living things: characteristics of 6–7, 11; classification of 12–15; and non-living things 6; and water 138
loam soil 233–6; improvements to 236
locomotion *see* movement
long sight, and its correction 281
louse *see* head louse
love vine (dodder) 148
LSD ('acid') 81
lubricants 224
lung cancer 102
lungs: and disease 102–3; as excretory organs 108, 113, 138; and exercise 250 *see also* respiratory system
lymphatic: glands 52, 53, 289; system 128
lymphocytes 128; and disease 53, 161; structure of 92

machines 226–31; efficiency of 230–1; and safety precautions 231
magic mushrooms 81
magnesium reactions: with acids 198; with oxygen 200
magnets **184–5**, 186, 187, 272
magnification using microscopes 12, 35
making slides 34
malaria 39, 148, **151**
male contraceptive 59
mathematical skills 31, 36–7
malnutrition 78
mangrove swamps 306
manipulative skills 30, 34
manure 239 *see also* fertilisers
marijuana 81; and lungs 102
marasmus 78
mass tables, for obesity 47
matter 11, 16–19, 22–3; and energy 11
maximum/minimum thermometer 130
measurement 30, 35; units of 8–9
mechanical: advantage 228; digestion 82, **84**, 86, 287; efficiency 230–1
medulla oblongata 125, 129
meiosis 60
Mendel 60
menopause 46
menstrual cycle 46, **48**; and contraceptive pill 57; and pregnancy 49, **62**, 63; and rhythm method 57
memory 28
mercury: barometer 308; as pollutant 146
metals, properties of 173, **198–201**; protection of 210, **214–15**; uses of 116, **202–3**
metals, reactivity: with acids 198; applications of 201, 215; with oxygen 200; series 201; with water 199
methane (biogas) 24, **153**
micro-organisms: and food storage 158–9; on skin 208, **289** *see also* bacteria; bread mould; protists; viruses
microphones **272**, 273
microscope 12, **35**; slides 34
microwave oven 115
milk 65, 222
milk teeth 83
millibar **309**, 310
mineral salts: and animals 79, 83, 129; and plants 72, 90; in soil 236
mirrors 265, 276–7
mitochondria 10, 106
mitosis 60
mixtures 23
molecules **18**, 23, 72, 258
momentum 290–1
monocotyledons 13

monosodium glutamate 77
moon 302; and living things 305; phases of 304; and tides 305; travel to 301
mosquitoes 39, **151**; control measures for 148, 151; and malaria 148, 151; types of 151
motor nerve cell **124**, 125, 288
mould and food spoilage 158 *see also* bread mould
mouth-to-mouth resuscitation 170 *see* artificial respiration
movement: of living things 7; of particles 16–17, **68–9**, 114–15; and skeleton 288 *see also* diffusion; osmosis
multiple sclerosis 289
muscles: as effectors 123; and exercise 250; and lactic acid 106
myxoedema 127
myxomatosis 155

narcotics 81
nasal cavity 99, **122**, 123
natural food products 76
natural resources 54–5, 190
nematode worms 14, 150; in soil 242
nerve cells: in reflex arc 124; structure of 288
nervous impulses **122**, 123, 124
nervous system **122–5**, 129; protection of 288, **289**
neutralisation 221; by lime 236; in mouth 209; in sugar test 75; and stain removal 221
neutrons 19
newtons (force unit) **227**, 230, 292, 309
nicotine 80; and athletes 253; and disease 102–3
night blindness 78
nitrates: plants' use of 72; in soil 243
nitrogen: in air 99; cycle **243**, 247; narcosis 257
nodules, on leguminous plants 239, **243**
noise 183, 248, **267** *see also* sound
non-biodegradable: insecticides 155; wastes **156–7**, 205
non-conductors 172 *see also* insulators
non-living things, characteristics of 6–7
non-metals 173 *see also* plastics
non-reducing sugars 74, **75**
nose *see* nasal cavity
nuclear: energy 21; fission 21; fusion **21**, 194
nucleus: of atom 19, 21; of living cell **10**, 60, 138
nutrition **6**, 66 *see also* digestion; photosynthesis
nylon 204

obesity 47; and diet 78, 79; and exercise 250
observation skills 30, 34
ocean currents 115; and El Niño 319

octane number 189
oestrogens **48–9**, 127; in contraceptive pills 57
Ohm's law 176–7
omnivores 244; in food chains 144
optic nerve 122, 278, 279
orbits **300**, 302, **303**, 304
organic compounds 24–5; and animals 12, 66
organic fertiliser 157, **236** *see also* compost; fertilisers
organs and systems 10–11
oscilloscope 267
osmoregulation: in fish 142; hormone controlling 127; in kidneys 112–13
osmosis **69**, 90; and food preservation 158; and osmoregulation 142
ovaries: in flowering plants 40, **43**; in mammals **50, 51**, 127
overloading 180–1
ovulation **48**, 51, 56
oxidising agents 198
oxides 200; of carbon 192, 193; of nitrogen 193
oxygen: in combustion 191; cycle 246–7; needed by decay bacteria 157, **158**; in gaseous exchange **98–101**, 108; and photosynthesis 72; preparation of **191**, 200; reaction with metals 200; and respiration 106–7; test for 200
oxyhaemoglobin 92
ozone hole 319

paints: coloured 282; use in preventing corrosion 214
Pap test (cervical smear) 51
pancreas 126, **127**, 129
parallel arrangements, in circuits 174–5
paralysis 289
parasites 148–152
parasitic worms 148, 150; adaptations of 150; spread of 150, 152, 153
Parkinson's disease 289
partially (semi-) permeable membranes 68–9
particles: arrangement in matter **17–18**, 23; movement 16–17, **68–9**, 114–15 *see also* diffusion; osmosis
Pascal (pressure unit) 309
passive immunity 161
passive smoking 102
peat 190
periods (menstruation) *see* menstrual cycle
peripheral nervous system 122–4
periscope 265
peristalsis 87
personal hygiene 153 *see also* hygiene
perspiration 109, 118, **119**, 134; and antiperspirants 208

pesticides: as contaminants 77, **147, 248**; handling of 155, 224
pests **154–5**, 156
petroleum gases 188, 189
pH: and enzymes 87, 88; scale 220; and soils **234**, 238; and toothpaste 209
phagocytes 92; and disease 160
phase change 17, **117**
phases of the moon 304; and organisms 305
phenobarbitone 81, 253
phloem 90–1
phosphorus: in diet 79; and teeth 83
photochemical reactions, in eye 278, **284–5** *see also* photography; photosynthesis
photography 278, **284–5**
photosynthesis 6, 12, **66–7**, 68, 71, 72, 195, 246
photovoltaic (solar) cell 197
pie chart 37
pigments 282–3
pill *see* contraceptive pill
pills, safe keeping of 162, 163
pinworm (threadworm, a nematode) 150
pitch *see* frequency
pituitary gland **126**, 127; and growth hormone 46, **127**; and sex hormones 48, **127**; and osmoregulation 127
placenta 63; as 'afterbirth' 64
planets 302–3
plankton in food chains 144
planning and designing 31, 32–3
plants: cells in 10; and animals 10, 11, 12; classification of 12–13; life cycle of 45
plaque: in arteries 96; on teeth 83, **209**
plasma 92, 93
Plasmodium 39, **151**
plastics 25, **204–5**
plate boundaries, in earth movement 314–15
platelets 92
Plimsoll line 255
plug, electrical 182
pneumoconiosis 103
poikilothermic animals 118
poisons: chemicals 163, **164–5**, 171; from heated plastics 205; pesticides 155, 224
poliomyelitis: disease 289; vaccine 161
pollen 40, **42**, 43, 103
pollination 42–3
pollutants 77, 248 *see also* pollution
pollution 248, 249; bauxite wastes 147; and fishing 145; from fuel burning **191**, 192–3; industrial **146–7**, 156; and lung cancer 102; mercury 146; motor vehicle 193; noise 267; and population growth 55; and ventilation 102, **135**; of water 146–7; and weather 309
poly(ethene)/polythene 25, 204
polymers 24, 25 *see also* plastics

Index 333

polystyrene, 116, 204; foam 116
polyvinylchloride (PVC) 204
population 245; control 56–9; growth 54–5
post-natal care 65
potential energy 20
power rating 175
practical: notebook 30, 33; skills 30–1, 34–5; work 30–5
predators, biological control of 155
prefixes for SI units 8–9
pregnancy 62–3; care in 64; signs of 62; teenage 49
pre-natal care 64
preserving food 77, **158–9**, 224
pressure: and ear 268; and diving 256–7; fluid 256, **309**; measurement of 308–9; units 309; and weather 310–12
pressure cooker 117
printing colours 283
producers, in food chains **144**, 244
profiles, of CXC examination 31
progesterone **48–9**, 127; in contraceptive pill 57
projector 277
prostate gland 50, **51**
proteins: digestion of 86–7; from fish 145; importance of **76**, 105; structure of 25, 74; tests for 75
protists **13**, 39, 287; *Plasmodium* 39, 151; and STDs 52
protoplasm 7, 10, **138**
protons 19
protozoans *see* protists
ptyalin *see* amylase
puberty 46, **48–9**
pulleys 228, **230**, 231
pulse, of blood circulation 94; after exercise 250
pupil, of eye 133, **278**, 279; reflex 280

quinine 148

radiant energy 194 *see also* solar energy
radiation 115, 116, **194**, 300, 320 *see also* infrared radiation; ultraviolet light
radio: frequencies **275**, 320; reception 275; waves 272–5
radioactivity 21
radiotherapy 47
rainfall 310–11; and flooding 312; in pollution 147; in water cycle 139
rats: and leptospirosis 151; as pests 154
rays, light 276–7; and eye 278, 280–1
reactivity series 201 *see also* metals, reactivity
receptors (senses) **122–3**, 124, 125
recessive genes 61

recycling: of basic materials 246–7, 249; of non-biodegradable wastes **156–7**, 205 *see also* carbon cycle; nitrogen cycle; oxygen cycle; water cycle
red blood cells 92, 93
reducing sugars 74, **75**
reduction 201
reflection, of light 276–7
reflex: action 124, **125**; arc 124
refrigeration 158, **159**
rehydration fluid 152, 153
relative humidity 134
reproduction: human 50–1, **60–3**; of living things 7, **38–9** *see also* asexual reproduction; sexual reproduction
reptiles 15, 118
resistance (electrical) 176–7, 180–1; variable 176–7
resistant strains 151, 155
respiration 6, **104–7**; and recycling 246–7
respiratory: diseases 102–3; substrates 78, **105**, 106; surfaces **100–1**, 287; system in fish 100, 142–3; system in mammals 98–103
response and stimuli 6, 122
results, recording and presenting 36–7
retina 122, **183**, **278**, **279**, 283
revision 33
rheostat 176
Rhesus factor, problems associated with 61
rhizomes 39, 73
rhythm method (contraception) 56
ribs: breathing 98; protective function 288
Richter scale 314
rickets 78
ring main 181
ringed worms 14 *see also* earthworms
ripple tank 264
road-making 237
rocket engine 301
rocky shore 232, **306–7**
rods (retina cells) 122, 278, 279
root: hairs 90; pressure 91
rotation of crops 239, **240**
roughage **74**, 84, 85
roundworms (nematode worms) 150
rubbish *see* waste materials
runner, of plant 39
rust (fungal disease) 148
rusting 210, 212–15; prevention of 214–15

saccharin 77
sacrificial anodes 215
safety: and earthquakes 315; and field trips 167; and machines 231; in the home **162–3**, 166, 180–1, 222–4; in the laboratory **164–6**, 180; in the workshop 167

salinity: of sandy soils **236**, 238; of sea 306
saliva 87, **88–9**
salivary amylase (ptyalin) 87, **88–9**
Salmonella **152**, 153
salt (sodium chloride): and circulation 97; and corrosion 213, 214; desalination 141; as excretory product 109; as food preservative 158, **159**
sandy soil 233–7; improvement of 236
sanitary control measures (sanitation) 153, **154**, 156–7
saponification 206
satellites **302**, 304; artificial 275, 311 *see also* moon
saturated fats 74; importance of **76**, 97
scalds 163, **170**
scales: on graphs 37; reading of 30, 35
scavengers, in food chains **144**, 244
schistosomiasis (bilharzia) 150, **152**; vector for 148, 152
school-based assessment (SBA) 30, 31, 34–5
science, reasons for studying 26–7
scientific method 26–7, **32–3**, 36
scouring powders 211
screw (machine) 228
scurvy **78**, 160
seashore 232, 264, **306–7**
sea water: and coasts 306–7, 318; and El Niño 319; and fish population **142**, 145; and food chains 144; levels 305, 307, 318; pollution 147; properties **142**, 306; waves 232, **264**
secondary sexual characteristics 48–9
section cutting, of plants 34
sedimentation of soil components **232**, 234
seeds: formation of 40, **43**; growth as seedlings 45; respiration in **104**, 107; structure and germination 44–5
seismic waves 314
self-pollination 43
semen 51
semi-circular canals 122, **268**, 269
semi-conductors 173
semi-(partially) permeable membrane 68–9
senses 122–3, 262–3
sensory cells *see* receptors
sensory nerve cells **124**, 125
series arrangements in circuits 174–5
serum 93; against Rhesus factor 61
sewage 233, **152–3**; uses of 153 *see also* waste materials
sex: determination 10, 60; hormones 48–9, **127**
sexual reproduction: of flowering plants 40–5; of mammals 50–1, **60–5**; of mould 38
sexually transmitted diseases (STDs) 52–3, 161
shock 170, **171** *see also* electric shock

shock wave 297
short sight, and correction of 280
ship protection (anticorrosion) 215
SI (Système International) units 8–9
sickle cell disease **61**, 160
sight *see* eyes
silver salts, in photography 284
skeleton 288; bird's 298
skin: of amphibian 100; and bacteria 289; and mammalian excretion 109; as receptor 123; and temperature control 113, **118–19**
skull 288
slides, making 34
slimming **78**, 105
smell 262–3; receptors 122, 123
smoking: and athletes 253; and cancer 102; and circulation 97, **103**; and pregnancy 64
smut (fungal disease) 148
snails 14; bilharzia vector 152; in soil 242
soaps **206–7**, 218; for personal use 208 *see also* detergents
soda acid fire extinguisher 169
sodium ions and hypertension 97
sodium hydrogen carbonate solution (bicarbonate): in carbon dioxide test 104, **246**; in deodorants 208; in mouthwash 209; in stain removal 221
sodium hydroxide (absorbing carbon dioxide) 67
soft water 218
soil 232–43; and added chemicals 147; components of 232–5; erosion and conservation 240–1; formation 232; improvement 236; nutrients 238; organisms 242–3; pH 234, 238; and plant growth 238–9; properties 238; and roads 237; water retention 235
soil air: importance of 233; measurement of 235
soil water: drainage and retention 235–6; importance of 233; uptake of 90
solar: cells 173, **195**, **197**; drier 197; energy 115, **194–7**, 302; still 196; water heating 195–7
solar system 302–3; model of atoms 19
solids 17; and heat transfer 114
solutions 16, 22, **216**
solvent(s): and cleaning 216–17; used as drugs 81; movement in osmosis 69; organic 217; water as 138, **216**
sound 262, 266–7; energy 290, 272; production **266–7**, 270; recording 270–1; tracks 270–1; transmission 272–5 *see also* frequency; hearing
space: bodies in 302–3; travel 300–1 *see also* planets; satellites; solar system
spectacles 280–1
spectrum, visible 132, **282**, 320

sperm, human **51**, 56
spermicides 56, 57
spinal cord 124, **288**
spores 38
sports equipment and playing surfaces 260–1
sprains 163, **171**
spread of disease 160; in sewage 152–3; in STDs 52–3
stability (balance) 294–5
stainless steel 213, **215**
stain removal 217, 221
starch 74; digestion of 86–7, 88–9; formation of 72; test for 66, **75**
states of matter 17, 117
state symbols 22
starvation 78
STDs *see* sexually transmitted diseases
steam turbine 21, **187**
steel 202; drums 202; stainless 213, 215
sterilisation (contraception) 58
stimulant ('pep') pills 81
stimuli 6, 122, 123, **124**
stings 163, **170**
stomata 70
storage organs: plant 73 *see also* liver
streamlining **297**, 299; in fish 142, 143
stress: and adrenaline 127; and asthma 103; and heart attacks 97
stretching *see* elasticity
stroke **96**, 289
subduction zone 314–15
sucrose (cane sugar) 74; digestion of **86**, 87; as preservative 158, **159**; test for 75
sun: effects on humans 171, 251; in solar system 302–3; and tides 305 *see also* solar energy
support: in animals 288; in plants 138
surface area to volume ratio 286–7
surface tension 255
suspensions 216
sweat: and excretion 109, **113**, 138; glands 109; and temperature regulation 118–19 *see also* perspiration
swim bladder 142
swimming, adaptations for in fish 142–3
switches, electric 172, 181
symbols for SI units 8–9
synapse **125**, 289
syphilis **52**, 53

tables of results 36
tapeworm 152
tarmac *see* road-making
tarnishing 210
tartrazine 77
taste buds 123

TB *see* tuberculosis
teenage pregnancies 49
teeth: cleanliness of 208; structure and functions of 82–3
telephone, structure of 272
television: sets 183; transmission 275, **285**
temperature: and enzymes 89; and global warming 318–19; and ignition 168; of inhaled and exhaled air 98, 99; judgement 130; measurement in humans 120; and respiration 99, **104–5**
temperature control: in animals **118–19**, 121, 171; of babies 65; process of 129; and thermostats 131
Temperature Humidity Index 135
termites 242, **243**
terracing, of soil 241
terrestrial habitats: land 244–5; soil 242–3
Terylene 204
testes 50–1; and cancer 51; and hormones 127
testosterone 48, 49, 127
tests: for chemicals 225; for food groups 75
test-tube babies 59
tetanus vaccine 161
tetraethyl lead **189**, 193
thalidomide 64
Thermit process 201
thermometers 120, 121, **130**
thermostats 131
thermosiphon system **195**, 196
thrombosis (coronary) 97
thrush (disease) 52
thunderstorms 173
thyroid gland 126, **127**
tidal waves **314**, 315
tides **305**, 306–7
tin 203
tissue culture 39
tissue fluid 128
tissues 10
toilet cleanliness 153, 194, 208
toilet flushing, and water use 138
tongue 122–3; genetic rolling ability 61
tooth decay **83**, 209
toothpaste 209; pH 209
touch 263
toxic chemicals 164, 165
toxins: on food 158; from germs 160
training, in athletes 252–3
tranquillisers 81
transformers 183, **187**
transpiration: for cooling 117, 118; for transport 90–1
transport, vehicles and 192–3
transport systems: need for 286–7; in plants **90–1**, 101; protection of 288–9

Index

trees 13, 117, **138**
trich (disease) 52
1,1,1-trichloroethane 217
tsunamis **315**, 316
tubal ligation 58
tuberculosis 53
tumours *see* cancer
turgidity 70, 138
twins 65
typhoid 152

ultrasound 64, **267**
ultraviolet light **133**, 183, 284
umbilical cord 63, 64
underwater: breathing 142; sports 256–7
units of measurement 8–9, 35
universal indicator: in neutralisation 221; with oxides and hydroxides 200; and soils 234
unsaturated fats 74; importance of 76, 97
uranium 21
urea: excretion of 111–12; formation of 108
urination: control of 110; and diabetes 127; in pregnancy **62**, 63
urine **111–12**, 113, 127; hygiene factors 152, **153**, 156
uterus 50, 62, **63**

vaccinations 152, **161**
vacuum flasks 104, **116**
vacuum pump 258, **308**
vagina 50, 51
vaginal: discharge 52; douches 208
valves: in heart 96; in veins 94
variables 33
variation, genetic 10, **60–1**
varicose veins 96
variety, of gametes 38, **60**
vascular bundles 90–1
vasectomy 58
vector **148**, 149, 160
vegetarians 76
vegetative reproduction 39; with food storage 72–3
vehicles, loading of 295
veins: in mammals **94–5**, 96; in plants 90–1
velocity 291
venereal diseases *see* sexually transmitted diseases (STDs)
ventilation 102, 103, **134–7**
vertebrae 288
vertebrates 15
vibration, sound 266–7, 272–3
villi 84, 128, 287
viruses 13; and antibiotics 160; and myxomatosis 155; and STDs 53

vitamins: in diet 76, 78; in supplements 224; storage of 129
vocal cords 262
volcanoes 314–15, **316–17**, 318
voltage **174–5**, 176–7, 182; and safety **173**, 180, 183
voltmeter 174, 175
voluntary actions 125
vomiting 85

warm-blooded animals *see* homoiothermic animals
washing soda 219
waste materials 156–7, 248; domestic 156; industrial **146**, 147, 156, 248; nuclear 21 *see also* sewage
water: conservation 138, **141**; as habitat 142–5, 306; hard and soft 207, **218–19**; for life **138**, 158; pH 220; polluted 146, 147; purification 140–1; safe 138; softeners 219; as solvent 138, **216**; sources 140, 146, 147 *see also* sea water; soil water; water waves
water balance of body 108, 112, **113**, 129
water cycle **139**, 246–7; and clouds 310
water waves: in sea 264, **305**; and hurricanes 312; and tsunamis 315, 316; and weathering 232
waterworks 140–1
wattage (power) **175**, 176, 178–9
waves: light 265; sound 267 *see also* water waves
weather 134, **310–11**, 312, 318–19; forecasting 311; satellites 302, **311**
weathering, of rocks 232; and waves 305
wedge (machine) 228
weight **292**, 300
weightlessness 300
wet and dry bulb hygrometer 130, **135**
wheel and axle 229
white blood cells 92; and disease 53, **160–1**, 289
wind pollination 42
windmills 197, **309**
winds: effect on sports 258; energy from 197; hurricanes **312**, 313, 318; land and sea breezes 115; and pollutant spread 309; and ventilation 134; weather factors 310–11
word equations 18, **22–23**, 198, 199, 200, 201, 218, 219, 221; photosynthesis 68
word processing 29
work **227**, 228–31
worms 14, **150**; parasitic 148, 150 *see also* flatworms; roundworms (nematode worms); ringed worms (earthworms)

xylem vessels 90–1

X-rays 284, 320; and cancer 47; and pregnancy 64

yeast 13; and anaerobic respiration 106, 107
yellow fever 151

zinc sheet *see* galvanised iron sheets
zones, on seashore 306–7
zygotes 38